MICHIGAN STATE UNIVERSITY
LIBRARY

JUL 07 2025

WITHDRAWN

RETURNING MATERIALS:
Place in book drop to
remove this checkout from
your record. FINES will
be charged if book is
returned after the date

The Chemistry of Weathering

NATO ASI Series

Advanced Science Institutes Series

A series presenting the results of activities sponsored by the NATO Science Committee, which aims at the dissemination of advanced scientific and technological knowledge, with a view to strengthening links between scientific communities.

The series is published by an international board of publishers in conjunction with the NATO Scientific Affairs Division

A	Life Sciences	Plenum Publishing Corporation
B	Physics	London and New York
C	Mathematical and Physical Sciences	D. Reidel Publishing Company Dordrecht, Boston and Lancaster
D	Behavioural and Social Sciences	Martinus Nijhoff Publishers
E	Engineering and Materials Sciences	The Hague, Boston and Lancaster
F	Computer and Systems Sciences	Springer-Verlag
G	Ecological Sciences	Berlin, Heidelberg, New York and Tokyo

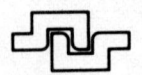

Series C: Mathematical and Physical Sciences Vol. 149

The Chemistry of Weathering

edited by

James Irving Drever
The University of Wyoming,
Department of Geology and Geophysics,
Laramie, Wyoming, U.S.A.

D. Reidel Publishing Company
Dordrecht / Boston / Lancaster

Published in cooperation with NATO Scientific Affairs Division

Proceedings of the NATO Advanced Research Workshop on
The Chemistry of Weathering
Rodez, France
July 2-6, 1984

Library of Congress Cataloging in Publication Data

NATO Advanced Research Workshop on the Chemistry of Weathering (1984: Rodez, France)
  The chemistry of weathering.

  (NATO ASI Series. Series C, Mathematical and physical sciences; vol. 149)
  "Published in cooperation with NATO Scientific Affairs Division."
  "Proceedings of the NATO Advanced Study Institute on the Chemistry of
Weathering, Rodez, France, July 2–6, 1984"-Verso t.p.
  Includes index.
  1. Weathering–Congresses. 2. Geochemistry–Congresses. I. Drever, James I. II. North Atlantic Treaty Organization. Scientific Affairs Division. III. Title. IV. Series: NATO ASI series. Series C, Mathematical and physical sciences; vol. 149.
  QE570.N38 1984   551.3'02   85-1843
  ISBN 90-277-1962-4

Published by D. Reidel Publishing Company
P.O. Box 17, 3300 AA Dordrecht, Holland

Sold and distributed in the U.S.A. and Canada
by Kluwer Academic Publishers,
190 Old Derby Street, Hingham, MA 02043, U.S.A.

In all other countries, sold and distributed
by Kluwer Academic Publishers Group,
P.O. Box 322, 3300 AH Dordrecht, Holland

D. Reidel Publishing Company is a member of the Kluwer Academic Publishers Group

2-0487-250 ts

All Rights Reserved
© 1985 by D. Reidel Publishing Company, Dordrecht, Holland.
No part of the material protected by this copyright notice may be reproduced or utilized in any form or by any means, electronic or mechanical, including photocopying, recording or by any information storage and retrieval system, without written permission from the copyright owner.

CONTENTS

PREFACE   vii

CHEMICAL MODELS OF WEATHERING IN SOILS   1
    Garrison Sposito

MULTICOMPONENT SOLID SOLUTIONS FOR CLAY MINERALS AND   19
    COMPUTER MODELING OF WEATHERING PROCESSES
    Bertrand Fritz

DISSOLUTION MECHANISMS OF PYROXENES AND OLIVINES   35
    DURING WEATHERING
    Jacques Schott and Robert A. Berner

THE EFFECTS OF COMPLEX-FORMING LIGANDS ON THE   55
    DISSOLUTION OF OXIDES AND ALUMINOSILICATES
    Werner Stumm, Gerhard Furrer, Erich Wieland
    and Bettina Zinder

KINETIC STUDY OF THE DISSOLUTION OF ALBITE WITH A   75
    CONTINUOUS FLOW-THROUGH FLUIDIZED BED REACTOR
    Roland Wollast and Lei Chou

INTERSTRATIFIED CLAY MINERALS AND WEATHERING PROCESSES   97
    M.J. Wilson and P.H. Nadeau

FORMATION OF SECONDARY IRON OXIDES IN VARIOUS ENVIRONMENTS   119
    U. Schwertmann

PHYSICAL CONDITIONS IN ALUNITE PRECIPITATION AS A   121
    SECONDARY MINERAL
    R. Rodriguez-Clemente and A. Hidalgo-Lopez

THIS PLANET IS ALIVE--WEATHERING AND BIOLOGY,   143
    A MULTI-FACETTED PROBLEM
    W.E. Krumbein and B.D. Dyer

SOLUBILIZATION, TRANSPORT, AND DEPOSITION OF MINERAL   161
    CATIONS BY MICROORGANISMS - EFFICIENT ROCK
    WEATHERING AGENTS
    Friedrich E.W. Eckhardt

CHEMICAL WEATHERING AND SOLUTION CHEMISTRY IN ACID   175
    FOREST SOILS: DIFFERENTIAL INFLUENCE OF SOIL
    TYPE, BIOTIC PROCESSES, AND H^+ DEPOSITION
    Christopher S. Cronan

PROTON CONSUMPTION RATES IN HOLOCENE AND PRESENT-DAY   197
    WEATHERING OF ACID FOREST SOILS
    Horst Fölster

EQUILIBRIUM AND DISEQUILIBRIUM BETWEEN PORE WATERS   211
    AND MINERALS IN THE WEATHERING ENVIRONMENT
    Wolfgang Ohse, Georg Matthess and Asaf Pekdeger

HYDROGEOCHEMICAL CONSTRAINTS ON MASS BALANCES IN   231
    FORESTED WATERSHEDS OF THE SOUTHERN APPALACHIANS
    Michael A. Velbel

PAST AND PRESENT SERPENTINISATION OF ULTRAMAFIC ROCKS;   249
    AN EXAMPLE FROM THE SEMAIL OPHIOLITE NAPPE OF
    NORTHERN OMAN
    Colin Neal and Gordon Stanger

MANGANESE CONCENTRATION THROUGH CHEMICAL WEATHERING OF   277
    METAMORPHIC ROCKS UNDER LATERITIC CONDITIONS
    Daniel Nahon, Anicet Beauvais and Jean-Jacques Trescases

RIVER CHEMISTRY, GEOLOGY, GEOMORPHOLOGY, AND SOILS IN   293
    THE AMAZON AND ORINOCO BASINS
    Robert F. Stallard

LIST OF WORKSHOP PARTICIPANTS   317

SUBJECT INDEX   321

PREFACE

Several important developments in our understanding of the chemistry of weathering have occurred in the last few years:

1. There has been a major breakthrough in our understanding of the mechanisms controlling the kinetics of silicate dissolution, and there have been major advances in computer modeling of weathering processes.
2. There has been a growing recognition of the importance of organic solutes in the weathering process, and hence of the inter-relationships between mineral weathering and the terrestrial ecosystem.
3. The impact of acid deposition ("acid rain") has been widely recognized. The processes by which acid deposition is neutralized are closely related to the processes of normal chemical weathering; an understanding of the chemistry of weathering is thus essential for predicting the effects of acid deposition.
4. More high-quality data have become available on the chemical dynamics of small watersheds and large river systems, which represent the integrated effects of chemical weathering.

We report here the results of a NATO Advanced Research Workshop which brought together experts in these and other aspects of weathering to exchange ideas and to try to integrate the different approaches. The organization of this volume follows the organization of the conference, with a natural progression from theoretical models and laboratory experiments to detailed field studies, and finally to large-scale drainage-basin studies. This volume cannot, unfortunately, include the extensive discussion that took place at the conference. There was a consensus at the end that thermodynamic and kinetic models could be used quite effectively to predict the chemistry of groundwaters and waters in the unsaturated zone below the soil, but we were still a long way from being able to model or predict the chemistry of weathering in the soil zone, where biological processes tend to dominate inorganic processes. In particular, it was felt that theoretical models could not, at present, predict the effects that acid deposition would have on weathering processes in the soil.

I am deeply grateful to my co-chairman, Yves Tardy, to the NATO Scientific Affairs Division for supporting the conference, and to the staff of the Hôtel Comtal for providing ideal arrangements at the conference site.

James I. Drever

CHEMICAL MODELS OF WEATHERING IN SOILS

Garrison Sposito

Department of Soil and Environmental Sciences
University of California, Riverside CA 92521

Chemical thermodynamics augmented by the Gay-Lussac-Ostwald step rule provides a theoretical approach to weathering phenomena in soils that requires few ad hoc constitutive assumptions. The efficacy of this approach is illustrated with several examples of natural and anthropogenically-induced weathering reactions involving kaolinite. In soils, kaolinite exhibits a continuum of structural disorder that is reflected by a corresponding spectrum of solubility product constants. The composition of the soil solution vis-à-vis this spectrum can be interpreted with the help of the step rule. The concept of a soil kaolinite continuum appears necessary to the evaluation of activity-ratio diagrams and other thermodynamic constructs.

INTRODUCTION

Soils are complex assemblies of matter whose properties evolve through the actions of biological, hydrological, and geological agents. The labile aqueous phase in soil, the soil solution, is the principal seat of this activity, which is mediated by the heterogeneity of the lithosphere and the transience of matter and energy flows from the atmosphere and biosphere. The chief local feature of soil solution lability is the continual transformation of dissolved constituents into different chemical species over a broad range of reaction time scales. This persistent but kinetically complex process, controlled by vicinal interactions involving solar photons and the biota, is the essential force that drives soil profile development and governs the patterns of chemical weathering.

The very complicated setting in which soil development takes place suggests that precedence be given to theoretical approaches requiring few ad hoc constitutive assumptions. The most prominent of these approaches is chemical thermodynamics (1-4). With reference to soils, a thermodynamic description is useful so long as the chemical phenomena of interest involve space and time scales that are incommensurable with those of a laboratory measurement (1,2). A soil sample from which a soil solution is extracted for study must occupy a volume palpably larger than that of a few solid grains and considerably smaller than that of a soil profile horizon. The chemical reactions investigated must either be at some stable point of development long before data are taken or be so unfavorable kinetically that the reactants can be assumed to be perfectly stable species to a high degree of approximation. These requirements often are met reasonably well for natural soils (1-3).

The focus of chemical thermodynamic models of weathering in soils is on solubility control, usually through the precipitation-dissolution reactions of hydrous oxides and aluminosilicates constrained by the Gibbs phase rule (1-3). As the very existence of the weathering process itself suggests, the assembly of solid phases reacting with the soil solution is not expected to represent more than a metastable system in general. That this fact does not pose an insuperable problem for chemical thermodynamics has been known for many years (5), but interpretive guidelines are necessary to avoid conceptual errors in the analysis of soil solution composition data (1). One of the more important of these guidelines is the Gay-Lussac-Ostwald (GLO) step rule (6). In the context of the Ion-Association model (2), this empirical rule can be summarized as follows:

(a) If the initial activity of an ion in the soil solution and attendant soil conditions make several solid-phase states potentially accessible to the ion, the solid phase which forms first will be the one for which the activity of the ion would be nearest below its initial value in the soil solution.

(b) Thereafter, other accessible solid phases will form in order of decreasing activity of the ion in the soil solution, and the rate of formation of each solid in the series will decrease as the corresponding ion activity decreases. In an open system (such as the soil solution tends to be), any one of the solid-phase steps may be maintained "indefinitely" on the time scale of weathering experiments.

The GLO step rule provides a conceptual framework for observed sequences of feldspar weathering by waters enriched in carbon dioxide (4). But the application of the rule is not limited to successions of different minerals. In soils, precipitated solid phases usually are of disordered structure relative to specimen minerals (7) and the GLO rule can be applied to a single soil mineral whose degree of crystallinity changes with time or with changing soil conditions. A good example is afforded by kaolinite [$Al_2Si_2O_5(OH)_4(s)$], which even in specimen minerals exhibits considerable variability in its Standard State chemical potential (8,9). It has been observed many times (9-13) that poorly crystalline kaolinite, with solubility characteristics approaching or equaling that of its disordered polymorph, halloysite (14), tends to form preferentially in soils. Conceivably, there exists a continuous spectrum of Standard State chemical potentials of soil kaolinites bounded by that of well crystallized, specimen kaolinite (-3799.7 kJ mol^{-1}) and that of specimen halloysite (-3780.5 kJ mol^{-1}) (15). According to the GLO rule, if part or all of this spectrum is accessible to Al^{3+}(aq) in a soil solution, there will be a succession of kaolinites in a soil profile, with the least crystalline forms appearing where weathering is least advanced or where conditions in the soil solution favoring kaolinite formation have developed most recently. Illustrations of this prediction in field studies have been reported by Calvert et al. (16) and Karathanasis and Hajek (9).

In this paper, several recent published investigations of soil weathering phenomena involving kaolinite will be interpreted with chemical thermodynamics incorporating the GLO step rule. The principal objective is to underscore the lability of soil kaolinite as a solubility-controlling solid phase of invariant composition. A secondary objective is to show the utility of the GLO step rule in the description of weathering phenomena induced by anthropogenic changes in soil profiles.

2. BEIDELLITE ⇌ KAOLINITE IN ALFISOLS

Beidellite is a smectite in which the charge deficit produced by isomorphic substitution in the tetrahedral sheet exceeds that in the octahedral sheet (17). A value of log $*K_{so}^o$ calculated for a Mg-beidellite with the solubility data published by Misra and Upchurch (18) is given in Table 1. This value and log $*K_{so}^o$ values for kaolinite, halloysite, and sepiolite were used to draw the boundary lines in Figure 1. The sepiolite log $*K_{so}^o$ value was calculated with the μ^o value for sepiolite [$Mg_2Si_3O_{7.5}(OH) \cdot 3H_2O(s)$] recommended by Bassett et al. (19). The region of Figure 1 bounded by vertical lines at pH_4SiO_4 = 3.2 and 4.9 represents the soil kaolinite continuum.

The data points in Figure 1 refer to the composition of 1:1 soil water extracts taken at different depths in Mexico silt loam, an Alfisol, seven years after $MgCO_3$ was added in the zone 0.4 to 0.6 m below the soil surface at the rate of 0.12 mol kg^{-1} (20-22). The sampling depths ranged between 0 and 0.8 m, with the soil extract pH value varying from 6.9 to 8.6 (22). The values of pH - 1/3pAl and pH - 1/3Fe(III) determined by Misra and Upchurch (18) for the Mexico soil were assumed to apply to the field soil samples as well. In general, the larger values

Table 1. Mineral dissolution reactions and log $^*K^o_{SO}$ values

Dissolution Reaction	log $^*K^o_{SO}$
$Al(OH)_3(s) + 3H^+(aq) =$ $Al^{3+}(aq) + 3H_2O(l)$	$8.05 \pm 0.73(15)$
$KAl_3(OH)_6(SO_4)_2(s) + 6H^+(aq)$ $= K^+(aq) + 3Al^{3+}(aq)$ $+ 2SO_4^{2-}(aq) + 6H_2O(l)$	$0.6 \pm 1.2(27)$
$AlOHSO_4 \cdot 5H_2O(s) + H^+(aq) = Al^{3+}(aq)$ $+ SO_4^{2-}(aq) + 6H_2O(l)$	$-3.8 \pm 0.2(27)$
$Al_4(OH)_{10}SO_4 \cdot 5H_2O(s) + 10H^+(aq) =$ $4Al^{3+}(aq) + SO_4^{2-}(aq)$ $+ 15H_2O(l)$	$22.5 \pm 0.2(27)$
$Al_2Si_2O_5(OH)_4(s) + 6H^+(aq) = 2Al^{3+}(aq)$ $+ 2H_4SiO_4^o(aq) + H_2O(l)$	$7.12 \pm 0.6(15)$
$Al_2Si_2O_5(OH)_4(s,dis) + 6H^+(aq)$ $= 2Al^{3+}(aq) + 2H_4SiO_4^o(aq)$ $+ H_2O(l)$	$10.48 \pm 0.6(15)$
$K_{0.24}Ca_{0.08}(Al(OH)_{2.61})_{1.45}[Si_{3.24}Al_{0.76}]$ $(Al_{1.56}Fe(III)_{0.24}Mg_{0.20})$ $O_{10}(OH)_2(s) + 12.83H^+(aq)$ $= 0.24K^+(aq) + 0.08Ca^{2+}(aq)$ $+ 3.77Al^{3+}(aq) + 0.24Fe^{3+}(aq)$ $+ 0.20Mg^{2+}(aq) + 3.24H_4SiO_4^o(aq)$ $+ 2.83H_2O(l)$	$20.1 \pm 1.9(10)$

Table 1. (Continued)

Dissolution Reaction	log $*K_{SO}^{O}$
$Al_{0.218}[Si_{3.55}Al_{0.45}]$ $(Al_{1.41}Fe(III)_{0.385}Mg_{0.205})$ $O_{10}(OH)_2(s) + 2.2H_2O(l)$ $+ 7.8H^+(aq) = 2.078Al^{3+}(aq)$ $+ 0.385Fe^{3+}(aq) + 0.205Mg^{2+}(aq)$ $+ 3.55H_4SiO_4^O(aq)$	7.17 ± 0.5^a
$Mg_{0.328}[Si_{3.55}Al_{0.45}]$ $(Al_{1.41}Fe(III)_{0.385}Mg_{0.205})$ $O_{10}(OH)_4(s) + 2.2H_2O(l)$ $+ 7.8H^+(aq) = 1.86Al^{3+}(aq)$ $+ 0.385Fe^{3+}(aq) + 0.533Mg^{2+}(aq)$ $+ 3.55H_4SiO_4^O(aq)$	$9.57 \pm 0.35 (18)$
$Mg_2Si_3O_{7.5}(OH)\cdot 3H_2O(s) + 1/2\,H_2O(l)$ $+ 4H^+(aq) = 2Mg^{2+}(aq)$ $+ 3H_4SiO_4^O(aq)$	$15.6 \pm 0.2 (19)$

[a] calculated with log $*K_{SO}^{O}$ for the dissolution of Mg-beidellite and an estimate (see Appendix) of μ^O [Al-beidellite] and μ^O [Mg-beidellite]

of pH - 1/2pMg plotted in Figure 1 correspond to samples at depths between 0.4 and 0.7 m in the soil.

3. BEIDELLITE ⇌ KAOLINITE IN ULTISOLS

Under acidic soil conditions, the weathering of smectite to kaolinite is expected to involve Al-beidellite, with disordered kaolinite precipitating from supersaturated soil solutions and transforming to more ordered forms as time passes (9). Figure 2 shows an activity-ratio diagram for Al^{3+}(aq) plotted with the

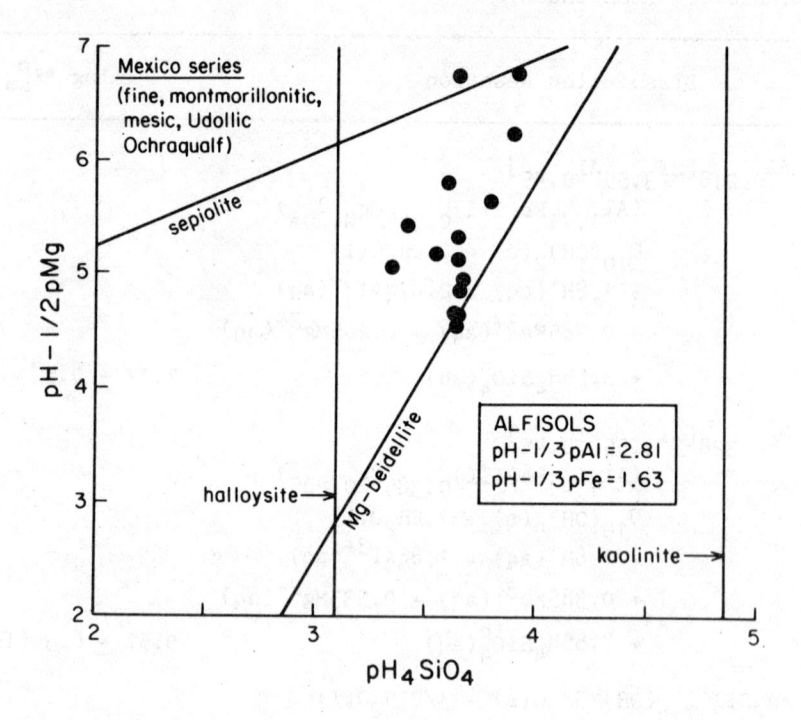

Figure 1. Solubility relationships in an Alfisol treated with $MgCO_3$.

log $*K_{SO}^O$ values for kaolinite, halloysite, and Al-beidellite in Table 1 for the conditions pH = 4.4, pFe(III) = 10.6, pMg = 4.3. (The log $*K_{SO}^O$ value for Al-beidellite was calculated with log $*K_{SO}^O$ for Mg-beidellite and estimates of $\mu°$ for the two beidellites based on a new version of the Mattigod-Sposito model (23), described in the Appendix.) The construction of activity-ratio diagrams is described by Sposito (1).

The data point in Figure 2 represents the mean and range of pAl and pH_4SiO_4 determined in centrifugates from the Bt and C horizons of four Ultisols by Karathanasis and Hajek (9). The soil solution compositions indicate that, at the pH, pFe(III), and pMg values in the soil centrifugates, an association between Al-beidellite and kaolinite of varying disorder is likely. This prediction was confirmed by X-ray diffraction analysis (9). The log $*K_{SO}^O$ values for the soil kaolinites ranged between 7.3 and and 10.4, spanning the kaolinite continuum completely, with the

CHEMICAL MODELS OF WEATHERING IN SOILS

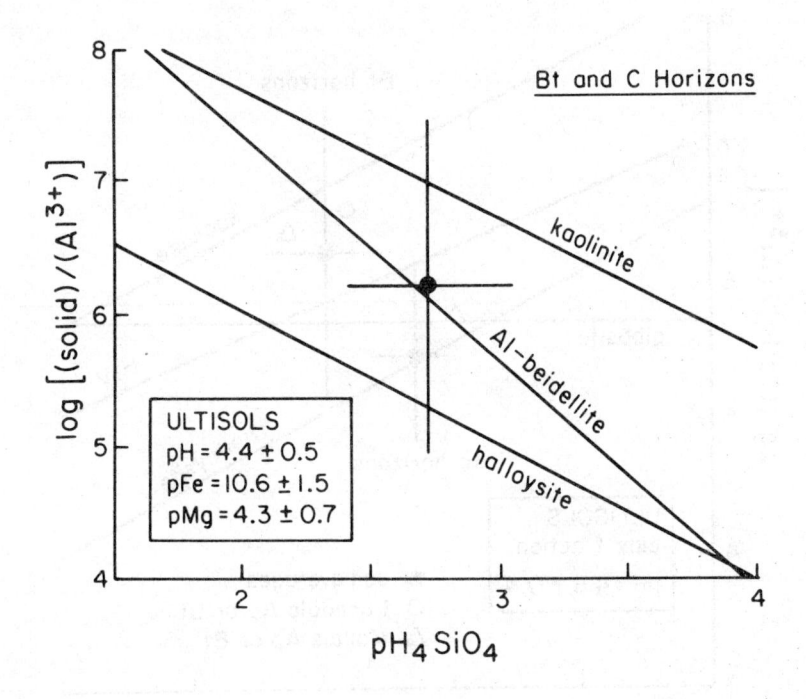

Figure 2. Activity-ratio diagram for $Al^{3+}(aq)$ in Bt and C horizons of four Alabama Ultisols.

larger values tending to be found in C horizons, where smectite predominated.

4. HYDROXY-INTERLAYER VERMICULITE ⇌ KAOLINITE ⇌ GIBBSITE IN ULTISOLS

Hydroxy-interlayer vermiculite (HIV) often is found in association with kaolinite and gibbsite in southern U.S. Ultisols (10). A value for log $*K_{so}^o$ of HIV based on solubility data for 28 soil samples (10) and an average HIV structural formula (24) is given in Table 1. Figure 3 shows an activity-ratio diagram for $Al^{3+}(aq)$ which incorporates the log $*K_{so}^o$ data for HIV and for the kaolinite continuum under the conditions pH = 4.6 and [0.24(pH - pK) + 0.16(pH - 1/2pCa) + 0.72 (pH - 1/3pFe(III)) + 0.4(pH - 1/2pMg)] = 1.65 ± 0.38. The filled circles in Figure 3 represent the mean and range of pAl and pH_4SiO_4 observed in centrifugates from surface (Ap) and subsurface (Bt) horizons of 16 Ultisols (10). These data points

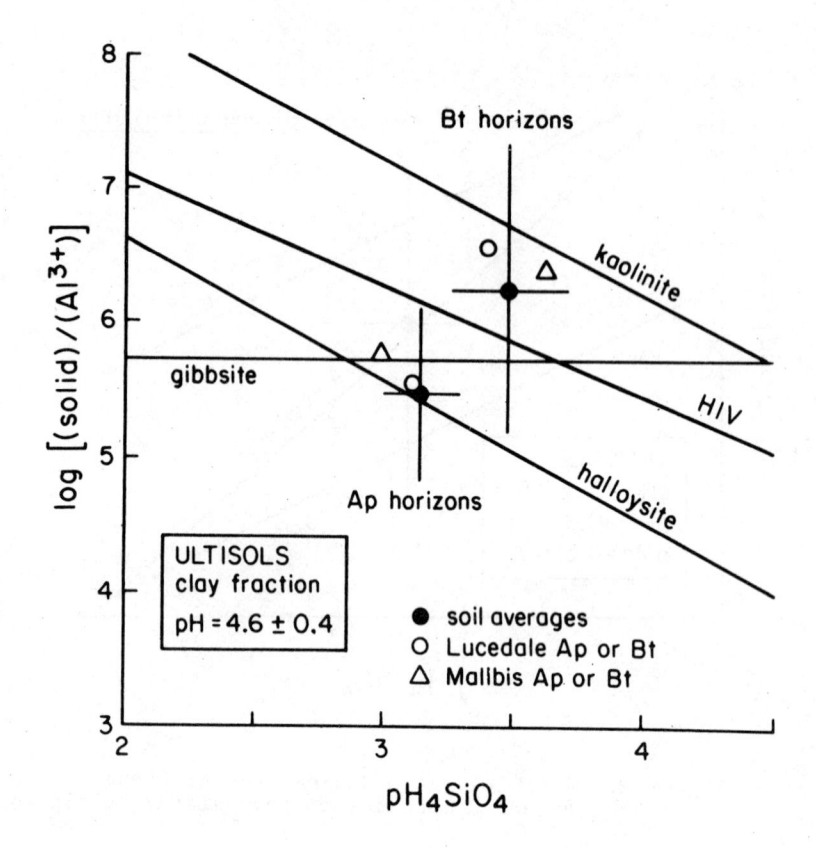

Figure 3. Activity-ratio diagram for Al^{3+}(aq) in Ap and Bt horizons of 16 Ultisols.

are consistent with the simultaneous presence of HIV, gibbsite, and kaolinite in the soils, which also was verified by X-ray diffraction analysis (10). There is a distinct tendency for kaolinite to be less ordered in the surface horizons than in the subsurface horizons; individual data points for the Lucedale (Rhodic Paleudult) and Malibis (Plinthic Paleudult) soils illustrate this trend in Figure 3.

X-ray diffraction analysis (10) shows that the HIV/kaolinite ratio approaches 2:1 in the Ap horizon of these soils, whereas in the Bt horizons kaolinite tends to be dominant. Gibbsite exhibits no particular trend with depth. These facts and the solubility data in Figure 3 suggest that HIV can persist only where it is in (metastable) equilibrium with disordered kaolinite.

The structural formulas of Al-vermiculite and Al-beidellite are not very different and hydroxy-interlayering of both minerals

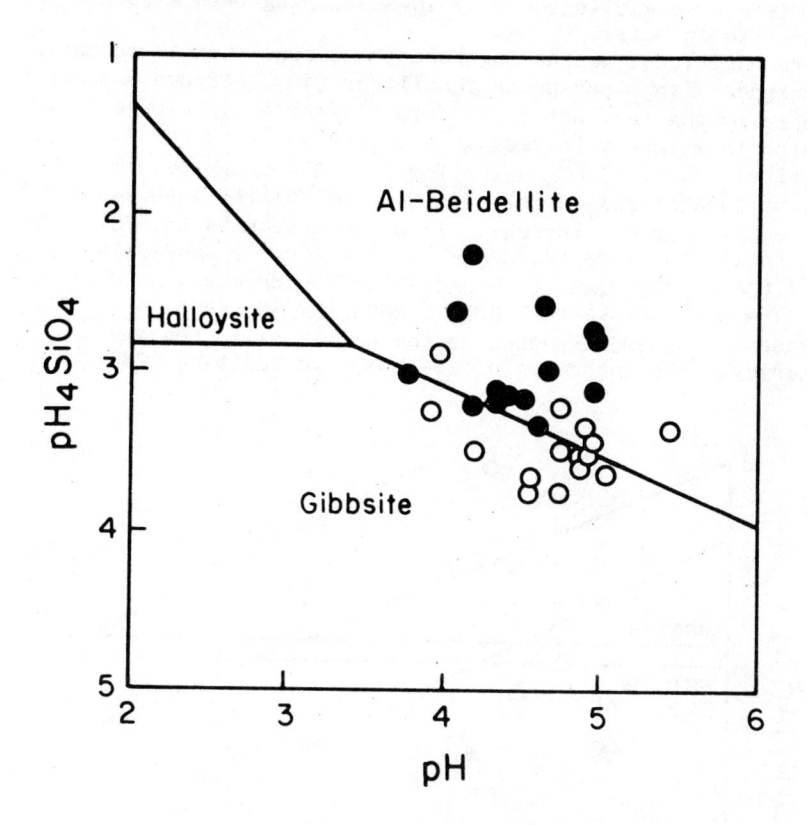

Figure 4. Predominance diagram in the system Al-SiO$_2$-H$_2$O. (● surface horizons, ○ subsurface horizons in Ultisols).

occurs under acidic soil conditions (9,10). Figure 4 shows a predominance diagram for Al-beidellite based on the data in Table 1 and the conditions imposed in Figures 2 and 3, except for the pH values. The open circles refer to subsurface horizon data summarized in Figures 2 and 3, whereas the filled circles refer to surface horizon data summarized in Figure 3. The tendency of the subsurface horizons to move away from Al-beidellite stability and that of the surface horizons to sustain higher silica activities are apparent. If well-crystallized kaolinite were included in the diagram, it would plot a boundary line with gibbsite at pH$_4$SiO$_4$ = 4.5 and aluminum beidellite would be relegated to a dog-ear in the upper right corner commencing at pH=4, pH$_4$SiO$_4$ = 1 and ending at pH$_4$SiO$_4$ = 3.2, pH = 6, with a slope $\simeq$ - 1. These characteristics emphasize the significant role that disordered kaolinites play in sustaining both gibbsite and Al-smectite in soils.

An additional weathering reaction occurs when these soils are amended with o-phosphate fertilizer (10). Figure 5 shows essentially the same activity-ratio diagram as in Figure 3 (the pH value is slightly larger) with a line included for "soil variscite". The log K_{SO}^O value for this Al-phosphate calculated by Karathanasis et al.(10) with solubility data was -29.8 (pK_{SO}^O = pAl + 2pOH + pH_2PO_4). In accordance with the GLO step rule, this value lies between -28.1 for freshly precipitated soil $Al(OH)_2H_2PO_4$(s,am) (25) and -30.5 for crystalline variscite (3). The data points in Figure 5 show the increase in pAl and decrease in pH_4SiO_4 expected as the aluminosilicates are weathered by the anthropogenic o-phosphate addition (25).

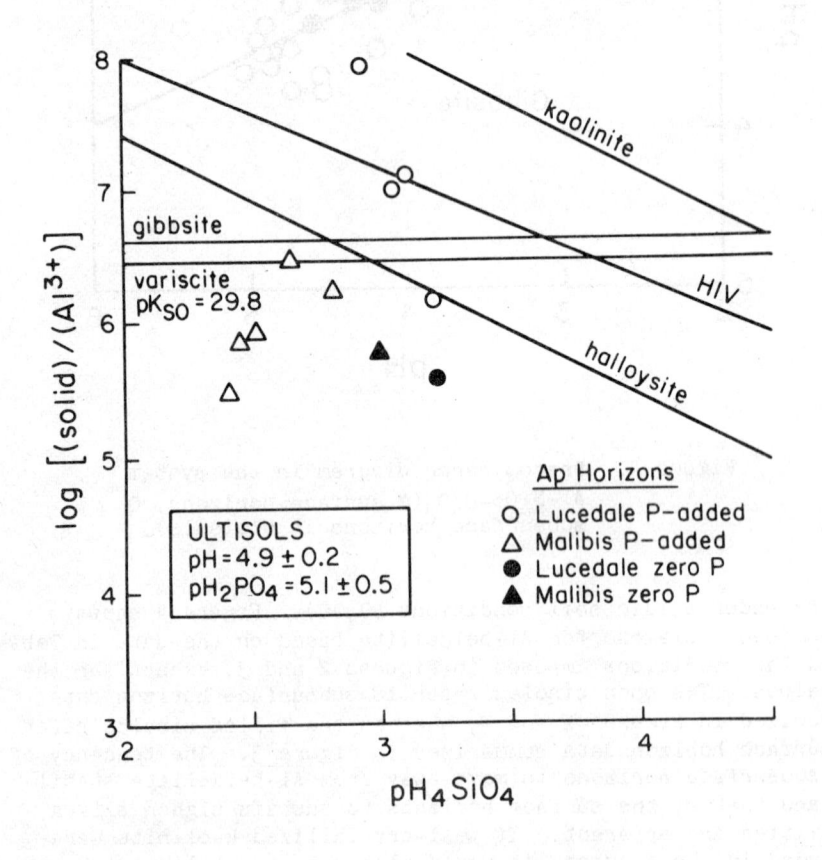

Figure 5. Activity-ratio diagram for Al^{3+}(aq) in Ultisols receiving o-phosphate amendments.

5. ANTHROPOGENIC ALUMINUM SULFATE IN OXISOLS

Oxisols in southern Brazil contain kaolinite and KCl-extractable aluminum in the clay fraction (26). The addition of gypsum to these soils produces a dramatic reduction in the extractable Al without a significant change in leachable Al or in the pH value (26). It is possible that this transformation involves the formation of an Al-sulfate solid and three candidates for the solid phase are listed in Table 1. The values of log *K^o_{so} given for alunite [$KAl_3(OH)_6(SO_4)_2(s)$], jurbanite [$AlOHSO_4 \cdot 5H_2O(s)$], and basaluminite [$Al_4(OH)_{10}SO_4 \cdot 5H_2O(s)$] were calculated from log K^o_{so} data compiled by Nordstrom (27). The least stable of these minerals for $pSO_4 < 4$ is basaluminite and it is this solid that might be expected to precipitate first according to the GLO step rule. The soil solution data points in Figure 6 support this prediction by clustering around the confluence of the halloysite-basaluminite lines appropriate to soil conditions after the gypsum treatment (26). In this case, the soil solution appears to be poised at the bottom of the kaolinite continuum.

6. APPENDIX

The Mattigod-Sposito model (1,23) permits the calculation of the absolute value of ΔG^o_r for the formation of a homoionic smectite and liquid water from a set of hydroxide components. This calculation, based on the concept that a layer silicate is a condensation polymer formed from hydroxides, requires an empirical regression equation containing C, the positive charge deficit produced by isomorphic substitution, R, the ionic radius of the exchangeable cation, and Z, the valence of the cation. The rationale behind the equation is that ΔG^o_r will have an important contribution from the exchangeable cation, which has been transferred from a higher-potential site in an hydroxide solid to a lower-potential site in the interlayer region to form the smectite (23). The number of these transfers per formula mass of the clay mineral is C and $|\Delta G^o_r|$ can be expected to increase with C. The effect of the transfer on $|\Delta G^o_r|$ should be smaller, however, for small, high-valence cations, since their ionic bonds are more localized and, therefore, less sensitive to the distribution of negative charge in a mineral. Thus $|\Delta G^o_r|$ increases with R and decreases with Z.

Given the validity of the physical arguments in support of the model, a simple relationship for $|\Delta G^o_r|$ can be imagined by combining R and Z into a single parameter of geochemical significance, the ionic potential. In the least complicated scenario, $|\Delta G^o_r|$ would be a linear function of the ratio of C to Z/R in accordance with the expected effects of these parameters on the Gibbs energy of smectite formation from hydroxides:

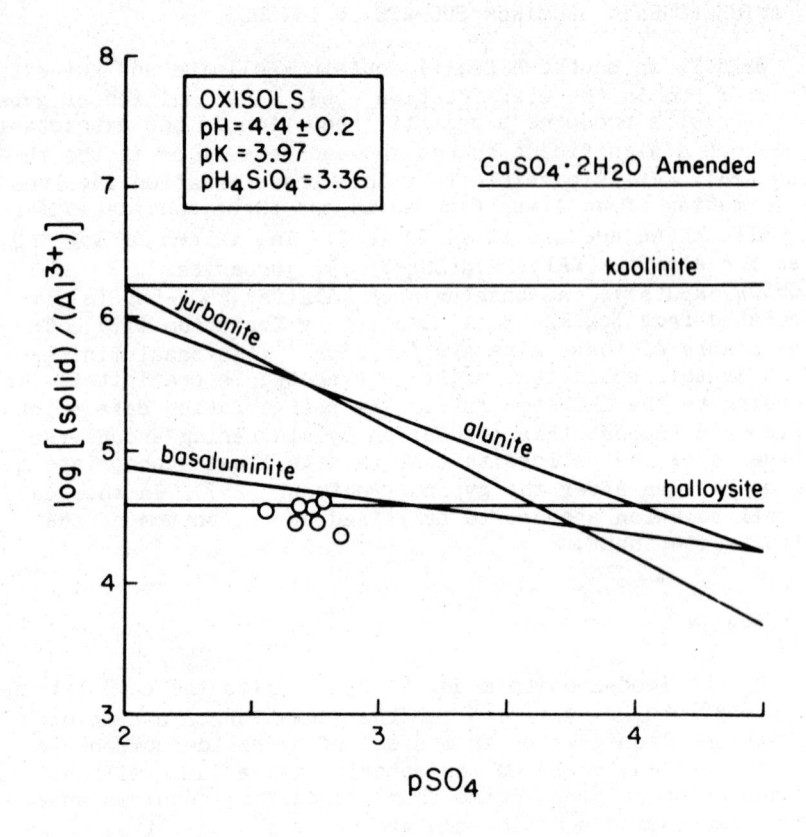

Figure 6. Activity-ratio diagram for Al³⁺(aq) in Oxisols treated with CaSO₄·2H₂O.

$$|\Delta G_r^o| = a + b(CR/Z) \qquad (6.1)$$

Figure 7 shows a test of Eq. 6.1 with the ΔG_r^o values for homoionic montmorillonites compiled in Table 5 of Mattigod and Sposito (23). The correlation is excellent ($r^2 = 0.977$). A similar relationship can be postulated for vermiculites since they are isostructural with smectites. Figure 8 shows a test of Eq. 6.1 for Transvaal vermiculite saturated with Li, Na, Mg, Ca, Sr, or Ba cations. The values of ΔG_r^o were calculated with the help of an estimate of μ^o for Na-vermiculite and ΔG_{ex}^o values for Na $\rightarrow$ M (M = Li, Mg, Ca, Sr, or Ba) exchange reactions, as described in the Data and Methods section of Mattigod and Sposito (23). The correlation once again is very good ($r^2 = 0.932$).

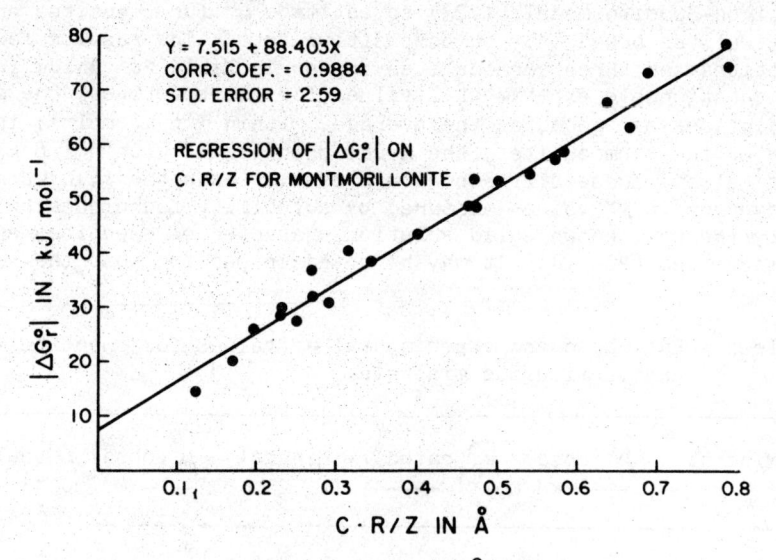

Figure 7. Regression of $|\Delta G_r^o|$ on CR/Z for montmorillonites.

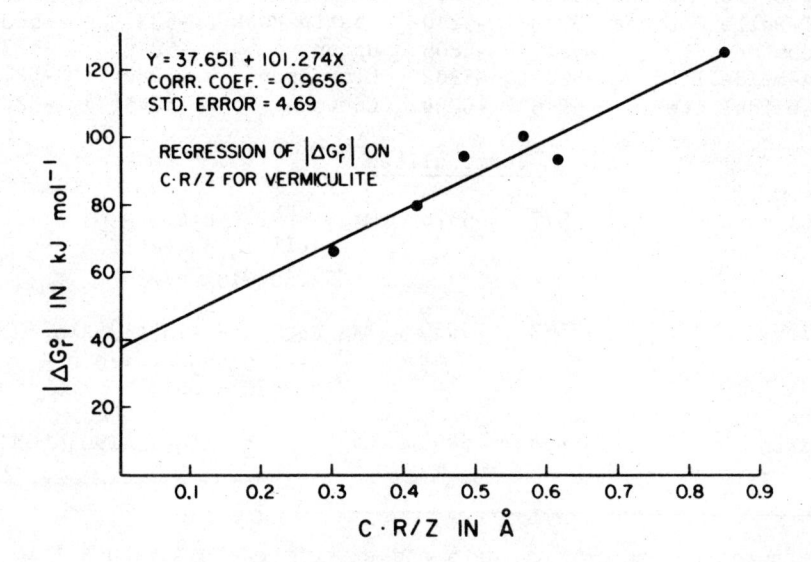

Figure 8. Regression of $|\Delta G_r^o|$ on CR/Z for vermiculites.

The new regression equations can be used as usual in the Mattigod-Sposito model (1,23) to estimate μ^o for smectites and vermiculites based only on composition data. The results for 14 smectites and three vermiculites appear in Table 2. Aside from two questionable experimental values for the smectites, the model predictions are good, on the average, within $\pm$ 6 kJ mol^{-1}; in the case of the vermiculites, the model predicts μ^o within $\pm$ 6 kJ mol^{-1} also. These differences are smaller than the $\pm$ 10 kJ mol^{-1} inaccuracy in μ^o values measured by solubility techniques brought on by the well known solid solution character of smectites and vermiculites (30-32). It may be noted in passing that the

Table 2. Estimated and experimental μ^o values for smectite and vermiculite minerals.

Mineral	μ^o(obsd)	μ^o(calcd)	Mineral	μ^o(obsd)	μ^o(calcd)
	---- kJ mol^{-1} -----			---- kJ mol^{-1} ----	

Smectites[a]

Mg-Aberdeen	-5219	-5217	Colony	-5262	-5261
Al-Aberdeen	-5200	-5196	Colony I	-5268	-5262
Mg-Belle Fourche	-5223	-5224	Colony II	-5262	-5267
Al-Belle Fourche	-5213	-5210	Castle Rock	-5337	-5337
Houston	-5215	-5205	Upton	-5218	-5238
Mg-beidellite	-5200	-5202	Clay Spur	-5226(?)	-5273
K-beidellite	-5215	-5234	Cheto	-5245(?)	-5284

Vermiculites[b]

Palabora	-5577	-5576	Mg$_{0.40}$[Si$_{3.145}$Al$_{0.855}$](Fe(III)$_{0.235}$Fe(II)$_{0.075}$Mg$_{2.60}$)O$_{10}$(OH)$_2$
Libby	-5537	-5550	Mg$_{0.43}$[Si$_{3.19}$Al$_{0.81}$](Al$_{0.08}$Fe(III)$_{0.295}$Fe(II)$_{0.015}$Mg$_{2.40}$)O$_{10}$(OH)$_2$
Harps	-5494	-5499	Ca$_{0.35}$[Si$_{3.3}$Al$_{0.7}$]Al$_2$O$_{10}$(OH)$_2$

[a] smectite composition data and experimental μ^o values from Mattigod and Sposito (23).

[b] Palabora and Libby data from Kittrick (28); Harps data from Henderson et al. (29).

y-intercepts in Figures 7 and 8 lead to estimates of -5237 and -5383 kJ mol^{-1} for the μ^o values of pyrophyllite and talc, respectively. These estimates compare well with the observed values, -5268 and -5543 kJ mol^{-1} (15).

ACKNOWLEDGEMENTS

Gratitude is expressed to Dr. Shas V. Mattigod for assistance with the calculations reported in the Appendix and to Professors H. C. Helgeson and J. I. Drever for helpful comments. This research was supported in part by the Kearney Foundation of Soil Science and by a travel grant from the University of California, Riverside.

REFERENCES

1. Sposito, G. 1981, "The Thermodynamics of Soil Solutions", Clarendon Press, Oxford.

2. Sposito, G. 1984, "Thermodynamics of the soil solution", in "Soil Physical Chemistry", D. L. Sparks, ed., C. R. C. Press, Boca Raton.

3. Lindsay, W. L. 1979, "Chemical Equilibria in Soils", John Wiley, New York.

4. Stumm, W. and Morgan, J. J. 1981, "Aquatic Chemistry", John Wiley, New York.

5. Lewis, G. N. and Randall, M. 1923, "Thermodynamics and the Free Energy of Chemical Substances", McGraw-Hill, New York.

6. Hemingway, B. S. 1982, "Gibbs free energies of formation of Bayerite, Nordstrandite, Al(OH)$^{2+}$, and Al(OH)$_2^+$, aluminum mobility, and the formation of bauxites and laterites," Advan. Phys. Geochem. 2, pp. 285-316.

7. Dixon, J. B. and Weed, S. B. 1977, "Minerals in Soil Environments", Soil Science Society of America, Madison.

8. Kittrick, J. A. 1980, "Gibbsite and kaolinite solubilities by immiscible displacement of equilibrium solutions," Soil Sci. Soc. Am. J. 44, pp. 139-142.

9. Karathansas, A. D. and Hajek, B. F. 1983, "Transformation of smectite to kaolinite in naturally acid soil systems: Structural and thermodynamic considerations," Soil Sci. Soc. Am. J. 47, pp. 158-163.

10. Karathanasis, A. D., Adams, F. and Hajek, B. F. 1983, "Stability relationships in kaolinite, gibbsite, and Al-hydroxy interlayered vermiculite soil," Soil Sci. Soc. Am. J. 47, pp. 1247-1251.

11. Bourrie, G. and Grimaldi, C. 1979, "Premiers résultats concernant la composition chimique des solutions issues de sols bruns acides sur granite en climat tempéré océanique: Données naturelles et expérimentales," in "Migrations Organo-Minérales dans les Sols Tempérés", Colloques internationaux du C.N.R.S., no. 303, pp. 41-48.

12. Singer, A. and Navrot, J. 1977, "Clay formation from basic volcanic rocks in a humid Mediterranean climate," Soil Sci. Soc. Am. J. 41, pp. 645-650.

13. Aurousseau, P., Curmi, P. and Charpentier, S. 1983, "Les vermiculites hydroxy-alumineuses dans des sols et les formations d'alteration du massif Armoricain: Approches minéralogique, microanalytique et thermodynamique," Geoderma 31, pp. 17-40.

14. Brindley, G. W. 1980, "Order-disorder in clay mineral structures," in "Crystal Structures of Clay Minerals and their X-ray Identification", G. W. Brindley and G. Brown, eds., Mineralogical Society, London, pp. 125-195.

15. Wagman, D. D., Evans, W. H., Parker, V. B., Schumm, R. H., Halow, I., Bailey, S. M., Churney, K. L. and Nuttall, R. L. 1982, "The NBS tables of chemical thermodynamic properties," J. Phys. Chem. Ref. Data 11, Supp. 2, pp. 1-392.

16. Calvert, C. S., Buol, S. W. and Weed, S. B. 1980, "Mineralogical characteristics and transformations of a vertical rock-saprolite-soil sequence in the North Carolina Piedmont: II. Feldspar alteration products-their transformations through the profile," Soil Sci. Soc. Am. J. 44, pp. 1104-1112.

17. Weaver, C. E. and Pollard, L. D. 1973, "The Chemistry of Clay Minerals", Elsevier, Amsterdam.

18. Misra, U. K. and Upchurch, W. J. 1976, "Free energy of formation of beidellite from apparent solubility measurements," Clays and Clay Min. 24, pp. 327-331.

19. Bassett, R. L., Kharaka, Y. K. and Langmuir, D. 1979, "Critical review of the equilibrium constants for kaolinite and sepiolite," in "Chemical Modeling in Aqueous Systems", E. A. Jenne, ed., Am. Chem. Soc., Washington, D.C., pp. 389-400.

20. Marshall, C. E., Chowdhury, M. Y. and Upchurch, W. J. 1973, "Lysimetric and chemical investigations of pedological changes. Part 2. Equilibration of profile samples with aqueous solutions," Soil Sci. 116, pp. 336-358.

21. Misra, U. K., Upchurch, W. J. and Marshall, C. E. 1976, "Lysimetric and chemical investigations of pedological changes: Part 3. Relative changes in the potassium- and magnesium-treated profiles and leaching losses," Soil Sci. 121, pp. 323-331.

22. Misra, U. K., Upchurch, W. J. and Marshall, C. E. 1976, "Lysimetric and chemical investigations of pedological changes: Part 4. Mineral equilibria in relation to potassium- and magnesium-enriched environment in the profile," Soil Sci. 122, pp. 25-35.

23. Mattigod, S. V. and Sposito, G. 1978, "Improved method for estimating the standard free energies for formation ($\Delta G_{f,298.15}^o$) of smectites," Geochim. et Cosmochim. Acta 42, pp. 1753-1762.

24. Kirkland, D. L. and Hajek, B. F. 1972, "Formula derivation of Al-interlayered vermiculite in selected soil clays," Soil Sci. 114, pp. 317-322.

25. Veith, J. A. and Sposito, G. 1977, "Reactions of aluminosilicates, aluminum hydrous oxides, and aluminum oxide with o-phosphate: The formation of X-ray amorphous analogs of variscite and montebrasite," Soil Sci. Soc. Am. J. 41, pp. 870-876.

26. Pavan, M. A., Bingham, F. T. and Pratt, P. R. 1984, "Redistribution of exchangeable calcium, magnesium, an aluminum following lime or gypsum applications to a Brazilian Oxisol, "Soil Sci. Soc. Am. J. 48, pp. 33-38.

27. Nordstrom, D. K. 1982, "The effect of sulfate on aluminum concentrations in natural waters: Some stability relations in the sytem $Al_2O_3-SO_3-H_2O$ at 298 K," Geochim. et Cosmochim. Acta 46, pp. 681-692.

28. Kittrick, J. A. 1973, "Mica-derived vermiculites as unstable intermediates," Clays and Clay Min. 21, pp. 479-488.

29. Henderson, J. A., Doner, H. E., Weaver, R. M., Syers, J. K. and Jackson, M. L. 1976, "Cation and silica relationships of mica weathering to vermiculite in calcareous Harps soil," Clays and Clay Min. 24, pp. 93-100.

30. Lippmann, F. 1977, "The solubility products of complex minerals, mixed crystals, and three-layer clay minerals," N. Jb. Miner. Abh. 130, pp. 243-263.

31. Lippmann, F. 1982, Proc. Int. Clay Conf. 1981, "The thermodynamic status of clay minerals," pp. 475-485.

32. Aagard, P. and Helgeson, H. C. 1983, "Activity/composition relations among silicates and aqueous solutions: II. Chemical and thermodynamic consequences of ideal mixing of atoms on homological sites in montmorillonites, illites, and mixed-layer clays," Clays and Clay Min. 31, pp. 207-217.

MULTICOMPONENT SOLID SOLUTIONS FOR CLAY MINERALS AND COMPUTER MODELING OF WEATHERING PROCESSES

Bertrand Fritz

Centre de Sédimentologie et de Géochimie de la Surface
C.N.R.S. / Institut de Géologie
1 rue Blessig - F 67084 STRASBOURG Cedex - France

ABSTRACT

Clays are very important secondary minerals in the weathering processes. They are non stoichiometric phases and react with aqueous solutions as high-order solid solutions. Their composition may change as a function of the composition of the aqueous phase, during or after their precipitation. This explains variations in clay composition from one geochemical environment to another, and also within the population of particles found in apparently homogeneous clay samples. Thermodynamics of equilibrium between multicomponent solid solutions and aqueous solutions is extremely complex and a general theory is not yet available with applicable data for the mixing energies of several cations in different structural sites. However, a computer program based on a very simple molecular mixing model has been elaborated and appears to be very helpful and extendable for the studies of clay mineral formation and interaction with natural solutions.

INTRODUCTION

In natural environments, water-rock interactions are responsible for most of the alterations of primary minerals. The chemical mass transfer induced by these processes modifies the composition of the aqueous phase and secondary minerals may be produced; among those, clay minerals are often dominant and play a very important role for almost all the major chemical elements involved. Unfortunately, clay minerals such as smectites, illites and chlorites, are all non stoichiometric phases and cannot be described by one characteristic formula as with salts and simple minerals. Furthermore, recent microprobe analyses have clearly shown that their chemical variability is not only obvious from one geochemical system to another (different montmorillonites in different weathering conditions, see Weaver and Pollard (33)), but also within one clay sample, from one particle to another, each sample appearing as a population of particles of more or less different composition (Velde (31); Carmouze et al. (4), Pedro et al. (21), Tardy et al. (28), Duplay (7)). For all these reasons, the geochemical approach of clay-water interactions is not easy, particularly if one tries to apply the now classical modeling approach (calculation of aqueous ion speciation, activities, saturation indices, and simulation of mass transfer due to irreversible dissolution reactions..., see Helgeson et al. (15), Fritz (8) and (9)). Thermodynamically, non stoichiometric minerals can be interpreted as solid solution phases:

- the chemical composition of any montmorillonite, as an example, is formally equivalent to a simple mixing of a series of stoichiometric end-members (pyrophyllite, muscovite, talc, phlogopite...).
- the chemical stability of such a phase with respect to the separate end-members must be defined in terms of enthalpies, entropies and volumes of mixing.

In practice, layer silicates present at least two or three possibilities for cation substitution within their structure (octahedral and/or tetrahedral sites) or between the layers (interlayer sites). Consequently the number of possible end-members is generally very high and their combination to produce one given clay is not unique. For that reason it is very difficult to solve the thermodynamic problem of their mixing, and clay minerals are only poorly taken into account in geochemical modeling; there are considered one by one using lists of different clays with a fixed composition and an experimental or estimated stability for each of them. This was done by Fritz (8) to model mass transfer in weathering profiles on granitic rocks, and appeared to be limited in applicability. If one uses a short data file for clays (one montmorillonite, one illite, one glauconite etc...) the simulation is only poorly representative of natural clays. On the other hand a large data file with provision for all types and compositions leads to a very "heavy" model, with unreasonable times and prices for the calculations. However, our tests of a very simply defined solid solution model have given very interesting results: simulated alteration sequences show secondary clay phases whose composition interact with the variations in the activities of aqueous species, due to different irreversible processes: dissolution of minerals, evaporation of water, or temperature variations.

I. CLAY MINERALS AND THERMODYNAMICS

1. From Binary to Multicomponent Solid Solutions

For a given total chemical composition, the difference between a solid solution of n end-members and the simple mixing of the same pure minerals is the mixing energy. For a solid solution containing m_i moles of each mineral M_i, the Gibbs free energy of formation is:

$$G_f^o = (\sum_{i=1,n} X_i G_{f,i}^o) + G_{f,mix}^o \quad (1)$$

If the mixing term is negative, the solid solution is more stable in a given solution than the separate pure end-members, and can be precipitated or preserved in a solution which is undersaturated with respect to all the end-members; an aluminous smectite, for example, can be stable when pyrophyllite and muscovite are not stable. Many works have been published by chemists and mineralogists on the theoretical problem of solid solution. Very few, however, concern the complex clay phases, particularly at low temperatures. General equations for solid solutions (essentially binary) were developed by Wagner et al. (32) and Prigogine and Defay (24). Solid solutions in rock-forming minerals were discussed by Saxena (26), Kerrick and Darken (17), and Grover (14) and Powell (23) developed a theoretical approach for multiple series. These authors give different equations for calculating mixing energies but generally not the thermodynamic constants (such as Margule's coefficients) one needs in order to apply practically these functions. This would need additional experimental calibrations or thermodynamic estimations in the range of temperature, pressure and aqueous chemistry considered. Data are available for binary systems such as magnesian calcites (De Boer (5); Thorstenson and Plummer (30); Lafon (18); Garrels and Wollast (13) and Berner (2) among others), other binary carbonates (Ca,Mn,Fe,Mg) studied by Lippmann (19) or Ba-Sr sulfates by Michard and Ouzounian (20).

The use of solid solutions in geochemical modeling has been introduced by Helgeson et al. (15) for binary ideal and regular solutions in the program PATHCALC. Pfeifer (22) used it for simulating the formation of alkaline solutions by weathering of ultramafic rocks. Different binary phases were tested: calcite-dolomite, forsterite-fayalite, talc-minnesotaite, and chrysotile-greenalite. In all these phases, the substitution of chemical elements concerned only one type of structural sites. Kerrick and Darken (17) showed the complexity of the problem if different kinds of sites may interact energetically in the same solid solution of oxides or chain and framework silicates. If the sites are energetically independent the total mixing energy is equal to the sum of the different mixing energies for the respective sites. However, if sites interact, cross terms must be introduced in the equations. This has been well documented by Helgeson et al. (16). The first applicable model was proposed by Stoessell (27) who used Kerrick and Darken's equations in a random site mixing model for illites containing only one cation in the interlayer site (potassium) and no ferric iron in the octahedral sites. In this model, the mixing of cations within one type of sites is ideal but the mixing of the molecular end-members (pyrophyllite, talc ...) is not ideal: the activities of these end-members are not equal to their mole fractions. As an example, the activity of pyrophyllite is related to the mole fraction of both pyrophyllite and muscovite:

$$A_{pyr} = X_{pyr}(X_{pyr}+X_{mus})^3(3 + X_{pyr})^4/4^4 \qquad (2)$$

As a conclusion of this approach, the author found an apparent excess of stability given by the model to the trioctahedral end-members, and discussed the possibility of extending the model to a regular site mixing. But, here again, this needs additional data to be able to calculate the excess of mixing energies and also to take into account ferric iron.

Finally, the latest attempt to give a theory for activity-composition relations among silicates and aqueous solutions is being developed now by Aagaard and Helgeson (1) who base their approach on random mixing approximations for cations in energetically equivalent sites in dioctahedral illites, montmorillonites and mixed-layer clays:
   (1) M(2) sites are supposed energetically equivalent and completely filled in the dioctahedral series.
   (2) M(1) sites are supposed essentially vacant.
   (3) Disorder is assumed in the tetrahedral sites.
The mineral phases are supposed to form rapidly at low temperatures, in a metastable state of disorder.

2. Clay Minerals and the Mass Action Law

The equilibrium condition between a solid solution and an aqueous solution is described by the mas action law applyed to all the end-members of the solid solution:

$$K_j = (Q_j/A_j) = (Q_j/X_j\lambda_j) \qquad (3)$$

with K_j = solubility product of the jth end-member
    Q_j = ion activity " " " ". " "
    λ_j = activity coefficient of the jth end-member
    X_j = mole fraction of the jth end-member
    A_j = activity of the jth end-member
Three questions appear immediately if one tries to apply these constraints: (i) how many end-members j do we have to consider?
(ii) what is the activity-mole fraction relation for each of these end-members? (iii) what is the solubility product K_j of each mineral j? The first question relates to the Gibbs phase rule. Let us

consider n moles of a 2:1 clay:

$$(Si_{4-x}Al_x)_T(Al_{2-y-\frac{2z}{3}} Fe^{III}_y Mg_z)O_{10}(OH)_2(K_a Na_b Ca_c Mg_d) \quad (4)$$

with x = a + b + 2c + 2d

without octahedral charge. This phase is entirely described by 7 variables (n,x,y,z,a,b,c) and we need 7 relations between the solid and liquid phases to define the composition of a clay phase in equilibrium with the aqueous ions. However, the number of possible stoichiometric end-members is 15:
- 3 end-members without tetrahedral substitution and with Al-Fe^{III}-Mg octahedral substitution;
- 12 end-members with Si-Al tetrahedral substitution, Al-Fe^{III}-Mg octahedral substitution and 4 possible cations in the interlayer sites (K,Na,Ca,Mg).

So, before discussing any equilibrium condition, ideal or non ideal behavior, we have to make a choice:
- Do we consider the *minimum number* of stoichiometric end-members sufficient to build by mixing *one* given clay formula (here 7 minerals)? Than we can write 7 *independent* mass action law equations, but we have a lot of possibilities for the choice of end-members.
- Do we consider the *total number* of possible end-members (here 15)? This set will then be convenient to produce *all* the clays corresponding to the general formula above, and is *unique*, but the 15 equations are *not independent*.

The first choice was made by Stoessell (27) for illites and one can see on figure 1 that, even for a very simple system (muscovite, pyrophyllite, phlogopite and talc) there are 4 different ternary solid solutions, and always two possibilities for any given composition. Stoessell chose pyrophyllite, muscovite and phlogopite for the Al-Mg system, with addition of annite for the ferrous iron content. Talc could have been chosen instead of pyrophyllite or phlogopite, and minnesotaite instead of annite or pyrophyllite. These possibilities have not been discussed by the author. Fritz (9) showed that the two possibilities concerning the Al-Mg substitution are not equivalent. With an assumption of ideal mixing in the random site mixing model, the equivalence needs a particular relation between the solubility products of the end-members:

$$\frac{K(muscovite)}{K(pyrophyllite)} = \frac{K(phlogopite)}{K(talc)} \quad (5)$$

which is not verified by available data like those given by Helgeson et al. (16) or Robie et al. (25). This relation means that the tetrahedral Si-Al substitution is energetically independent from the octahedral Al-Mg substitution.

The number of possible choices for the necessary independent end-members among all the possible ones, increases rapidly with the number of substituing cations and makes the choice very difficult,

General formula : $(Si_{4-x}Al_x)_T(Al_{2-\frac{2y}{3}}Mg_y)O_{10}(OH)_2K_x$

Muscovite — Phlogopite
$Si_3AlAl_2O_{10}(OH)_2K$ — $Si_3AlMg_3O_{10}(OH)_2K$

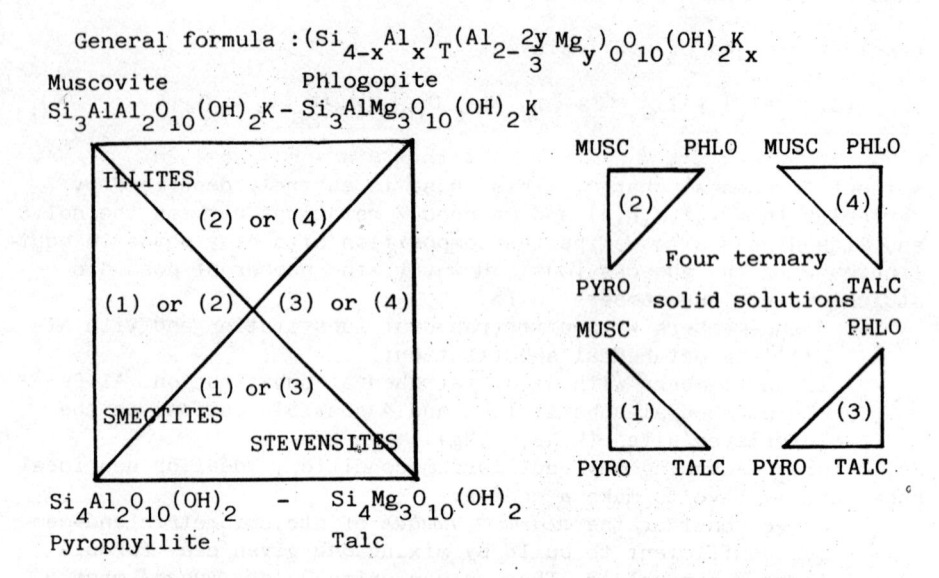

Fig. 1. Possible ternary solid solutions in the pyrophyllite-muscovite-talc-phlogopite system.

while the consequences of this choice are not the same with the various options. For that reason we used the second possibility, considering the unique solid solution with all the end-members. This has been made with a first ideal approximation, but may be extended, taking into account thermodynamically described departures from ideality.

The equilibrium condition is written for all the possible end-members and the mole fraction X_j of one end-member j is calculated from the saturation state of this mineral in the solution:

$$K_j = \frac{Q_j}{A_j} = \frac{Q_j}{X_j} \cdot \frac{1}{\lambda_j} \tag{6}$$

with $\lambda_j = 1$, in a first approximation (ideal mixing)
These n equations are not independent because the Q_j are related in the aqueous solution. For example:

$$\frac{Q(pyrophyllite)}{Q(talc)} = \frac{(Al^{3+})^2}{(Mg^{2+})^3} = \frac{Q(muscovite)}{Q(phlogopite)} \tag{7}$$

which combines with equation (6) to give:

$$\frac{X_{pyro} \cdot X_{phlo}}{X_{musc} \cdot X_{talc}} \cdot \frac{\lambda_{pyro} \cdot \lambda_{phlo}}{\lambda_{musc} \cdot \lambda_{talc}} = \frac{K_{musc} \cdot K_{talc}}{K_{pyro} \cdot K_{phlo}} \tag{8}$$

For the simple Al-Mg system described on figure 1, this equation adds to the mass balance equations:

$(Si) = 4-x = 4(Xpyr + Xtalc) + 3(Xmusc + Xphlo)$ (9)
$(Al)octahedral = 2 - 2(y/3) = 2(Xpyr + X musc)$ (10)
$(Mg)octahedral = y = 3(Xtalc + Xphlo)$ (11)

This four equations-four unknowns system is always soluble and gives one solution and only one (Fritz (9) and Tardy et al. (28)). This result can be generalized to a multicomponent solid solution of n end-members. For a given set of thermodynamic data (K_j) and defined functions for activities (A_j, ideal or not):
 -any distribution of mole fractions X_j deduced from saturation tests of the end-members in the aqueous solution ($X_j = Q_j / \lambda_j K_j$ and $\Sigma X_j = 1$ as saturation condition) defines only one clay.
 -one clay formula corresponds to only one distribution of the n mole fractions X_j consistent with the chosen solid solution model.
This double relation means mathematically that the system is not undefined even if one uses more equations than the minimum number really necessary. Two small computer programs are used in order to calculate the equilibrium conditions:
 -ARSAT calculates clay formulae in equilibrium with given aqueous solutions.
 -CISSFIT calculates activities of aqueous species in equilibrium with a given clay.
At present these models use only the ideal mixing assumption and all the minerals constituing the solid solution by mixing have a mole fraction equal to their saturation index in the aqueous solution. The solubility products used in this multicomponent solid solution model have been published by Tardy and Fritz (29). They have been estimated partially by a trial and error method, using the computer programs mentioned above, to fit at low temperature with experimental data obtained for different type of phyllosilicates.

3. Multicomponent Solid Solutions and Mass Transfer Calculations

In computer programs for mass transfer calculations, the detection of a new mineral phase at equilibrium is followed by an increase in the number of unknowns and equations. For a monomineralic phase one considers only one new constraint (equilibrium condition) and one unknown (the number of moles of this mineral per kg H_2O). For a solid solution phase with n end-members, n equations are added to the system (equilibrium conditions discussed above), corresponding to the n new unknowns (numbers of moles of the n end-members per kg H_2O). At a given stage of a simulated process, a solid solution is entirely known if one can calculate these "molalities" x_j of the end-members. Mole fractions X_j are then calculated from the total number of moles of minerals produced in the solid solution:

$$x = \Sigma x_j \qquad (12)$$
$$X_j = x_j/x \quad \text{for } j=1,n \qquad (13)$$

These molar variables were used by Helgeson et al. (15) to introduce solid solutions in the PATHCALC model. This choice, convenient for binary solutions, leads to extreme difficulties in the calculations of multicomponent solutions when some end-members are minor in the mixing (x_j very small) and particularly in the initial stages of precipitation of the complex phase (all x_j very small). Under the same conditions, the mole fractions X_j and the total number of moles x have generally more significant values and where selected as calculation variables in the programs DISSOL, EVAPOR and THERMAL (9). In practice, the considered unknowns are x and $(n-1)$ values of X_j. The last mole fraction X_n is of course related to the others ($X_j = 1 - \Sigma X_k$ for all k different of j) and all the molar variables can be calculated easily ($x_j = x \cdot X_j$). This change of variables simplifies the equations necessary for the derivation of the mole fractions in the following equations:

$$\frac{dLn\, X_j}{d\xi} = \frac{x'_j}{x_j} - \frac{X_j}{x_j}(x'_j) \qquad (14)$$

$$\frac{d^2 Ln\, X_j}{d\xi^2} = \frac{x''_j}{x_j} - \frac{X_j}{x_j}(x''_j) - \frac{x'^2_j}{x_j^2} + \frac{X_j}{x_j^2}(x'_j)^2 \qquad (15)$$

which become:

$$\frac{dLn\, X_j}{d\xi} = \frac{1}{X_j} \cdot \frac{dX_j}{d\xi} = \frac{X'_j}{X_j} \qquad (16)$$

$$\frac{d^2 Ln\, X_j}{d\xi^2} = \frac{1}{X_j} \cdot \frac{d^2 X_j}{d\xi^2} - \frac{1}{X_j^2} \cdot \frac{dX_j}{d\xi} \cdot \frac{dX_j}{d\xi} = \frac{X''_j}{X_j} - \frac{X'^2_j}{X_j^2} \qquad (17)$$

These relations involve only the mole fractions and their derivatives, and are applicable even for early precipitation stages (X_j are known, but x_j are still equal to zero). They allow simulation of the production of a mineral phase whose composition changes during the precipitation, following equilibrium conditions with a changing aqueous composition: *all the end-members try to participate in the solid solution phase (mole fractions X_j) to the extent that their respective saturation indices ($A_j = Q_j/K_j$) increase in the aqueous solution.*

II. APPLICATIONS

1. Clay Minerals Produced in Simulated Weathering of Granitic Rocks

The computer program DISSOL, with an option for multicomponent solid solution, has been applied to the calculation of weathering reactions of granitic rocks. These simulations give three major results:
- the continuous evolution of the chemical composition of the aqueous phase with an increasing extent of weathering;
- the stability sequence for primary minerals which defines the relative "solubilities" of the rock-forming minerals under weathering conditions;
- the stability sequence for secondary minerals, which shows the progressive production of minerals by alteration, and particularly clay minerals.

Results concerning the aqueous chemistry will not be discussed here. They give us a description of the total range of possible chemical composition of natural waters due to weathering mass transfer (see Fritz (8) and (9), Fritz and Tardy (29)), in areas where human activity and industrial pollution do not interfere with natural processes.

2. Physico-Chemical Conditions for the Simulation

The granite considered represents the so called "Granite des Crêtes" from the Vosges Massif (northeast France) described in the simulation by six major constituents (table 1).

Table 1. Chemical and mineralogical composition of the granite used in the simulated alteration.

Mineral	weight %	mole %	Oxides	weight %
Quartz	17.17	53.9	SiO_2	66.05
Microcline	26.58	18.0	Al_2O_3	12.60
Albite	21.83	15.7	FeO	4.69
Anorthite	6.49	4.4	MgO	4.17
Annite	11.13	4.1	CaO	3.63
Tremolite	16.80	3.9	K_2O	5.52
total =	100.00	100.0	Na_2O	2.58
			H_2O	0.76
			total =	100.00

The initial reacting solution is assumed to be a very pure rain water, with negligible alkalinity (HCO_3^- dominant and equal to H^+) and cation molalities (Na,K,Ca,Mg). This solution is equilibrated with soil or underground conditions for CO_2 (fCO_2 equal to 0.00316 instead of 0.000316 commonly in atmosphere). This gives an initial pH equal to 5.15 at 25°C. Redox conditions are oxidizing (Eh = 550 mV for pH = 5.15 and $logfO_2$ = -25.). These conditions may represent subsurface conditions for weathering processes.

3. Predicted sequences for Mineral Stabilities

The dissolution of granite is almost immediately incongruent, due to early precipitation of iron and aluminium oxides; micromoles of mineral dissolution (per kg H_2O) are sufficient. This initial stage of alteration is very probable in all weathering environments: the solution is progressively enriched in silica and some cations (Na,Ca,Mg,K), while the dissolved aluminium and iron ions are quantitatively fixed in the secondary phases (iron oxyhydroxides such as goethite FeOOH and gibbsite $Al(OH)_3$). The cation concentration increases go together with an alkalinity increase (essentially bicarbonate and some carbonate), while the pH goes up to the 6.5-7.5 range. This is a general sequence of weathering that can be summarized for the mass transfer to the solution by:

$$Rock + x\ CO_2 + y\ H_2O = x\ HCO_3^- + x\ cation\ equivalents + a\ SiO_2 + b\ FeOOH + c\ Al(OH)_3$$

The carbon dioxide is here the major alteration agent due to its acid effect ($CO_2 + H_2O = H^+ + HCO_3^-$), and also explains the alkalinity in common natural waters. This carbon dioxide comes either directly from the atmosphere, or from the oxidation of organic matter in soils. This is illustrated in waters sampled in granitic areas from springs or rivers:
- in small drainage basins in the Vosges Massif (France) as shown by Bourrié (3),Fritz (10) and Fritz et al. (11);
- in large basins like in the Lake Chad system for the Chari River as discussed by Gac (12).

The primary minerals are progressively stabilized with respect to the solution and are no longer dissolved in the simulation. The order of saturation is the following: quartz, microcline, tremolite albite (figure 2). The silica concentration in the solution can be used as a guide to place a given natural water in this theoretical sequence. The saturation of quartz corresponds to an early stage of alteration (SiO_2 = 6 mg/l at 25°C). The saturation point of microcline (SiO_2 = 30 to 50 mg/l at 25°C) corresponds to a more intense alteration. Natural waters mentioned above range between these values: 10 mg/l in the Vosges Massif (at about 10°C mean annual temperature) and 20 to 40 mg/l in Chad (at 25-30 °C). After the microcline saturation point silica concentration is stabilized

```
Primary minerals reaching saturation
    Quartz     -----------------------|
    Microcline ---------------------------------|
    Tremolite  ------------------------------------|
    Albite     -------------------------------------------|

Secondary minerals precipitated
            Goethite    |---------------|
            Gibbsite    |--|
            Kaolinite       |------------|
            Smectites                    |---------------- - - -
            Calcite                          |------------ - - -
Ph       5.15                     6.0    7.0     7.75
        |    |    |    |    |    |    |    |    |
        -9.  -8.  -7.  -6.  -5.  -4.  -3.  -3.  -2.
                        Log (Reaction Progress  )
```

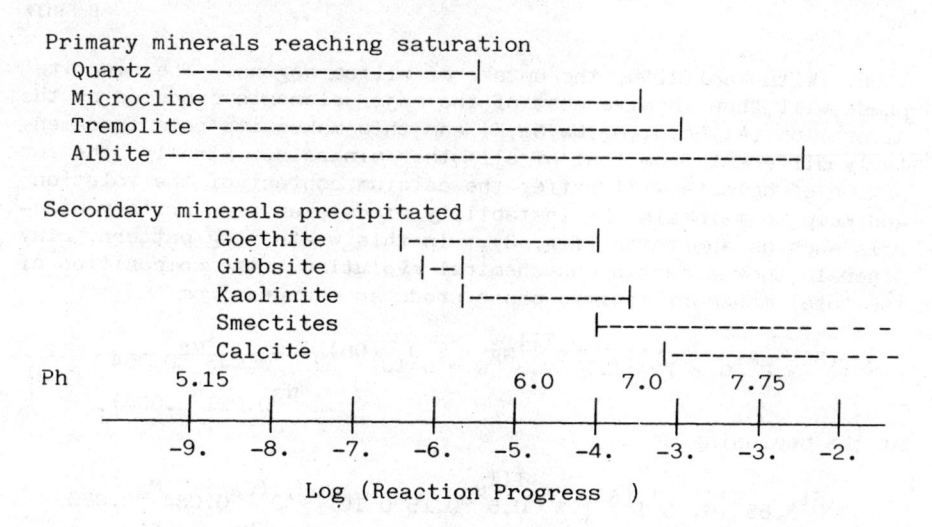

Fig. 2. Simulated sequences or primary minerals and secondary minerals saturation with corresponding pH values.

because less silica is now being dissolved, and silica is also precipitated in kaolinite and finally in smectites. The saturation point of albite would correspond to very intense alteration of the rock in an almost closed system. At this point anorthite and annite woult even not be stable. This explains the important instability of these minerals under common weathering conditions; Ca-rich plagioclases and biotites observed in thin sections of weathered granites are always much more altered than all other constituents. Reducing conditions could stabilize biotites but we found in the simulation that this would need Eh values as low as -200 to -400 mV. Such conditions are not often possible in subsurface conditions at low temperatures (0 to 30 °C).

This thermodynamic sequence of stability of rock-forming minerals explains what has been often observed in weathered granites by petrologists: quartz and microcline are generally well preserved, or slightly altered, amphiboles and albite or sodic plagioclases are moderately altered, anorthites and calcic plagioclases are highly altered. Minerals produced by these alterations are also ordered in a sequence by the simulation:
   -gibbsite + goethite
       -goethite + kaolinite
           -kaolinite + smectites
               -smectites + calcite...

In this sequence chlorite-type clays do not appear, but they were tested and saturation was not reached. The beginning of the sequence is essentially an uptake af aluminium and iron produced by alteration, which are very "insoluble" in the alteration solution. With kaolinite, the uptake of silica begins. The smectite phase will then involve most of the major elements produced by the alteration (Al,Fe,Si,Mg,Ca,Na,K) and this makes its role fundamentally different from that of all other minerals. Finally, the formation of calcite will buffer the calcium content of the solution and help to maintain the instability of calcium-rich primary minerals such as anorthite (fig. 3). In this weathering pattern, clay minerals show a continuous chemical evolution. the composition of the total montmorillonitic phase produced changes from

$$(Si_{3.8}Al_{0.2})_T(Al_{1.7}Fe^{III}_{0.3}Mg_{0.0})O_{10}(OH)_2(Ca_{0.043}Mg_{0.054} Na_{0.001}K_{0.006})_I$$

at the beginning to

$$(Si_{3.65}Al_{0.35})_T(Al_{1.3}Fe^{III}_{0.6}Mg_{0.15})O_{10}(OH)_2(Ca_{0.084}Mg_{0.080} Na_{0.004}K_{0.018})_I$$

at the saturation point of albite. The distribution of cation in the different type of sites is not constant; it varies with the variations of the corresponding cation/proton activity rations in solution ($(Al^{3+})/(H^+)^3$; $(K^+)/(H^+)$ etc...) and with the activity

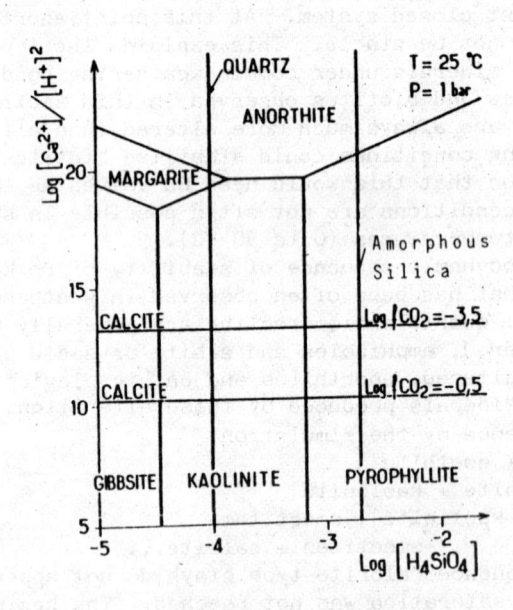

Fig. 3 The calcite barrier to anorthite stabilization under low temperature subsurface conditions.

of aqueous silica. It is important to observe here that this model does not predict "crazy" clays like pure pyrophyllite, muscovite or talc, but smectites with realistic compositions; the evolution of this clay phase is continuous; variations are progressive and limited. For a given chemistry of the system, the solid solution model gives directly the composition of a clay consistent with the system and then adjusts it slowly corresponding to variations in the system. This seems to be a very good representation of an important chemical buffering effect of the clay on the system. These results obtained in a closed system with a fixed bulk composition do not mean that drastic variations are impossible, but those need major variations of the activity of at least one cation or silica in the solution. This is possible for elements present in the solution only as traces (e.g. ferric iron, even ferrous iron and aluminium) which are, however, major elements in some primary minerals being altered. Also local availability of one of these elements does not automatically imply local high aqueous activity unless particular physico-chemical conditions are realized (pH, Eh, pO_2, pCO_2, water activity). This problem of particular conditions in microenvironments is being investigated now in the laboratory.

4. Influence of Temperature

The same simulations have been conducted for different temperatures between 0 and 200°C. The results have been discussed in a study of hydrothermal alteration and diagenetic transformation by Fritz (9). It is interesting to recall here the general trend obtained for clay mineral composition with increasing temperature:
  (1) The Si-Al substitution in tetrahedral sites increases. Montmorillonites are formed at low temperature (Si_4 to $Si_{3.6}Al_{0.4}$) and illites appear more stable at higher temperature ($Si_{3.5}Al_{0.5}$ to $Si_{3.1}Al_{0.9}$)
  (2) Aluminium is dominant in the octahedral sites under oxidizing conditions, while iron is present as Fe^{3+} in goethite and in the clay. Under reducing conditions ferrous iron may be present also in the octahedral sites but will follow magnesium into the chlorite phase, whose stability increases rapidly from 25 to 100°C. At low temperature kaolinite can remain stable in the smectite stability field but the quantity produced is not important. At higher temperatures, kaolinite is less stable than the illite.
  (3) The proportion of potassium in the interlayer sites increases rapidly. At low temperatures (below 60°C) calcium and magnesium are the dominant exchangeable cations of montmorillonites produced in alteration profiles. Sodium may be fixed too, but only in restricted environments (see Droubi (6)) or in sea water. In contrast, at higher temperature, potassium becomes dominant, followed by sodium

calcium and magnesium. Furthermore, this occurs with almost all type of waters, except brines for which calculation of ion activities require a new approach.
A comparison of clays obtained by simulated alteration of the same granite (table 1) at different temperatures is given in table 2.

Table 2. Different clay compositions obtained by simulated alteration of the same granitic rock at 25, 60 and 150°C.

T(°C)	Tetrahedra Si	Al	Octahedra Al	Fe^{III}	Mg	Interlayer sites Ca	Mg	Na	K
25	3.65	0.35	1.30	0.60	0.15	0.084	0.080	0.004	0.018
60	3.60	0.40	1.61	0.35	0.06	0.080	0.030	0.070	0.110
150	3.50	0.50	1.70	0.28	0.03	0.015	0.020	0.040	0.410

These calculations show that the model can be used as a guide to distinguish between clays produced by low temperature processes and clays related to hydrothermal events.

CONCLUSION

Clay minerals are complex non-stoichiometric mineral phases. Therefore the clay-aqueous solution interaction is not easy to understand in terms of thermodynamic equilibrium conditions. At present, the theoretical problem of energetic interaction between different sites for different cation substitution is not solved. Theoretical approaches have been proposed in the literature, but none is enough developed to be used in water-rock interaction models in order to simulate the precipitation of a multicomponent solid solution. However, our tests of a very simply defined model of solid solution, applied in the program DISSOL, give very encouraging results, even if the model is too simple thermodynamically. At present, ideal or regular solid solution models for multicomponent phases are probably only analog models rather than being realistic thermodynamic models, but they allow the introduction into water-rock interaction modeling, of mineral phases with much more realistic composition, which vary with the composition of the solution. This is fundamental for the study of chemical constraints in alteration processes. Furthermore, with these models "the door is open" for the introduction of accurate thermodynamic data for non-stoichiometric minerals.

ACKNOWLEDGMENTS

The author is deeply indebted to Alain Clément for the transformation of the program DISSOL with a multicomponent solid solution option. The numerous suggestions of Doctor James Drever were very helpful for improvement of the English.

REFERENCES

(1) Aagaard P. and Helgeson H.C.,1982,Amer.J.of Sci.,282,pp.237-285.
(2) Berner R.A.,1978,Amer.J.of Sci.,278,pp;1475-1477.
(3) Bourrié G.,1976,Sci.Géol.,Mém.,Strasbourg,52,pp.1-174.
(4) Carmouze J.P.,Pedro G; and Berrier J.,1977,C.R.Acad.of Sci., Paris,284,D,pp.615-618.
(5) De Boer R.B.,1977,Geochim.Cosmochim.Acta,41,pp.265-270.
(6) Droubi A.,1976,Sci.Géol.,Mém.,46,pp.1-177.
(7) Duplay J.,1982,*Thèse 3ème cycle,University of Strasbourg*,pp.1-110.
(8) Fritz B.,1975,Sci.Géol.,Mém.,Strasbourg,41,pp.1-152.
(9) Fritz B.,1981,Sci.Géol.,Mém.,Strasbourg,65,pp.1-197.
(10) Fritz B.,1982,Rech.Géogr.à Strasbourg,19-20-21,pp.179-184.
(11) Fritz B,Massabuau J.C. and Ambroise B.,1984,IAHS-AISH Publ., 150,pp.249-264.
(12) Gac J.Y.,1979,Trav.Doc. ORSTOM,Paris,123,pp.1-251.
(13) Garrels R.M. and Wollast R.,1978,Amer.J.of.Sci.,278,pp.1469-1474.
(14) Grover J.,1976,In *Thermodynamics in Geology*,D.Reidel Publ.Co., Fraser D.G.(Ed.),pp.67-98.
(15) Helgeson H.C.,Brown T.H.,Nigrini A. and Jones T.A.,1970,Geochim. Cosmochim.Acta,34,pp.569-592.
(16) Helgeson H.C.,Delany J.M.,Nesbitt H.W. and Bird D.K.,1978,Amer. J.of Sci.,278-A,pp.1-229.
(17) Kerrick D.M. and Darken L.S.,1975,Geochim.Cosmochim.Acta,39, pp.1431-1442.
(18) Lafon M.,1978,Amer.J.of Sci.,278,pp.1455-1468.
(19) Lippmann F.,1980,N.Jb.Miner.Abh.,139,1,pp.1-25.
(20) Michard G. and Ouzounizan G.,1978,C.R.Acad.Sci.,Paris,287,C, pp.183-189.
(21) Pedro G.,Carmouze J.P. and Velde B.,1978,Chem.Geol.,23,pp.139-149.
(22) Pfeifer H.R.,1977,Schweiz.Miner.Petr.,Mitt.,57,pp.361-396.
(23) Powell R.,1976,In *Thermodynamics in Geology*,D.Reidel Publ.Co., Fraser D.G.(Ed.),pp.57-66.
(24) Prigogine I. and Defay R.,1954,*Chemical Thermodynamics*,Longmans,Green and Co.(Ed.),London.
(25) Robie R.A.,Hemingway B.S. and Fisher J.R.,1978,Geol.Surv.Bull., 1452,pp.1-456.

(26) Saxena S.K.,1973,*Thermodynamics of rock-forming crystalline solutions*,Springer Verlag (Ed.),New York.
(27) Stoessell R.K.,1979,Geochim.Cosmochim.Acta,43,pp.1151-1159.
(28) Tardy Y.,Duplay J. and Fritz B.,In *Developments in Sedimentology*,Van Olphen and Veniale (Ed.),35,pp.441-450,1981.
(29) Tardy Y. and Fritz B.,Clay Minerals,16,pp.361-373,1981.
(30) Thorstenson D.C. and Plummer L.N.,1978,Amer.J.of Sci.,277, pp.1203-1223.
(31) Velde B.,1977,Developments in Sedimentology,21,Elsevier.
(32) Wagner C.,Mellgren S. and Westbrook J.H.,1952,*Thermodynamics of alloys*,Addison Wesley Publ.Co.(Ed.),Reading,Mass.,pp.1-161.
(33) Weaver C.E. and Pollard L.D.,Developments in Sedimentology, 15,Elsevier.pp.1-213.

DISSOLUTION MECHANISMS OF PYROXENES AND OLIVINES DURING WEATHERING

Jacques Schott[1] and Robert A. Berner[2]

1. Laboratoire de Minéralogie et Cristallographie, Université Paul-Sabatier, 31062 Toulouse Cédex, France.

2. Department of Geology and Geophysics, Yale University, New Haven, CT 06511, U.S.A.

ABSTRACT

The mechanisms of dissolution of both iron-free and iron-rich pyroxenes and olivines have been studied by means of chemical analyses of reacting solutions and by the use of X-ray photoelectron spectroscopy.

Iron-free minerals (all conditions) and iron-rich minerals in the absence of oxygen underwent congruent dissolution after the initial formation of a thin (< 10Å) protonated layer depleted in Fe, Ca or Mg relative to Si. These results, combined with findings of high activation energies, rule out a control of dissolution by a solution-transport mechanism. Dissolution is strictly a surface process and occurs by a H^+ (or H_3O^+) exchange reaction at the surface.

Under oxygenated conditions dissolution of iron-rich minerals results in the formation of two surface layers. The outer layer is hydrated ferric oxide. The inner layer is probably a Fe^{+3} - Mg silicate and it is protective toward silica release. At pH 6 or lower this layer does not continue to grow with very long times of reaction. However, at higher pH values, such a protective layer might be stabilized and continue to grow and this might help explain the observations of, for example, layers of biopyribole type minerals on the surfaces of partly weathered pyroxenes.

INTRODUCTION

In this paper we address the important problem of whether or not Ca-Mg-Fe silicate dissolution during weathering proceeds via the formation of a diffusion-inhibiting protective surface layer. Results from recent studies both on feldspars and on Ca-Fe-Mg silicates [Lagache (8); Petrovic et al. (13); Holdren and Berner (6); Schott et al. (17); Schott and Berner (18)] suggest that during dissolution the rate limiting step is a surface-controlled reaction resulting consequently in a constant rate of dissolution for each of these minerals. On the other hand, previous studies on Fe-Mg-silicates [notably that of Luce et al. (10)] suggest that dissolution proceeds incongruently with the consequent formation of a residual protective layer of altered composition on the surface of the mineral. As a result so-called parabolic kinetics is found where the rate of dissolution decreases with time due to the growth in thickness of such surface layers. The layer may consist of:
- a Fe-Mg-Ca depleted, protonated version of the underlying silicate [as assumed by Luce et al. (10) for enstatite];
- a newly formed Fe-Mg-silicate of somewhat different structure type [Huguenin, (7); Eggleton and Boland (3)];
- a precipitate of hydrous ferric oxide [Siever and Woodford (19)].

In all cases, the layer is assumed to be protective in the sense that diffusion of dissolved species to and from the unaltered primary mineral surface is inhibited by the surface layer. In the case of iron-rich silicates, two more questions should be added: does oxidation of iron takes place within the pre-existing silicate structure or is iron released in the ferrous form and oxidized only when in the aqueous state? Last but not the least, is it possible to treat the observed experimental data within the framework of kinetic theories?

Using our recent experimental data [Schott et al. (17); Schott and Berner (18)], the aim of this work is to answer some of the above questions and more particularly to test the hypotheses of incongruent dissolution, altered surface composition and parabolic kinetics. In this work our previous studies on the dissolution in the laboratory of both iron-free and iron-rich silicates are coupled with our observations on partially weathered grains of pyroxenes and amphiboles from soils [Berner and Schott (2)]. Also, in order to get less equivocal results, classical solution chemical analyses are combined with careful studies of mineral surfaces (composition, oxidation state and structural form of iron, search for OH bonds etc...). By coupling this information with morphological studies of the silicate surfaces, it is possible to propose a model describing the first stages of Mg-Fe silicate dissolution both in the laboratory and during wheathering in soils.

METHODS

Details of methods of preparing and purifying mineral samples and solutions and analyses of products can be found in Schott et al. (17) and Schott and Berner (18). They will not be repeated here. A large effort was made to be sure that surfaces of minerals were free of surface defects and of adhering fine particles produced by grinding. Also, mineral purity was assured by hand-picking and chemical removal of traces of accompanying phases, and checks were made for purity using X-ray diffraction, optical petrography, and scanning electron microscopy. Checks for possible alteration of surfaces, due to chemical pre-treatment, were done using X-ray photoelectron spectroscopy, and after trial and error, those treatments were adopted which produced negligible alteration.

X-ray photoelectron spectroscopy (XPS) was used to examine the surface chemistry of reacted minerals, both in terms of XPS peak areas, which reflect the relative amounts of each element, and peak positions which reflect relative binding energies. Binding energies can be used to deduce important data on oxidation states, bonding environment, etc., and, thus, are very useful in the study of the role of surface chemistry during mineral dissolution. To check our XPS results, chemical analyses of the reacting solutions were also conducted by the usual means (see the above references for details).

RESULTS AND DISCUSSION

Silica release to solution

Iron-free minerals (all conditions), iron minerals dissolved at pH 1, and iron minerals dissolved at pH 6 under anoxic conditions. The course of dissolution of diopside, enstatite, augite and bronzite are given in Fig. 1a. It can be seen that after a minor, rapid initial reaction, the rate of silica release is constant during experiments (for more than 600 hours). This agrees well with other studies [Holdren and Berner (6); Grandstaff (5)], which also found that linear kinetics of dissolution resulted if samples were first pretreated to remove fine particles adhering to the mineral grains.

The effect of H^+ activity on the linear rate of silica release for enstatite, diopside and augite can be seen in Table 1. The results of Table 1 suggest an H^+ reaction with the solid surface as being the rate limiting step as proposed by Fogler et al. (4) for feldspar dissolution in HF-HCl solutions. More direct evidence of this H^+ reaction will be given later.

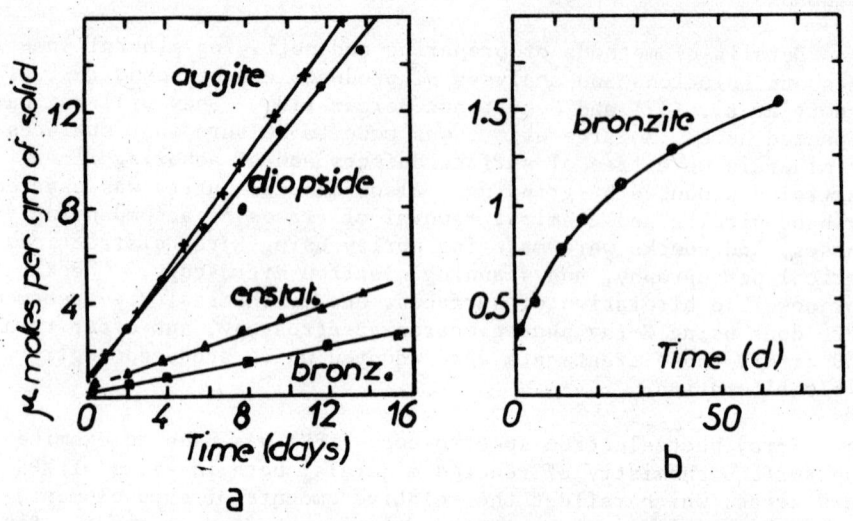

Fig.1.- Plot of amount of silica released vs time for the dissolution of etched enstatite, bronzite ($pO_2 = 0$), diopside and augite at pH 6 (a) and for bronzite in oxic conditions at pH 6 (b). T=50°C for enstatite, diopside and augite; 20°C for bronzite. After Schott et al. (17) and Schott and Berner (18)

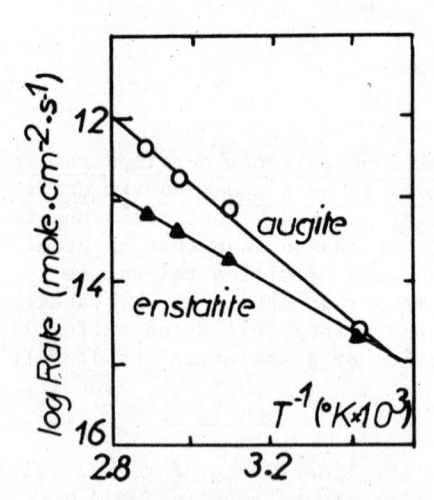

Fig.2.- Reciprocal temperature plot for silica release from etched enstatite and augite at pH 6. After Schott et al. (17)

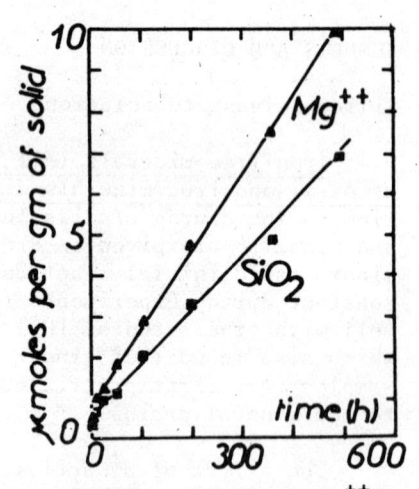

Fig.3.- Plot of released Mg^{++} and SiO_2 vs time for the dissolution of etched enstatite at pH 6, T = 20°C. After Schott et al. (17)

Table 1.— Dependence of the linear rate of silica release (r) on H^+ activity (T=20°C) [After Schott et al. (17), plus unpublished data for augite].

	Diopside
$r_{SiO_2} = a_{H^+}^{0.7}$	(pH range 2-6)
$r_{SiO_2} = a_{H^+}^{0.75}$	(pH range 2-6), T=50°C
	Enstatite
$r_{SiO_2} = a_{H^+}^{0.8}$	(pH range 1-6)
	Bronzite [after (5)]
$r_{SiO_2} = a_{H^+}^{0.6}$	(pH range 1-6)
	Augite
$r_{SiO_2} = a_{H^+}^{0.7}$	(pH range 1-6)

A plot of the logarithm of the initial rate of silica release against the reciprocal temperature T for etched augite, diopside and enstatite, at pH 6, over the temperature range 20-75°C, is shown in Fig. 2. From these data the apparent activation energy ΔE has been calculated. Resulting values of ΔE are: augite 78 kJ·mol^{-1}, enstatite 49 kJ·mol^{-1} and diopside 50-150 kJ·mol^{-1}. The range given for diopside dissolution probably reflects different rate-controlling mechanisms at different pH values.

These high values for activation energies rule out a control of dissolution by a solution-transport mechanism. The reactions at the surface, responsible for higher activation energy, are rate controlling as confirmed by the formation of etch pits on the surface of both naturally-weathered and laboratory-dissolved mineral grains. Note also that these activation energies, while higher than those based on transport control, are notably lower than those expected from breaking bonds in crystals (bulk diffusion activation energies range from 80 to 400 kJ·mol^{-1}). As we will see later, the catalytic effects of adsorption on surfaces reduce the activation energies to this intermediate energy range.

Table 2.- Surface composition results, in terms of XPS peak area ratios, for the dissolution of enstatite and diopside. T = 20°C [After Schott et al. (17)]

pH	Time (days)	Mg_{2s}/Si_{2p}	$Ca_{2p_{3/2}}/Si_{2p}$
	Enstatite		
	starting material	0.61	–
6	22	0.56	–
6	28	0.54	–
1	28	0.31	–
	Diopside		
	starting material	0.32	1.90
6	2	0.33	1.62
6	22	0.32	1.62
1	2	0.28*	1.28*
1	22	0.23*	1.23*

* – solution satured with respect to amorphous silica

Table 3.- Surface composition results, in terms of XPS peak area ratios, for the dissolution of bronzites T = 22°C [After Schott and Berner (18)]

pH	O_2	Time (days)	Mg_{2s}/Si_{2p}	$Fe_{2p_{3/2}}/Si_{2p}$	O_{1s}/Si_{2p}
		$Mg_{1.77}Fe_{0.23}Si_2O_6$ (Webster)			
		starting material	0.56	1.12	8.3
6	oxic	60	0.53	1.60	9.3
6	oxic	60 (after soneration)	0.53	1.20	8.7
6	anoxic	25	0.53	0.91	8.3
1	oxic	28	0.42	0.83	8.4
		$Mg_{1.70}Fe_{0.30}Si_2O_6$ (Gem quality)			
		starting material	0.53	1.40	8.1
6	oxic	21	0.54	2.00	8.7
6	anoxic	21	0.51	1.00	8.0
1	oxic	21	0.41	0.90	8.1

Iron minerals dissolved at pH 6 under oxic conditions. At pH 6 under oxygenated conditions, the rate of silica release is not constant (Fig. 1b). The shape of the curve suggests that the dissolution follows a more-or-less parabolic rate law. It is worth noting also that the dissolution rate is significantly lower than that found for anoxic conditions. This is obviously connected to the oxidation of iron and this result will be discussed in detail later in this paper.

Cation Release to Solution

Mg-Ca. In Tables 2 and 3 are shown XPS results, in terms of peak area ratios of Ca or Mg to Si, for the surface composition of enstatite, diopside, and bronzite from experiments conducted at various pH values. At pH 6 there is a small or negligible depletion of Mg, relative to Si over reaction periods ranging up to 60 days. This indicates essentially congruent dissolution of Mg relative to Si. By contrast, also at pH 6, Ca depletion from diopside is much greater. However, on extended reaction the Ca/Si ratio remains constant which means that there is initial incongruent dissolution of Ca relative to Si, followed by congruent dissolution. At pH 1 Mg and Ca are both released incongruently relative to Si but with Ca being released to a greater extent.

These results are confirmed by chemical analyses of the reacting solutions as shown in Figures 3-5. In particular, note that the XPS results for Ca in diopside are confirmed. In Figure 4 initial incongruent release of Ca is followed after about 100 hours by congruent release relative to both Si and Mg.

Fe. In Table 3 and 4 data for the XPS peak area ratio of Fe to Si in bronzite and fayalite are given. Under anoxic conditions or acidic (pH=1) conditions the behaviour of Fe is similar to that of Ca in diopside; there is incongruent release of Fe relative to Si, as shown by lowered ratios on the mineral surfaces of Fe/Si. Also, for fayalite at pH 1 there is very great preferential Fe release. By contrast, in the presence of oxygen the surface ratio of Fe/Si rises during reaction and this is matched by a rise in the oxygen content of the surface. The probable explanation for this is the formation of a precipitate layer of hydrous ferric oxide (see below).

The XPS results for iron are confirmed by chemical analyses of reacting solutions. Under anoxic conditions or at pH 1 there is incongruent loss of Fe relative to Si and under oxic conditions there is essentially no release of Fe to solution.

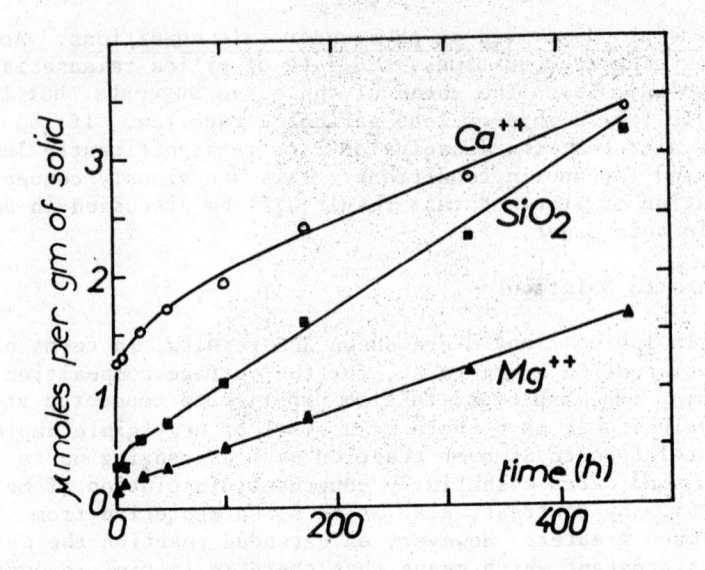

Fig.4.- Plot of released Ca^{++}, Mg^{++} and SiO_2 vs time for the dissolution of etched diopside at pH 6, T = 20°C. [After Schott et al. (17)]

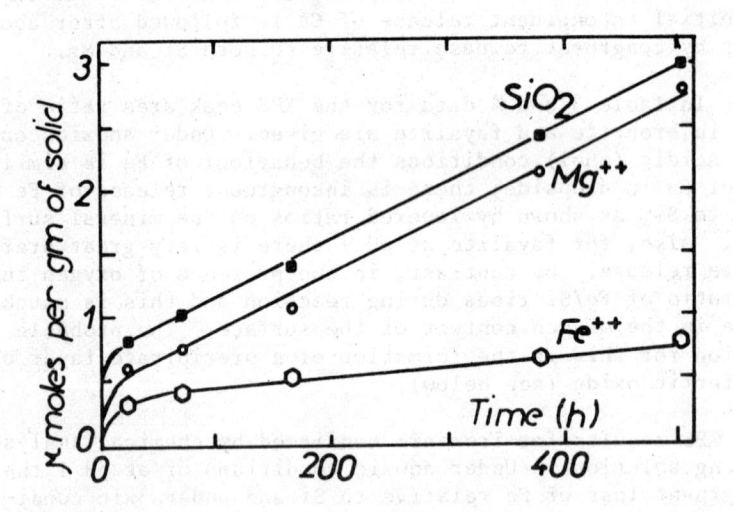

Fig.5.- Plot of released Mg^{++}, Fe^{++} and SiO_2 as a function of time for the dissolution of etched Webster bronzite at pH 6, pO_2 = 0 atm. (anoxic conditions). [After Schott and Berner (18)].

Table 4.- Summary of surface composition results, in terms of XPS peak area ratios, for the dissolution of fayalite. (T = 22°C). [After Schott and Berner (18)]

pH	O_2	Time (days)	$Fe_{2p_{3/2}}/Si_{2p}$	O_{1s}/Si_{2p}
	starting material		15.4	10.5
6	oxic	30	19.4	-
6	oxic	30 (fine fraction)	23.0	25.4
6	oxic	30 (after ultrasonic cleaning)	16.5	12.9
6	anoxic	21	12.9*	10.5*
1	oxic	14	1.6	7.8
1.5	oxic	1	9.0	10.4

* - Solution saturated with respect to amorphous silica.

Table 5.- Site occupancy and Madelung site energies of some pyroxenes [data from Ohashi and Burnham (11)] and in fayalite [data from Raymond (15)]. [After Schott and Berner (18)]

Mineral	Site	Occupancy	Energy (kJ mol^{-1})
Diopside	M1	Mg	-4346
	M2	Ca	-3157
Orthoenstatite	M1	Mg	-4318
	M2	Mg	-3940
Orthoferrosilite	M1	Fe	-4133
	M2	Fe	-3806
Fayalite	M1	Fe	-3652
	M2	Fe	-3980

Initial cation release and the structure of pyroxenes and olivines

The preferential release of Fe and Ca from bronzite and fayalite, and diopside respectively can be explained in terms of bonding energies. In the pyroxenes and olivines cations are located in two different types of structurally different sites, called M1 and M2, and these sites differ in the energy by which they hold various cations. Data for site occupancy and respective Madelung energies are shown in Table 5. Note that, except in fayalite, cations are more weakly bonded in the M2 sites than in the M1 sites. In the enstatite-fayalite series, Fe is preferentially located in the less stable sites. This helps to explain why there is preferential Fe release during (anoxic) dissolution. In diopside all Ca is in the M2 sites and this, again, helps to explain its incongruent release. The small loss of Mg from enstatite is roughly half the loss of Ca from diopside. This is what one would expect if the loss of Mg from enstatite was confined to M2 sites, since in enstatite Mg occupies both M1 and M2 sites.

Cation exchange and surface reaction

If dissolution is strictly a surface process, dissolution of M2 sites could occur by a H^+ (or H_3O^+) exchange reaction at the surface:

$$M1M2Si_2O_6 + 2H^+ \rightleftharpoons M2 + M1H_2Si_2O_6 \qquad (1)$$

Following the hypothesis of Lasaga (9) for feldspar dissolution, reaction (1) may be viewed as chemisorption of H^+ followed by reaction with a surface Ca^{++} (or Fe^{++}) and the detachment of Ca^{++} (or Fe^{++}).

The forward rate of reaction (1) may be written

$$r_+ = \frac{dn_{M2}}{dt} = k_+ X_{M2} \Theta_H^2 \qquad (2)$$

with Θ_H expressed by the Langmuir isotherm

$$K = \frac{\Theta_H}{m_H \Theta_E} \qquad (3)$$

Θ_H and Θ_E are respectively the fraction of adsorption sites occupied by H^+ and vacant; $X_{M2} = 1 - X_H$ = fraction of surface M2 sites occupied by M2 cations; K = equilibrium constant for the adsorption reaction and m_H = concentration of H^+ in the solution.

DISSOLUTION MECHANISMS OF PYROXENES AND OLIVINES

X_{M2} and X_H can be expressed as

$$X_{M2} = \frac{n_{M2px}}{n_{px}} = 1 - X_H = 1 - \frac{n_{Hpx}}{n_{px}} \qquad (4)$$

where n_{M2px} and n_{Hpx} are the number of moles of surface pyroxene M2 sites with M2 cations and with hydrogen respectively, and $n_{px} = n_{M2px} + n_{Hpx}$.

X_{M2} and X_H can be determined from the composition of the reacting solution as demonstrated by Murphy (unpublished work). If it is assumed that the exchange reaction occurs only at surface M2 sites, then the number of moles of protonated pyroxene created equals the number of moles of cations from M2 sites released to solution. On the other hand, as during the silicate hydrolysis reaction one protonated exchanged pyroxene is removed from the surface with each M1 site (see eq. 1), the number of moles of protonated pyroxene at the surface is given by:

$$n_{Hpx} = n_{M2_S} - n_{M1_S} \qquad (5)$$

where n_{M2_S} and n_{M1_S} are the number of moles of cations from M2 sites and M1 sites released to solution.

Combining equation (5) with eq. (2-4) yields

$$r_+ = \frac{dn_{M2}}{dt} = k_+ \Theta_H^2 \left[1 - \frac{(n_{M2_S} - n_{M1_S})}{n_{px}}\right]$$

$$= k_+ K^2 \Theta_E^2 m_H^2 - \frac{k_+ K^2 \Theta_E^2 m_H^2}{n_{px}} (n_{M2_S} - n_{M1_S}) \qquad (6)$$

At constant pH, equation (6) implies a linear relation between the rate of the exchange reaction and $n_{M2_S} - n_{M1_S}$ which is confirmed by our experiments (see Fig. 6).

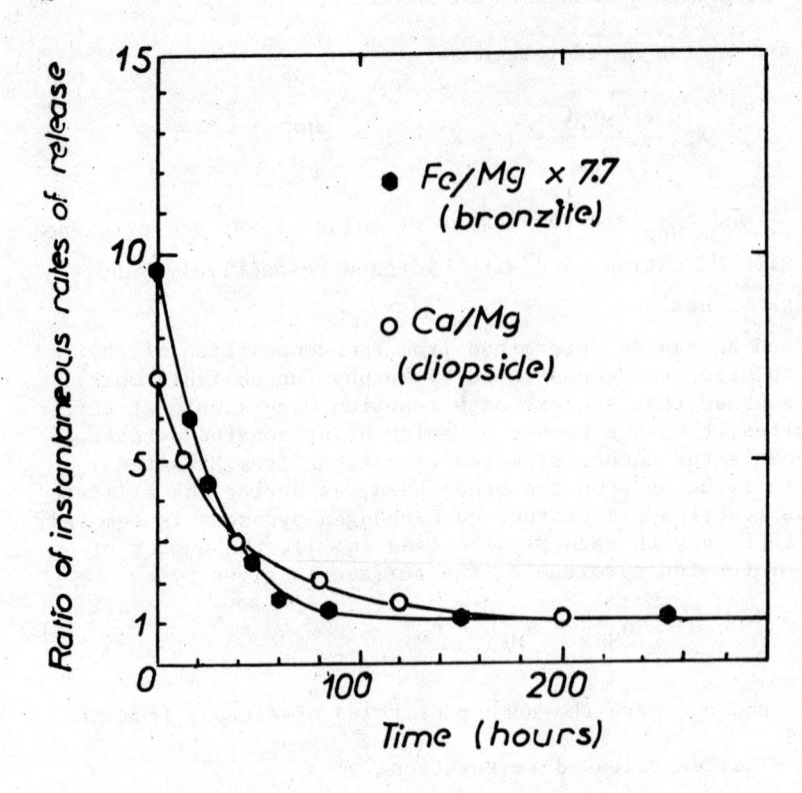

Fig.7.- Plot of ratio of instantaneous rates of cations release vs time for the dissolution of etched diopside (Pitcairn) and bronzite (Webster) at pH = 6, T = 20°C (pO_2 = 0 for bronzite).

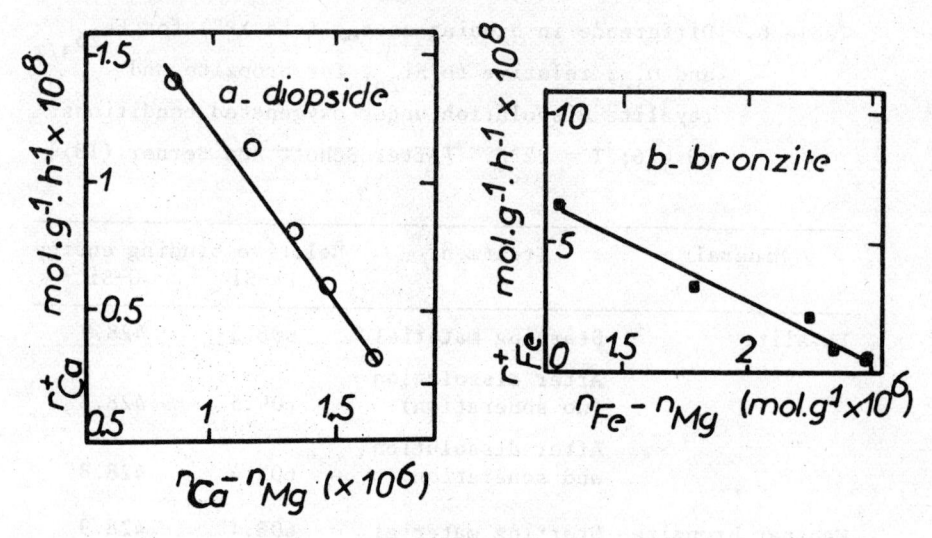

Fig.6.- Plot of released M2 cations, as a function of $n_{M2_S} - n_{M1_S}$, for the dissolution of diopside (a) and bronzite (b).

The altered layers

In Figure 7 is shown incongruent, followed by congruent dissolution of Ca in diopside and Fe in bronzite (under anoxic conditions). Congruency is reached after about 100 hours. This change can be attributed to the attainment of a cation-depleted surface layer of constant thickness. After attainment of congruency, dissolution at the outer surface of the layer occurs at the same rate as cation loss at the inner surface and, as a result, a steady-state thickness is maintained, an idea already proposed for Mg silicates by Luce et al. (10). However, our thickness estimates of this layer, based on XPS data [Schott et al. (17) ; Schott and Berner (18)] suggest that the layers are no more than 10-20Å thick. If so, then the idea of the layers being protective, in the sense of inhibiting diffusion, makes no sense. The concept of a classical diffusion gradient breaks down when the scale is only that of one or two unit cells. Apparently, the cation-depleted layer is so unstable chemically, that it continues to break down and dissolve before it can become appreciably thick. Therefore, it should not serve in a protective sense like the protective corrosion layers that form on metals. Because of the measured dependence of the rate of dissolution on H^+ concentration, and the arguments presented above, it is likely that cation depletion results from the substitution of H^+ for cations, but the exact nature of this surface layer is still a matter demanding further attention.

In the case of bronzite and fayalite dissolution under oxygenated conditions, the situation is considerably different.

Table 6.- Difference in binding energy (via XPS) for $Fe_{2p_{3/2}}$ and O_{1s}, relative to Si_{2p}, for bronzite and fayalite dissolution under oxygenated conditions. pH = 6; T = 22°C. [After Schott and Berner (18)]

Mineral	Treatment	Relative binding energy Fe-Si	O-Si
Fayalite	Starting material	608.5	428.9
"	After dissolution (no soneration)	609.5	428.3
"	After dissolution and soneration	609.4	428.8
Webster bronzite	Starting material	608.1	428.9
" "	After dissolution and soneration	609.3	428.9
Gem quality bronzite	Starting material	608.0	428.8
" "	After dissolution and soneration	609.2	428.9

Table 7.- Difference in binding energy (via XPS) for $Fe_{2p_{3/2}}$ and O_{1s}, relative to Si_{2p}, for some selected minerals (Si_{2p} energies for non-silicate minerals obtained from small amounts of Forsterite added as a standard). [After Schott and Berner (18)].

Mineral	Relative binding energy Fe-Si	O-Si
Fayalite	608.5	428.9
Bronzite	608.3	428.9
Augite	608.7	428.9
Biotite	608.2	428.9
Nontronite	609.9	428.9
Magnetite	608.0*	--
Hematite	608.4	427.0
Goethite	609.4	428.0

* - This peak is the sum of Fe^{3+} (608.5 eV) and Fe^{2+} (607.1 eV).

This is due to the oxidation of iron. The increase of Fe/Si and O/Si ratios shown in Tables 3 and 4 for runs at pH 6 can be explained in terms of the precipitation of hydrous ferric oxides on the mineral surfaces after release of Fe^{+2} to solution and reaction with dissolved O_2. Since this material is readily removed by ultrasonic cleaning as indicated by consequent drops in both Fe/Si and O/Si (see Tables 3 and 4), it probably is non-protective due to its lack of strong adherence. It is also acid-soluble as evidenced by a lack of its presence at low pH. Such a precipitated layer would be expected to be porous and non-protective from the arguments presented by other workers [see Ryzhak et al. (16); Baird et al. (1); and Petrovic (13) for an extended discussion]. In our experiments it could not have exceeded about 50Å in thickness based on the degree of release of accompanying cations to solution plus surface area measurements, as well as on SEM observations.

Evidence that the precipitated Fe-oxide layer is most likely a hydrous ferric oxide is given in Tables 6 and 7. The binding energies of iron and oxygen relative to silicon for non-sonically cleaned fayalite, shown in Table 6, are most similar to the values listed in Table 7 for goethite and not for hematite and magnetite. The data for Fe alone suggest oxidation from Fe^{+2} to Fe^{+3} in agreement with our conclusion based on XPS peak areas.

After sonic cleaning and removal of the non-adherent hydrous ferric oxide layer, the binding energy data show that the surfaces of bronzite and fayalite subjected to dissolution in the presence of O_2 still show an energy shift indicative of oxidation to Fe^{+3}, but lack of a shift in the O_{1s} binding energy (see Table 6). This result best matches that for nontronite (a hydrous ferric silicate) in Table 7. Thus, it appears that there are two layers resulting from dissolution under oxic conditions. The outer layer is hydrated ferric oxide, as discussed above. The inner layer is probably a Fe^{+3}-Mg silicate and it is not removed by ultrasonic cleaning; thus, it may be protective toward silica release and this helps explain the parabolic rate data shown in figure 1b. Proof of actual parabolic behaviour is shown by a linear square-root-of-time plot in Figure 8. Growth of the Fe^{+3}-Mg silicate layer could bring about decreasing rates of dissolution with time due to the inhibition of diffusion through this layer, thus, resulting in parabolic behaviour.

The exact chemical nature of the inner Fe^{+3}-Mg silicate layer is not known but additional XPS data suggest the presence of hydrogen. Data for the width of the O_{1s} peak at half-maximum peak height is shown in Table 8 for reacted and unreacted bronzite and fayalite and for hydrous-plus-anhydrous Fe, Mg, and Si oxide mineral pairs. It is apparent that the addition of H to each mineral causes an increase in peak-width because of the

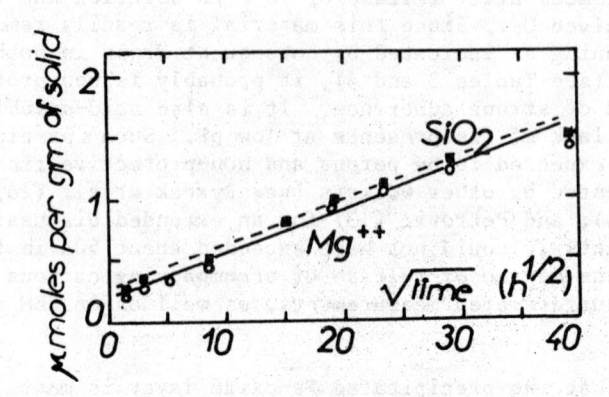

Fig. 8.- Dissolved Mg and SiO₂ plotted as a function of the square root of time for the dissolution of Webster bronzite, at pH 6, P_{O_2} = 0.2 atm. After Schott and Berner (18).

introduction of higher energy O-H bonds to the already present Si-O, Fe-O, and Mg-O bonds. (This agrees with the observation of shorter O-H bond lengths than the O-Si bond lengths found in pyroxenes and olivines, for example see Pauling (12)). The increase in peak-width of the surface of fayalite upon reaction with solution may be due to similar formation of O-H bonds. However, bronzite shows no increase in peak-width upon reaction (see Table 8). The probable reason for this is that bronzite, to start with, exhibits a broadened peak ($d_{1/2H}$ = 3.1 vs 2.0 eV in fayalite) due to the presence of both bridging (between Si) and non-bridging oxygens in the silicate chain. Yin et al. (20) have shown that the presence of the two types of oxygen produces peak broadening over that in olivine and that the two different oxygen types can be deduced via a peak deconvolution technique. Thus the addition of H to the bronzite surface may not be discernible because the O_{1s} peak is already broadened.

With our present techniques we have no way of telling how thick the Fe^{+3}-Mg silicate layer became during the experiments. The Fe, Mg, and O contents relative to Si, as determinated via XPS peak area ratios (see Tables 3 and 4), are essentially the same in the layer and in the underlying unaltered material for each mineral (small differences could easily be due to incomplete

Table 8.— Half-peak-height width ($d_{1/2H}$) of the O_{1s} peak as determined by XPS. [After Schott and Berner (18)]

Mineral	$d_{1/2H}$(eV)
Unreacted fayalite	2.0
Fayalite after reaction at pH 1 or anoxically at pH 6	3.0
Fayalite after reaction oxically at pH 6 and soneration	2.9
Unreacted bronzite	3.1
Bronzite after reaction at pH 1 or anoxically at pH 6	2.9
Bronzite after reaction oxically at pH 6 and soneration	3.2
SiO_2 (quartz)	2.1
$SiO_2 \cdot nH_2O$	2.9
MgO (periclase)	2.0
$Mg(OH)_2$ (brucite)	3.0
FeOOH (hematite)	1.7
$HFeO_2$ (goetite)	2.9

removal, via sonification, of residual hydrous ferric oxide). Thus, there is no way of calculating the thickness of material added or removed. Only the oxidation state of iron has increased from Fe^{+2} to Fe^{+3}.

Regardless of thickness, however, one can say that for charge balance to be maintained, the oxidation of iron must be balanced by some other process. For bronzite, which contains only about 10 mol % Fe, the charge balance mechanism may be the loss of a small percentage of Mg (the major cation) which is actually seen in Table 3. However, for fayalite, which contains essentially no Mg, this mechanism is not available. Also, with very long times of reaction the composition of the surface of Fe-silicates does not change. This is demonstrated by the fact that the surfaces of partly weathered pyroxene grains, taken from soils also exhibit essentially the same Fe/Si, O/Si, and Mg/Si ratios as the underlying unaltered mineral [Berner and Schott (18)]. Again, only oxidation of Fe^{+2} to Fe^{+3} is observed. Since the O_{1s}

line broadening discussed above suggests the production of O-H bonds, perharps the missing charge is supplied by the addition of OH$^-$ to the mineral surface.

Our results apply only to pH values of 6 or lower. (This also includes the soils studied). At such pH values no known hydrous Fe^{+3}-Mg-silicate is stable relative to dissolution. This provides some evidence against the idea of the formation during weathering of a thick, protective surface layer. However, at higher pH values, such a layer might be stabilized and continue to grow and this might help explain the observations, for example Eggleton and Boland (3), of layers of biopyribole type minerals, in topotactic relationship, on the surfaces of partly weathered pyroxenes.

REFERENCES

1. Baird T., Fryer J.R. and Galbraith S.T. 1977, *Iron oxide and oxyhydroxide preparations and their reactions with silica.* In Dev. Electron. Microsc. Anal., pp. 211-214, Bristol.
2. Berner R.A. and Schott J. 1982, *Mechanisms of pyroxene and amphibole weathering II. Observations of soil grains.* Amer. J. Sci. 207, pp. 1214-1231.
3. Eggleton R.A. and Boland J.N. 1982, *The weathering of enstatite.* Clays and Clay Mineral 30, pp. 11-20.
4. Fogler H.S., Lund K. and McCune C.C. 1975, *Acidizations III. The kinetics of the dissolution of sodium and potassium feldspar in HF/HCl acid mixtures.* Chem. Eng. Sci. 30, pp. 1325-1332.
5. Grandstaff D.E. 1980, *The dissolution rate of forsteritic olivine from Hawaiian beach sand.* Proc. 3rd Internat. Symp. Water-Rock Interaction, pp. 72-74.
6. Holdren G.R. Jr and Berner R.A. 1979, *Mechanism of feldspar weathering-I. Experimental studies.* Geochim. Cosmochim. Acta 43, pp. 1161-1171.
7. Huguenin R.L. 1974, *The formation of goethite and hydrated clays mineral on Mars.* J. Geophys. Res. 79, pp. 3895-3905.
8. Lagache M. 1976, *New data on the kinetics of the dissolution of alkali feldspars at 200°C in CO$_2$ charged water.* Geochim. Cosmochim. Acta 40, pp. 157-161.
9. Lasaga A.C. 1981, *Transition State Theory.* In kinetics of Geochemical Processes, A.C. Lasaga & R.J. Kirpatrick, Editors, pp. 135-169, Mineralogical Society of America.
10. Luce R.W., Bartlett R.W. and Parks G.A. 1972, *Dissolution kinetics of magnesium silicates.* Geochim. Cosmochim. Acta 36, pp. 36-50.
11. Ohashi Y. and Burnham C.W. 1972, *Electrostatic and repulsive energies of the M1 and M2 cation sites in pyroxenes.* J. Geophys. Res. 77, pp. 5761-5766.

12. Pauling L. 1960, *The nature of the chemical bond and the structure of molecules and crystals : an introduction to modern structural chemistry.* Cornell University Press.
13. Petrovic R. 1976, *Rate control in feldspar dissolution-II. The protective effect of precipitates.* Geochim. Cosmochim. Acta 40, pp. 1509-1521.
14. Petrovic R., Berner R.A. and Goldhaber M.B. 1976, *Rate Control in dissolution of alkali feldspar-I. Study of residual feldspar grains by X-ray photoelectron spectroscopy.* Geochim. Cosmochim. Acta. 40, pp. 537-548.
15. Raymond M. 1971, *Madelung constants for several silicates.* Carnegie Inst. Yearb. 70, pp. 225-227.
16. Ryzhak I.A., Krivoruchko O.P., Buyanov R.A., Kefeli L.M. and Ostan'kovich A.A. 1969, *Genesis of iron (III) hydroxide and iron oxide.* Kinet. Katal. 10, pp. 372-385 (Russ).
17. Schott J., Berner R.A. and Sjöberg E.L. 1981, *Mechanisms of pyroxene and amphibole weathering-I. Experimental studies of iron-free minerals.* Geochim. Cosmochim. Acta 45, pp. 2133-2135.
18. Schott J. and Berner R.A. 1983, *X-ray photoelectron studies of the mechanism of iron silicate dissolution during weathering.* Geochim. Cosmochim. Acta 47, pp. 2233-2240.
19. Siever R. and Woodford N. 1979, *Dissolution kinetics and the weathering of mafic minerals.* Geochim. Cosmochim. Acta 43, pp. 717-724.
20. Yin L.I., Ghose S. and Adler I. 1971, *Core binding energy difference between bridging and nonbridging oxygen atoms in a silicate chain.* Science 173, pp. 633-635.

THE EFFECTS OF COMPLEX-FORMING LIGANDS ON THE DISSOLUTION OF OXIDES AND ALUMINOSILICATES.

Werner Stumm, Gerhard Furrer, Erich Wieland and Bettina Zinder

Institute for Water Resources and Water Pollution Control (EAWAG), Dübendorf,
Swiss Federal Institute of Technology, Zurich, Switzerland.

INTRODUCTION

The composition of natural waters is controlled to a significant extent by processes at the solid(particle)/water interface, above all by the dissolution and precipitation of minerals. The weathering of rocks (aluminosilicates) is often an incongruent dissolution, i.e., the dissolution is accompanied (or followed) by the formation (precipitation) of a new phase; e.g. the dissolution of a feldspar

$$NaAlSi_3O_8(s) + H^+ + 7H_2O \rightarrow \begin{cases} Al(OH)_3(s) + Na^+ + 3H_4SiO_4 \\ 1/2\, Al_2Si_2O_5(OH)_4(s) + Na^+ + 2H_4SiO_4 \end{cases} \quad (1)$$

is accompanied by the precipitation of $Al(OH)_3$ or kaolinite. The sequence of the various processes may be quite involved: mass transport, surface detachment and surface attachment are important reactions; one of the elementary steps in these reactions may become rate-determining (1-9). From a chemical point of view, both the dissolution and the formation of a new solid phase, are characterized by a change in the coordination of the reactive partners.

Coordinative Properties of Surfaces

We believe that the detachment and attachment of reactants at the solid surfaces, as they occur in dissolution and nucleation, respectively, should be interpreted as chemical processes, or more specifically, as surface coordination reactions.

The hydrous solid surfaces contain functional groups, e.g., hydroxo groups, >Me-OH, at the surfaces of oxides of Si, Al and Fe. These groups have coordinative properties similar to those of their counterparts in soluble compounds; despite some geometric restrictions imposed by the solid nature of surfaces, an oxygen donor group in a hydrous oxide surface tends to form coordinative bonds in a similar way as oxygen donor groups in solutes. Thus, the interaction of hydrous oxide surfaces H^+, OH^- and metal ions and with anions and weak acids are surface coordination (complex formation and ligand exchange) reactions (Fig. 1) (10-14).

Fig. 1

An oxide surface, covered in the presence of water with amphoteric surface hydroxyl groups,>Me-OH, can be looked at as a polymeric oxoacid or base. The surface OH group has a complex forming O-donor atom that coordinates with H^+ and metal ions. The underlying central ion in the surface layer of the oxide--acting as a Lewis acid--can exchange its structural OH-ions against other ligands (anions or weak acids), (15) p. 625-640.

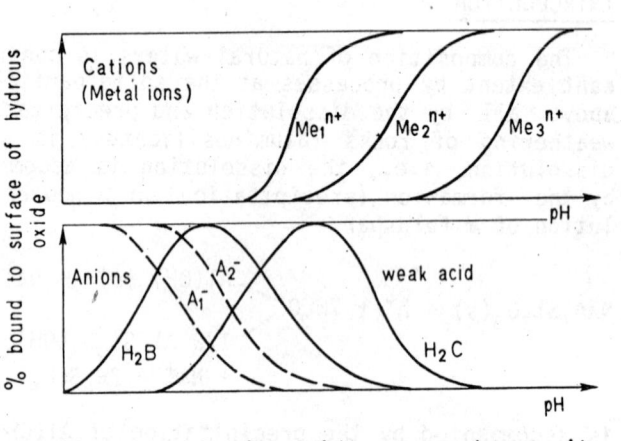

The extent of surface coordination and its pH dependence can be quantified by mass action equation (surface equilibrium constants) and can be explained by considering the affinity of the surface sites for metal ion or ligand and the pH dependence of the activity of surface sites and ligands. The tendency to form surface complexes may be compared with that to form corresponding solute complexes.

DISSOLUTION OF OXIDES

Surface Coordination Equilibria

Acid-base:
>Me—OH $\rightleftarrows$ Me—O$^-$ + H$^+$
>Me-OH + H$^+$ $\rightleftarrows$ >Me—OH$_2^+$

Metal ions:
>MeOH + M^{2+} $\rightleftarrows$ >MeOM$^+$ + H$^+$
(>MeOH)$_2$ + M^{2+} $\rightleftarrows$ (>MeO)$_2$M + 2H$^+$

Anions:
>MeOH + L^{2-} $\rightleftarrows$ >MeL$^-$ + OH$^-$
(>MeOH)$_2$ + L^{2-} $\rightleftarrows$ (>Me)$_2$L + 2OH$^-$

Acids:
>MeOH + H$_2$L $\rightleftarrows$ >MeHL + H$_2$O

EPR and ENDOR spectroscopic measurements by Motschi have shown that these surface complexes are typically inner-sphere coordination compounds (Fig. 2) (15-17). Thus, such compounds ought to be treated as chemical species which enter directly into mass action equation and rate laws.

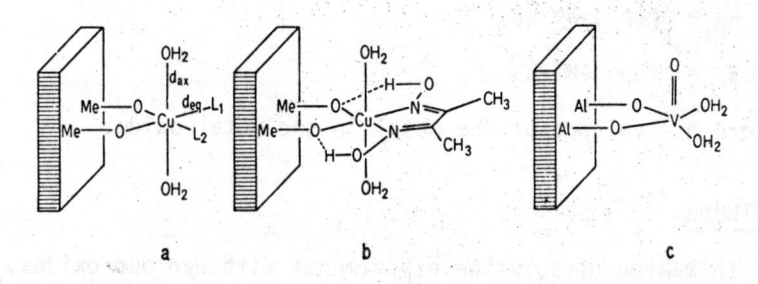

Fig. 2 Suggested structure of the Cu(II) surface complexes, derived from EPR and ENDOR results, (16,17).

a) General structure for the ($\equiv$AlO)$_2$CuL$_1$L$_2$(H$_2$O)$_2$ complex. The interatomic distances are estimated to be:

d_{eq} = 2.0(1)Å, d_{ax} = 2.6(1)Å for L$_1$=L$_2$ = H$_2$O and

d_{eq} = 2.0(1)Å, d_{ax} = 2.8(1)Å for L$_1$L$_2$ = bipy

b) Structure of the Cu-dimethylglyoxime surface complex involving hydrogen bonds between oxime and surface oxygens, (16).

c) Possible structure for the δ-Al$_2$O$_3$ surface bond VO^{2+}, based on ^{27}Al ENDOR study (17).

The equatorial positions of the VO^{2+} fragment are occupied by surface functional groups.

In order to abstract from the complexity of incongruent systems, we investigated first separately in relatively simple systems the kinetics of congruent oxide dissolution and of heterogeneous nucleation at oxide surfaces. Specifically, we evaluated how changes in surface coordination (surface protonation, surface complex formation) affect the surface detachment and surface attachment, respectively.

In a few case studies we have evaluated the effects of H^+ and various complex-forming anions on the dissolution kinetics of few oxides. The results can be generalized into a simple rate law that shows that the dissolution rate, R, depends on the relative concentration of protonated surface groups and on the relative concentration of anionic surface complexes (11,18):

$$R_H = \text{prop } \{>\text{Me-OH}_2^+\}^n$$

$$R_L = \text{prop } \{>\text{Me-L}\} \tag{2}$$

where n = valence of the metal in the metal oxide.

METHODS

In making dissolution experiments with hydrous oxides, special attention must be devoted to the properties of the surface. If heterogeneities of surface properties (different phases, different particle size, different surface energies) exist, parabolic dissolution rates are typically observed. Linear rate laws are usually obtained if the pretreatment renders the surface properties sufficiently homogeneous.

Extensive details on the experimental methods will be given in Furrer and Stumm (19) and Zinder and Stumm (20).

Dispersions of δ-Al_2O_3, α-FeOOH and BeO were treated with dilute HF (1% by weight) and subsequently washed with H_2O. The dispersions were then "conditioned" by preexposing them for three days in a milieu similar to that used in the subsequent dissolution experiments.

During the dissolution, pH [calibrated as -log H^+ concentration at a given ionic strength (0.1 M $NaNO_3$)] was kept constant with the help of an automatic titrator. This titrator compensates the hydrogen ions used in the dissolution reaction.

Progress in dissolution (during 30 to 50 hours) was measured by following the concentration of dissolved metal ion, as determined by flameless atomic absorption spectroscopy. Dissolution rate was computed from the linear [Me(aq)] vs time plots. The

DISSOLUTION OF OXIDES

extent of ligand adsorption was measured in case of oxalate, malonate, succinate and citrate by using C-14 labelled compounds and in case of salicylate, phthalate and benzoate by UV spectrophotometry (measurement of residual conc. in solution), respectively. Surface complex formation equilibria were formulated and defined as in earlier publications (12, 13).

RESULTS

Effect of H⁺ and Surface Protonation, respectively

Representative results, on the dissolution of δ-Al$_2$O$_3$ obtained at various pH values (25 °C) are given in Fig. 3. Phenomenologically the hydrogen ion dependence was observed and the rate was found to be proportional to $[H^+]^{0.4}$, this condition is valid for pH>3.5 (Fig. 3b). This suggests that the surface is saturated with protons at pH 3.5 and below.

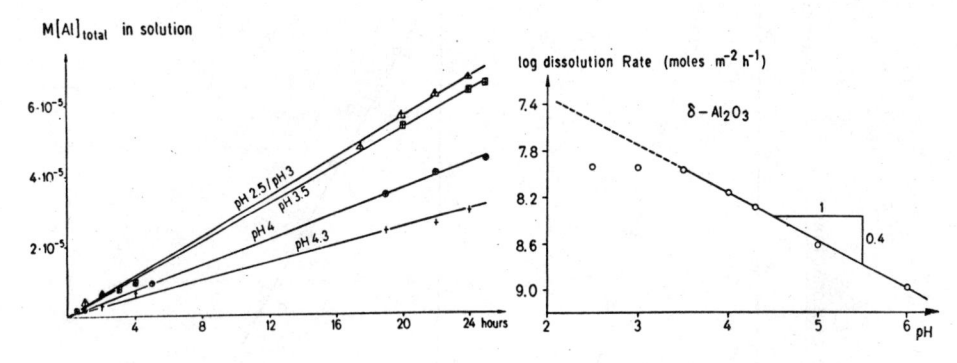

Fig. 3a

Effect of hydrogen ion concentration on the dissolution rate of δ-Al$_2$O$_3$, (2.2 g δ-Al$_2$O$_3$/l) Linear rate laws are obtained for dispersion that have been pretreated with HF and washed. The dissolution rate is controlled chemically by reaction at the surface.

Fig. 3b

Phenomenological interpretation of hydrogen ion dependence of the dissolution rate of δ-Al$_2$O$_3$:

$$R_H = k_H \cdot [H^+]^{0.4}$$

(applicable to 3.5<pH<6).

The results can be reinterpreted in terms of the concentration of surface protonated groups $\{\exists\text{-OH}_2^+\}$ (Fig. 4). The dependence of the concentration of protonated surface groups on solution pH can be evaluated from surface equilibrium constants which have been determined in separate experiments from acidimetric titration curves (Fig. 4) (14,21).

The constants have to be adjusted for the surface charge. The dissolution rate of a trivalent and bivalent metal oxide, $\delta\text{-Al}_2\text{O}_3$ and BeO, are compared in Fig. 5.

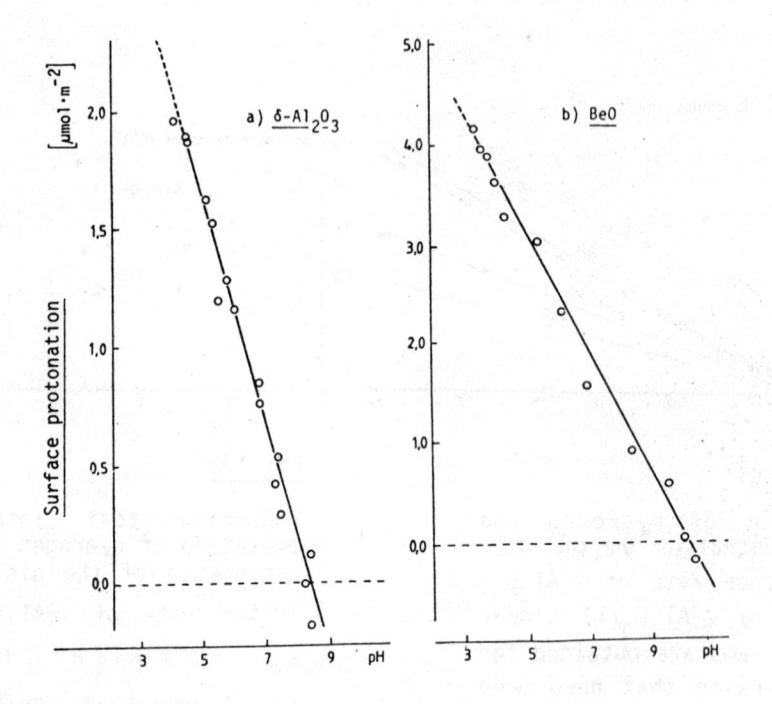

Fig. 4

The surface protonation of $\delta\text{-Al}_2\text{O}_3$ and BeO as a function of the solution pH.

DISSOLUTION OF OXIDES

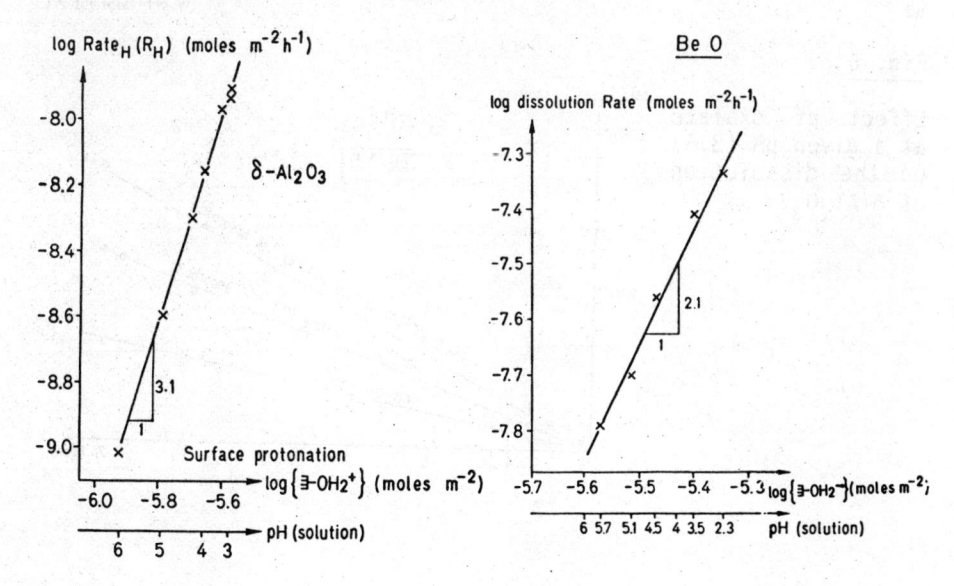

Fig. 5a

Effect of pH on the dissolution rate of δ-Al$_2$O$_3$. This dependence (Fig. 3b) can be reinterpreted in terms of a dependence on the concentration of protonated surface groups $\{\text{э-OH}_2^+\}$. The rate depends on $\{\text{э-OH}_2^+\}^3$.

Fig. 5b

Rate of dissolution of BeO as a function of the concentration of protonated surface groups. The rate depends on $\{\text{э-OH}_2^+\}^2$.

Effect of Ligands

The effect of an organic ligand, oxalate, on the dissolution of δ-Al$_2$O$_3$ at a given pH is exemplified in Fig. 6. Oxalate enhances the dissolution rate (18).

There is a linear dependence of the rate on the concentration of the ligand-surface complexes. Their concentration is known from surface equilibrium constants, e.g. $\{>\text{MeL}\} / \{>\text{MeOH}\} \cdot [\text{L}^{2-}]$, that have been determined previously (Sigg and Stumm (12) and

Fig. 6

Effect pf oxalate at a given pH (3.5) on the dissolution of δ-Al$_2$O$_3$.

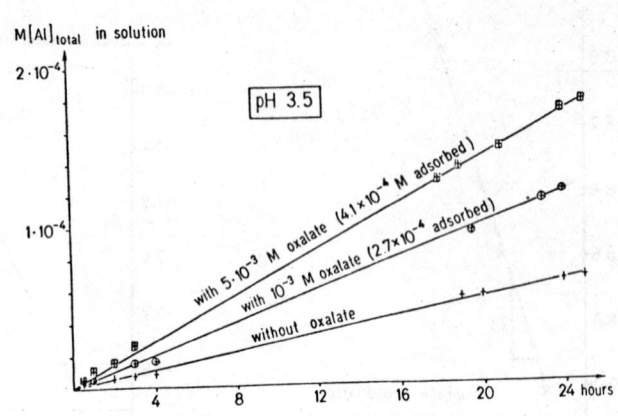

Kummert and Stumm (13)). The effect of the ligand becomes superimposed on that of surface protonation; i.e, the dissolution rate can be considered to be composed of two (additive) rates,

i) a surface protonation $\{\exists\text{-OH}_2^+\}$ dependent rate R_H, and
ii) a rate R_L which depends on the concentration of ligand surface complexes, $\{\exists\text{-L}\}$,

$$R = R_H + R_L = k_H \cdot \{\exists\text{-OH}_2^+\}^n + k_L \cdot \{\exists\text{-L}\} \tag{3}$$

The effect of various organic complex formers on the dissolution rate, R_L of δ-Al$_2$O$_3$ are compared in Fig. 7. As table 2 illustrates, the most rapid dissolution rate is obtained when bidentate mononuclear surface chelates are formed. Most efficient are five- and six-ring chelates; seven-ring chelates are somewhat less effective. Monodentates are typically more inert with regard to dissolution reactions. Fluoride is an exception; it accelerates the dissolution rate markedly. (Zutic and Stumm, to be published).

A similar enhancement of the dissolution rate by surface complexing oxalate is observed on the dissolution of Fe(OH)$_3$ (amorphous), α-Fe$_2$O$_3$ (hematite) and α-FeOOH (goethite), (Fig. 8).

DISSOLUTION OF OXIDES

Fig. 7

The dissolution rate on the presence of organic ligand anions (pH 2.5 - 6) can be interpreted as a linear dependence on the surface concentrations of deprotonated ligands, $\{\text{\textbardbl}-L\}$. R_L' [nmoles·m^{-2}·h^{-1}] is that portion of the rate which is dependent on surface complexes only. In case of citrate and salicylate at pH 4.5 corrections accounting for the protonation of the surface complexes were made.

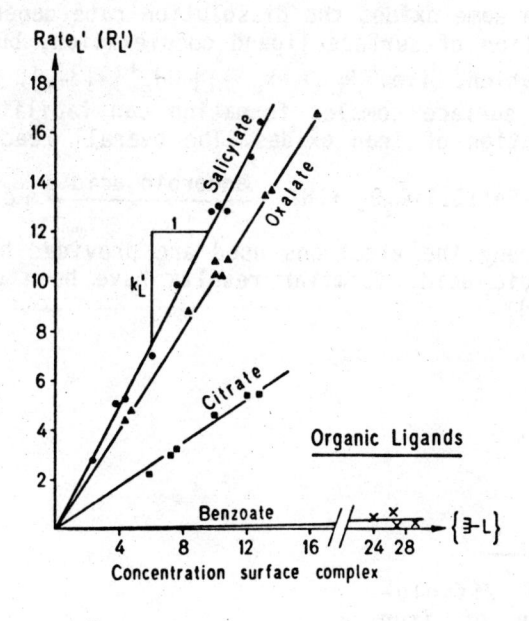

Fig. 8

Oxalate enhances the dissolution rate of amorphous Fe(OH)$_3$, α-Fe$_2$O$_3$ (hematite) and α-FeOOH (goethite) (with exclusion of light).

The curves do not intersect at zero because, as a consequence of the preconditioning, some dissolved Fe is present at t=0.

In the absence of oxalate, rates of dissolution of these oxides are—within the time period of observation—negligible.

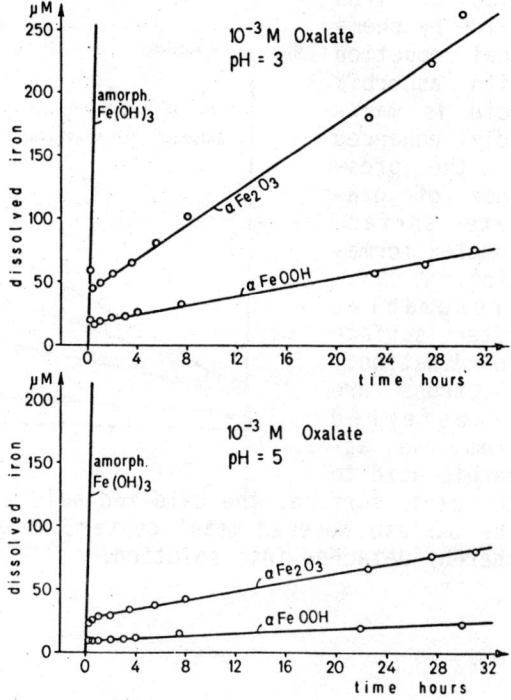

With some oxides the dissolution rate depends not only on concentration of surface ligand complexation, but also on surface protonation, i.e. $R_L = k_L' \cdot \{]-OH_2^+\} \cdot \{]-L\}$. As Fig. 9 illustrates the surface complex formation can facilitate the reductive dissolution of iron oxides. The overall reaction can be written as

$$>Fe(III)=C_2O_4 + H^+ \xrightarrow{ascorbic\ acid} Fe^{2+}(aq) + C_2O_4H^- + > \quad (4)$$

whereby the electrons used are provided by the oxidation of ascorbic acid. (Similar results have been reported by Schwertmann (22)).

Fig. 9

The dissolution of iron oxide by chemical reduction with ascorbic acid is markedly enhanced in the presence of oxalate surface complex formation. Presumably, after surface coordination, electrons are transferred from the ascorbic acid to the oxide surface, the oxidized acid subsequently being desorbed. The oxalate covered metal center, more amenable to reduction, is thereby detached into solution.

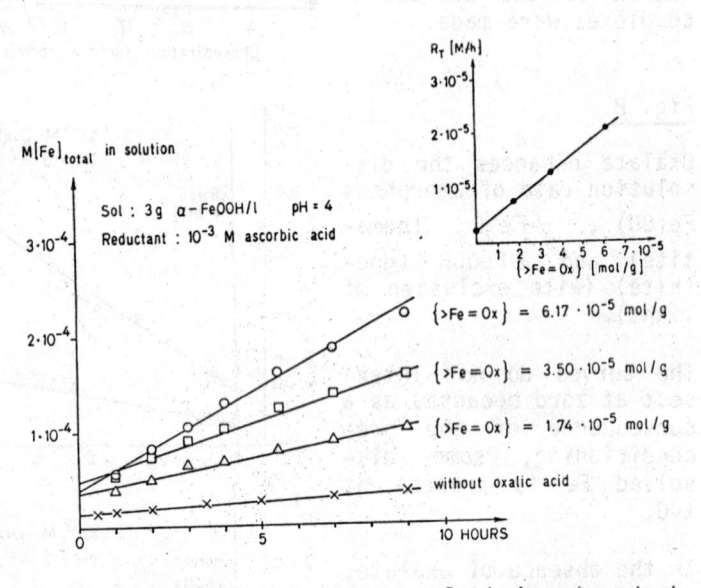

Sol: 3g α-FeOOH/l pH = 4
Reductant: 10^{-3} M ascorbic acid

$\{>Fe=Ox\} = 6.17 \cdot 10^{-5}$ mol/g
$\{>Fe=Ox\} = 3.50 \cdot 10^{-5}$ mol/g
$\{>Fe=Ox\} = 1.74 \cdot 10^{-5}$ mol/g
without oxalic acid

INTERPRETATION OF THE RESULTS

The effect of surface coordination on the dissolution of oxides:

As pointed out before, the dissolution of a metal or metalloid oxide means a change of the ligands of the metal or metalloid ions. As shown by Valverde and Wagner (23) one has in essence the scheme

$$Me^{z+} \text{ (crystal)} = Me^{z+} \text{ (aq)} \tag{5}$$

$$O^{2-} \text{ (crystal)} + 2H^+ \text{ (aq)} = H_2O \tag{6}$$

In comparing dissolution with the vaporization of a metal into a vacuum, Valverde and Wagner assume that the rate-determining step is essentially characterized by the release of ions from kinks into the adjacent solutions.

Table 1 illustrates in a simplified way, how acids and ligands affect the oxide surface groups and polarize the Me-O bonding. In a ligand exchange reaction, the nucleophilic ligand, L, binds to a Me center and replaces an OH group. This also polarizes the particular Me-O bonds (11).

Table 1

OXIDE DISSOLUTION   (Simple Hypothesis on Rate-Determining Step)

1.) Surface coordination reactions

i) $\equiv$OH + H$^+$ $\xrightarrow{\text{fast}}$ $\equiv$OH$_2^+$

ii) $\equiv$OH + HL $\xrightarrow{\text{fast}}$ $\equiv$L + H$_2$O

iii) $\equiv$OH / $\equiv$OH + H$_2$B $\xrightarrow{\text{fast}}$ $\equiv\equiv$B + 2H$_2$O   "blocking"

2.) Surface controlled dissolution reactions

i) $\equiv$OH$_2^+$ $\xrightarrow{\text{slow}}$ Me(OH$_2$)$_x^{z+}$ + $\equiv$

ii) $\equiv$L $\xrightarrow{\text{slow}}$ MeL (aq) + $\equiv$

Dissolution Rate

Related to $\{\equiv\text{OH}_2^+\}^m$, $\{\equiv\text{L}\}$

Far away from dissolution equilibrium, back reaction is negligible

Table 2: Effect of surface complex formation on dissolution rate of $\delta\text{-Al}_2\text{O}_3$

		k_L
five ring chelates	: oxalate	$10.8 \cdot 10^{-3}$ h^{-1}
six ring chelates	: malonate	$6.9 \cdot 10^{-3}$ h^{-1}
	salicylate	$12.5 \cdot 10^{-3}$ h^{-1}
seven ring chelates	: succinate	$2.4 \cdot 10^{-3}$ h^{-1}
	phthalate	$3.0 \cdot 10^{-3}$ h^{-1}
monodentate ligands	: benzoate	$\ll 10^{-3}$ h^{-1}

Oxalate, malonate, salicylate, succinate, phthalate form bidentate (mononuclear) surface complexes, e.g.

$$>\!\!Al\!\!<^{OH_2}_{OH} + C_2O_4^{2-} + H^+ \rightleftharpoons\, >\!\!Al\!\!<^{O-C=O}_{O-C=O} + 2H_2O$$

$K^s = 10^{11} M^{-2}$

where K^s is defined as

$K^s = \{>\!AlOx\}/\{>\!AlOHOH_2\}[Ox^{2-}][H^+]$

concentration in { } are surface concentrations in moles kg^{-1} or moles m^{-2}, and concentrations in [ ] are in moles dm^{-3}, where benzoate forms monodentate complexes.

The weakening of the Me-O bonds by the protonation of surface OH groups and the formation of (inner sphere) surface complexes with suitable ligands enhance the subsequent detachment of a Me·H$_2$O or a MeL group. This detachment is rate-determining. Under alkaline conditions, the deprotonation or additional hydroxylation of the surface OH-groups may also facilitate the dissolution of an oxide.

Derivation of the rate laws

Fig. 10 illustrates schematically the individual steps in the surface reaction controlled dissolution of hydrous oxides or hydroxides. In accordance with (5) and (6), a number of protons equivalent to the valency of the Me-center, is needed to detach a Me(aq) group into the aqueous solution. The stepwise protonation occurs relatively fast. The slow - and rate-determining - step is the detachment of the Me(aq) group (reaction c in Fig. 10). In a bivalent metal oxide two neighboring surface groups need to become protonated (formula c in Fig. 10). The rate of detachment (step c) thus depends on the surface-concentration of surface metal centers that contain two neighboring protonated surface hydroxo groups.

The probability to find two neighboring protonated groups is proportional to Θ_H^2 where Θ_H is the degree of surface protonation

$$\Theta_H = \frac{\{]\text{-OH}_2^+\}}{\{]\text{-OH}\} + \{]\text{-OH}_2^+\}} \qquad (7)$$

For a three-valent metal oxide, three neighboring surface sites need to be protonated. The probability to find three neighboring protonated groups is proportional to Θ_H^3.

Thus the rate determining step for the dissolution of a metal oxide in slightly acidic solutions (reaction c in Fig. 10) is given by

$$R_H = k \cdot \Theta_H^n = k' \cdot \{]\text{-OH}_2^+\}^n \qquad (8)$$

where n corresponds to the valency of the metal in the metal oxide; i.e.,

$$R_H = k \cdot \Theta_H^2 = k' \cdot \{]\text{-OH}_2^+\}^2 \text{ for BeO, and}$$

$$R_H = k \cdot \Theta_H^3 = k' \cdot \{]\text{-OH}_2^+\}^3 \text{ for Al}_2O_3 \qquad (9)$$

This rate determining detachment occurs probably primarily at kink or step sites. This rate law is in agreement with the results given in Figures 4 and 5.

Figure 11 illustrates the effect of ligands on the dissolution rate. The detachment of the MeL group is the determining step:

$$R_L = k_L \cdot \Theta_L = k_L' \cdot \{]\text{-L}\} \qquad (10)$$

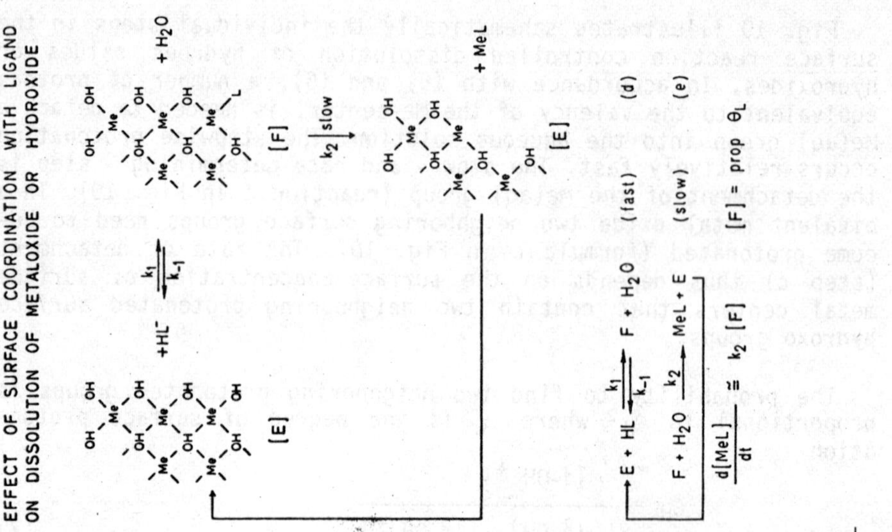

EFFECT OF SURFACE COORDINATION WITH LIGAND ON DISSOLUTION OF METALOXIDE OR HYDROXIDE

[E] $\xrightleftharpoons[k_{-1}]{k_1}$ +HL⁻ [E] $\xrightarrow[\text{slow}]{k_2}$ +MeL [F] + H₂O

E + HL $\xrightleftharpoons[k_{-1}]{k_1}$ F + H₂O (fast)

F + H₂O $\xrightarrow{k_2}$ MeL + E (slow)

$$\frac{d[MeL]}{dt} \cong k_2 [F]$$

[F] = prop θ_L

Fig. 11

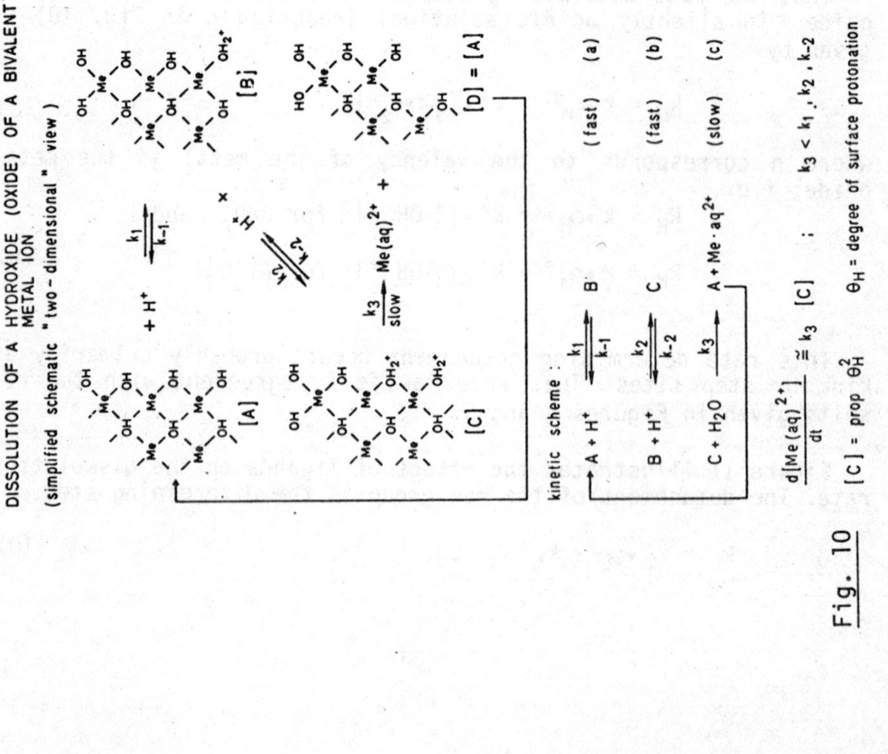

DISSOLUTION OF A HYDROXIDE (OXIDE) OF A BIVALENT METAL ION
(simplified schematic "two-dimensional" view)

[A] +H⁺ $\xrightleftharpoons[k_{-1}]{k_1}$ [B]

[C] $\xrightleftharpoons[k_{-2}]{k_2}$ +H⁺

[C] $\xrightarrow[\text{slow}]{k_3}$ Me(aq)²⁺ + [D] = [A]

kinetic scheme:

A + H⁺ $\xrightleftharpoons[k_{-1}]{k_1}$ B (fast) (a)

B + H⁺ $\xrightleftharpoons[k_{-2}]{k_2}$ C (fast) (b)

C + H₂O $\xrightarrow{k_3}$ A + Me·aq²⁺ (slow) (c)

$$\frac{d[Me(aq)^{2+}]}{dt} \cong k_3 [C] \quad ; \quad k_3 < k_1, k_2, k_{-2}$$

[C] = prop θ_H^2 θ_H = degree of surface protonation

Fig. 10

DISSOLUTION OF OXIDES

The presence of water soluble anions of organic acids, such as malic, malonic, oxalic, acetic, succinic, tartaric, vanillic and p-hydroxy benzoic acids has been demonstrated in top soils in concentration as high as 10^{-5}M to 10^{-4}M, oxalate being the most abundant. Many of these complex formers are released from organisms and plants. These substances are also present in surface waters (24); oxalate also occurs in rainwater in concentrations as high as 10^{-6}M.

The ligands form complexes with Al^{3+} and Fe^{3+} (and other metal ions) in solution and on the surface of the oxides of Al(III) and Fe(III). The formation of oxalate surface complexes for exemple can be given as

$$>Al\overset{-OH_2}{-OH} + C_2O_4^{2-} + H^+ \rightleftharpoons >Al\overset{-O-C(=O)}{-O-C(=O)} + 2H_2O \qquad (11)$$

$$K^s = 10^{11} M^{-2}$$

The complexing in solution tends to raise the Al and Fe concentrations in solution. It is often assumed that the effect of ligands in enhancing the weathering rates is caused by this complex formation in solution which could allow for significant mass transfers in solution. Complex forming ligands increase the rate of dissolution, however, primarily by surface complex formation (cf. Eq. 11); furthermore they extend the domain of congruent dissolution of minerals; i.e. higher concentrations of soluble Al(III) or Fe(III) can be built up before a new phase is formed.

The dependence of the dissolution rate of δ-Al_2O_3 upon $\{\overset{}{\text{\}}}$-$OH_2^+\}^3$ corresponds to a fractional order dependence on $[H^+]$, i.e., $[H^+]^{0.4}$ for pH ⩾ 3.5 (see Fig. 3b), because surface protonation is not linearly related to $[H^+]$ (Fig. 4). Rate laws with a fractional order dependence on $[H^+]$ have been reported from most oxides (Grauer and Stumm (25)) and for many minerals (Table 3). Our rate law (Eq. 9) is compatible with these reaction orders and reflects a chemical mechanistic interpretation of the dissolution process.

The weakening of the Fe(III)-O bonds by surface-coordinated ligands makes the Fe(III) center apparently more amenable to reduction (Eq. 4, Fig. 9).

PERSPECTIVES ON HETEROGENEOUS NUCLEATION

The free energy of the formation of a nucleus consists essentially of free energy gained from bonds (related to the oversaturation ratio) and of work required to create a surface. This latter interfacial energy is minimized if the nucleus is formed in contact with some other solid phase, especially if there is some similarity in the atomic structure of the two phases.

Heterogeneous nucleation, rather than nucleation from homogeneous solution, is the predominant formation process for crystals in natural waters and the formation of new solid phases in weathering reactions. The structure of the template (the surface on which the surface is formed) is essential in favoring or directing the nucleation of certain crystal structures. But in addition to the matching of the structures, the specific adsorption, i.e. the chemical bonding of nucleus constituents to the surface of the solid substrate is a prerequisit for the nucleation. For example, in the nucleation of an ion lattice $\{AB\}_n$ the following steps may occur (perhaps preferably near a kink of the solid substrate):

$$A(H_2O)_n + S \underset{}{\overset{k_1}{\rightleftharpoons}} SA(H_2O)_m + (n-m) H_2O \tag{12}$$

$$B(H_2O)_x + S \underset{}{\overset{k_2}{\rightleftharpoons}} SB(H_2O) + (x-y) H_2O \tag{13}$$

$$SA(H_2O)_m + SB(H_2O)_y \underset{k_{-3}}{\overset{k_3}{\rightleftharpoons}} \{AB\}_n + 2S + (m+y)H_2O \tag{14}$$

where S is a surface site, e.g., a hydrous oxide or organic surface with an oxygen donor atom such as MeOH which forms surface complexes with A and B (in the latter case by ligand exchange); the surface binding is accompanied with at least a partial dehydration of the ions A and B (m n, x y). If the last step is rate determining the rate of AB nucleus formation is given by

$$\frac{d\{AB\}_n}{dt} = k_3 \cdot \Theta_A \cdot \Theta_B \tag{15}$$

i.e., the rate of nucleus formation is proportional to the product of the relative coverage of sites with A and with B, where

$$\left(\Theta_A = \frac{\text{sites occupied by A}}{\text{all sites available}}\right).$$

The product of $\Theta_A \cdot \Theta_B$ relates to the probability to find neighboring sites occupied by A and B. Thus, the rate of nucleation appears to be related to a two-dimensional solubility product $\Theta_A \cdot \Theta_B = Q$ (26).

For heterogenous nucleation of a non-symmetric crystal, e.g. CaF_2, the rate of nucleus formation depends on the concentration of the surface complexes formed

$$\frac{d\{CaF_2\}_n}{dt} = k \cdot \Theta_{Ca} \cdot \Theta_F^2 \tag{16}$$

CHEMICAL WEATHERING OF ALUMINOSILICATES

The concepts given here can be extended to the weathering of aluminum silicates. Hydrogen ions and ligands may associate with the $\delta\text{-}Al_2O_3$ or silicate layers, e.g. if the attachment of ligands to the Al-centers enhances the dissolution rate, this may serve as an indication that chemical processes at the $\delta\text{-}Al_2O_3$ surface are rate determining for the dissolution of the aluminum silicate. The results found in our laboratory studies appear to be in qualitative agreement with some observations made on the chemical weathering of rocks in crystalline areas.

Chemical weathering is a key process serving to neutralize the internal production and the anthropogenic input of acids to a watershed (27,28). The rate of chemical weathering in soils and sediments is obviously a function of mineralogy, temperature, flow-rate, surface area, ligand and CO_2 concentration in soil water. It is also a strong function of hydrogen ion activity when the pH of soil water goes below 5.0.

The rate of weathering of $\delta\text{-}Al_2O_3$ has been shown (Eq. 3) to be proportional to the degree of surface protonation and to the surface complexation:

$$R = k_H \cdot \{\exists\text{-}OH_2^+\}^3 + k_L \cdot \{\exists\text{>}L\} \tag{17}$$

Table 3 Hydrolysis Reaction Order for the Dissolution of Minerals.

Mineral	Formula	Solution	Reaction Order
Dolomite	$(Ca,Mg)CO_3$	HCl	$[H^+]^{0.5}$
Bronzite	$(Mg,Ca)SiO_3$	HCl	$[H^+]^{0.5}$
Enstatite	$MgSiO_3$	HCl	$[H^+]^{0.8}$
Diopside	$CaMgSi_2O_6$	HCl	$[H^+]^{0.7}$
K-feldspar	$KAlSi_3O_8$	Buffer	$[H^+]^{0.33}$
Iron hydroxide	$Fe(OH)_3$-Gel	Various acids	$[H^+]^{0.48}$
Aluminum oxide	$\delta\text{-}Al_2O_3$	HCl	$[H^+]^{0.4}$

Equation 17 results in a dissolution rate that is proportional to $[H^+]$ to the 0.4 power (see Fig. 3b). Many common minerals have been reported to undergo fractional order dissolution in acids (Table 3) (28).

A general expression for chemical weathering that corresponds to Eqs (3) and (17) is

$$W = k_H \cdot [H^+]^n + k_o \qquad (18)$$

where W = chemical weathering rate (rate of H^+ ion consumption) [eq·ha^{-1}·yr^{-1}]

k_H = rate constant [eq·ha^{-1}·yr^{-1}]

n = fractional order constant

k_o = rate constant in the absence of free acidity (due to CO_2 and ligand weathering [eq·ha^{-1}·yr^{-1}].

This fractional order dependence on $[H^+]$ is typically observed in crystalline lake areas receiving acid atmospheric depositions (29, 29). Since the increase in weathering rate upon increase in $[H^+]$ is not linear--a tenfold increase in $[H^+]$ nearly doubles the weathering rate--increased acidification resulting from atmospheric input cannot be fully compensated by increase in the neutralization rate in the watershed.

REFERENCES

(1) AAGARD P. and HELGESON H.C. (1982) "Thermodynamic and kinetic constraints on reaction rates among minerals and aqueous solutions." 1. Theoretical considerations. Amer. J. Sci. 282, pp. 237-285.
(2) BERNER R.A. (1977) "Mechanism of feldspar weathering: some observational evidence." Geology 5, pp. 369-372.
(3) BERNER R.A. (1978) "Rate control of mineral dissolution under earth surface conditions." Amer. J. Sci. 278, pp. 1235-1252.
(4) BERNER R.A. (1981) "Kinetics of weathering and diagenesis. In: Kinetics of Geochemical Processes" (eds. A.C. LASAGA and R.J. KIRKPATRICK), Rev. Mineral. 8, pp. 111-134.
(5) BOYLE J.R. and VOIGT K.G. (1973) "Biological weathering of silicate minerals. Plants and Soil." 38, pp. 191-201.
(6) DIBBLE W.E. and TILLER W.A. (1981) "Non-equilibrium water/rock interactions. I. Model for interface-controlled reactions." Geochim. Cosmochim. Acta 40, pp. 191-202.
(7) PACES T. (1978) "Reversible control of aqueous aluminium and silica during the irreversible evolution of natural waters." Geochim. Cosmochim. Acta 42, pp. 1487-1493.
(8) PETROVIC R. (1976). "Rate control in feldspar dissolution. II. The protective effect of precipitates." Geochim. Cosmochim. Acta 40, pp. 1509-1521.
(9) WOLLAST R. (1976) "Kinetics of alteration of K-feldspar in buffered solutions at low temperature." Geochim. Cosmochim Acta 31, pp. 635-648.
(10) STUMM W. KUMMERT R. and SIGG L. (1980) "A ligand exchange model for the adsorption of inorganic and organic ligands at hydrous oxide interfaces." Croat. Chem. Acta 53, pp. 291-312.
(11) STUMM W. FURRER G. and KUNZ B. (1983) "The role of surface coordination in precipitation and dissolution of mineral phase." Croat. Chem. Acta 58, pp. 593-611.
(12) SIGG L. and STUMM W. (1981) "The interactions of anions and weak acids with the hydrous goethite surface. Colloids and Surfaces. 2, pp. 101-117.
(13) KUMMERT R. and Stumm W. (1980) "The surface complexation of organic acid and hydrous γ-Al_2O_3." J. Colloid. Interface Sci. 75, 373-385.
(14) SCHINDLER P.W. (1981) "Surface complexes at oxide-water interfaces." In: Adsorption at the Solid-Liquid Interface (eds. M.A. ANDERSON and A. RUBIN), pp. 1-49. Ann Arbor Press.
(15) STUMM W. and MORGAN J.J. (1981), "Aquatic Chemistry" 2nd ed. Wiley-Interscience New York.

(16) RUDIN M. and MOTSCHI H. (1984) "A molecular model for the structure of copper complexes on hydrous oxide surfaces." An ENDOR study. J. Coll. and Interf. Sci. 98, pp. 385-393.
(17) MOTSCHI H. and RUDIN M. (1984) "^{27}Al ENDOR study of VO^{2+} adsorbed on δ-alumina. Direct evidence for innersphere coordination with surface functional groups." Coll. and Polymer Sci. 262 (in press).
(18) FURRER G. and STUMM W. (1983) "The role of surface coordination in the dissolution of δ-Al_2O_3 in dilute acids." Chimia 37, pp. 338-341.
(19) FURRER G. and STUMM W. (to be submitted).
(20) ZINDER B. and STUMM W. (to be submitted).
(21) HOHL H. and STUMM W. (1976) "Interaction of Pb^{2+} with hydrous Al_2O_3." J. Coll. and Interf. Sci. 55, pp. 281-288.
(22) SCHWERTMANN U. (1964) "The differentiation of iron oxide in soils by a photochemical extraction with acid ammonium oxalate." Z. Pflanzenernähr. Bodenkunde 105, pp. 146-163.
(23) VALVERDE N. and WAGNER C. (1976) "Considerations on the kinetics of the dissolution of metal oxides in acid solutions." Ber. Bunsen-Gesellschaft 80, pp. 330-340.
(24) GRAUSTEIN W.C. (1977) "Calcium oxalate; occurrence in soils and effect on nutrient and geochemical cycles." Science 198, pp. 1252-1254.
(25) GRAUER R. and STUMM W. (1982) "Die Koordinationschemie oxidischer Grenzflächen und ihre Auswirkung auf die Auflösungskinetik oxidischer Festphasen in wässrigen Lösungen." Colloid and Polymer Sci. 260, pp. 959-970.
(26) HOHL H. et al. (to be published)
(27) SCHNOOR J. L. and STUMM W. (1984) "Acidification of Aquatic and Terrestrial Ecosystems." In: Chemical Processes in Lakes (ed. STUMM W.), Wiley Interscience, New York (in press).
(28) SCHNOOR J. L and STUMM W. "Chemical Weathering" (to be submitted)
(29) SCHNOOR J.G., SIGG L., STUMM W. and ZOBRIST J. (1983) "Acid precipitation and its influence in Swiss lakes." EAWAG News 15, pp. 6-12.

KINETIC STUDY OF THE DISSOLUTION OF ALBITE WITH A CONTINUOUS FLOW-THROUGH FLUIDIZED BED REACTOR

Roland WOLLAST and Lei CHOU[*]

University of Brussels, BELGIUM
Northwestern University, USA[*]

ABSTRACT

The dissolution kinetics of albite were studied using a continuous flow reactor based on the fluidized bed technique. The influence of pH and concentration of dissolved Na, Al and Si was investigated. The results indicate that the mechanism involves three successive steps: 1) rapid exchange of Na$^+$ with H$^+$, 2) build-up of a residual layer depleted in Na and also depleted in Al under acidic conditions, and 3) a steady-state and congruent dissolution stage where the rate is controlled by a surface reaction between the residual layer and the solution. It might be possible to describe the kinetics of this steady-state stage in terms of activated surface complexes, but their nature is certainly more complicated than previously considered.

INTRODUCTION

We intend to present here a synthesis of the results of a study of weathering of albite conducted with a continuous flow reactor, based on the fluidized bed technique. This new experimental approach allowed us to obtain precise and simultaneous measurements of the rate of release of Na, Al and Si during different stages of feldspar weathering. Furthermore, the influence of various parameters such as pH, concentration of dissolved Na, Al and Si, on the reaction rate has been investigated and quantified. The large amount of information provided by these experiments will be used to test different mechanisms proposed earlier in the literature. More precisely, we would like to show here that it is possible to reconcile broadly the apparent contradictions among the previous interpretations and to propose a sequence of reaction steps which are consistent with the experimental observations. We will briefly summarize first the main weathering mechanisms previously described in the literature. We will then successively analyze and discuss the various steps of the reaction according to our experimental results.

LITERATURE REVIEW

When freshly ground feldspars are exposed to aqueous solutions, one first observes, at the very beginning of the reaction, a large amount of alkali ions released to solution and an increase in pH, suggesting that the main process occurring during this initial stage is the exchange of alkali ions on the feldspar surface for protons. Tamm (28) first demonstrated that this exchange was reversible for K-feldspar and Nash and Marshall (20) extended this concept for a whole range of feldspars. Busenberg and Clemency (4) estimated that the depth to which cations were exchanged in microcline, oligoclase and anorthite after a few minutes of reaction, was equal to or slightly more than one unit cell. However, they observed that this exchange reaction proceeds continuously at a decreasing rate, involving deeper layers. Garrels and Howard (11) tried to interpret more quantitatively the exchange reaction in terms of a chemical equilibrium between the aqueous solution and the feldspar considered as an ion exchanger. They concluded from their experimental results that the reaction may be expressed by a constant that includes the activities of the dissolved species and the concentration of the surface sites occupied by potassium.

The rapid exchange process is followed by the dissolution reaction itself, during which various elements of the feldspar framework are released to the solution. The numerous hypotheses presented in the literature concerning the dissolution mechanisms and the rate limiting steps during the weathering process can be classified into two broad categories. In the first category, the dissolution rate is controlled by the formation of a residual layer at the surface of the reacting mineral, through which the reactants and products of weathering must diffuse. In the second category, the dissolution rate is continuously controlled by the reaction of the unaltered feldspar with the solution at the solid-solution interface. We will consider these two models in more detail.

(a). Residual Layer Model

The hypothesis of the formation of a residual layer on the surface of the reacting feldspars is mainly based on the fact that the ratio of alkali ions to silicon to aluminum observed in solution during the reaction does not correspond to the stoichiometric ratio in fresh feldspar. This lack of congruency may be explained by differential leaching of the various components, leaving a residual layer at the surface of the mineral (4, 7, 8, 19 and 29). The rate of the reaction is then controlled by diffusion of the feldspar components through this protective layer. The dissolution of feldspars was usually studied with batch reactors by following the composition of the solution as a function of time. The evolution of the concentrations with time indicates that the dissolution kinetics, observed after the rapid initial exchange reaction, follow satisfactorily a parabolic law (4, 15, 18, 19 and 29). This behavior supports the hypothesis of a reaction stage during which the rate is governed by diffusion through a growing layer (19, 21 and 29).

For longer reaction times, the change of concentration is no longer parabolic and can be better approximated by a linear increase with time. Wollast (29) and Helgeson (13) suggested that the departure from the parabolic law was due to secondary reactions occurring between the residual layer and dissolved compounds, as soon as the solution becomes oversaturated with respect to some alumino-silicates. However, Correns and von Engelhardt (8) observed that during this stage the feldspar dissolution approached congruency and concluded that this step corresponded to a steady-state where the layer reached a constant thickness due to the dissolution of its outer surface. Luce et al. (19) also observed short-term incongruency and long-term near congruency for the dissolution of magnesium silicates.

Paces (21) outlined a more detailed description of the weathering model of albite and formulated it mathematically. In his model, similar to the picture given by Correns and von Engelhardt (8) he assumes that sodium is rapidly removed from the feldspar framework and its rate of release is controlled by diffusion through the depleted layer. The residual aluminosilicate layer collapses due to excess strain and dissolves at the solid-solution interface. At the beginning of the reaction, the diffusion of sodium is faster than the structural collapse of the residual layer, and the concentration gradient becomes progressively smaller as the dissolution proceeds. Consequently, the rate of diffusion becomes slower and finally equals the rate of retreat of the outer layer. At this stage the thickness of the layer becomes constant and the dissolution congruent.

(b). Surface Reaction Model

During her hydrothermal studies on the weathering of feldspars, Lagache (17 and 18) found that fresh crystals and previously altered ones dissolve the same way in a given aqueous solution. Furthermore she observed that the rate of dissolution was roughly proportional to the square root of the surface area, instead of simply proportional to the surface area as predicted by various diffusion models. She suggested that the rate was thus rather controlled by the reaction of the solution with the unaltered feldspar, at the solid-solution interface.

Petrović et al.(24), Holdren and Berner (16) and Schott et al. (25) examined the surface of weathered feldspar and other silicates using X-ray photoelectron spectroscopy (XPS) and were unable to detect the existence of a leached layer thicker than a few angstroms. They also concluded that the reaction was thus not controlled by diffusion, although it is not evident that sample preparation for XPS measurements does not alter the surface of the reacted mineral.

If the surface of the weathered feldspar has the same composition as the bulk mineral, then the dissolution step must be congruent and the departure from the stoichiometric ratio observed for the dissolved components must be due to secondary reactions, such as precipitation of alumino-silicates distinct from the feldspar surface (15, 17 and 23). In batch experiments, precipitation of Al-hydroxide and/or of alumino-silicates is likely to occur due to the oversaturation realized during these experiments (3, 13 and 14).

Holdren and Berner (16) explained the high initial rate of dissolution, corresponding to the parabolic stage described above, by the preferential dissolution of fine particles attached to the surface of the larger grains. They found that pretreatment of the sample with HF would suppress this effect and that only the linear part of the dissolution kinetics was then observed.

Attempts to reconcile parabolic dissolution kinetics with surface controlled reaction mechanisms have been also based on the formation of etch pits (12) which have been observed in naturally weathered or artificially etched feldspars (16). Dibble and Tiller (9) discussed in more detail the time dependence of interface reaction velocities and showed that linear or logarithmic rate laws applied when uniform detachment or a layer source generation mechanism such as screw dislocations controlled the dissolution rate. A parabolic time dependence could result from a change in surface detachment parameters if they were functions of the square root of time, but these authors were not able to elucidate the origin of such a dependence. Alternatively, Holdren and Adams (15) tried to explain the apparent parabolic law by secondary precipitation distinct from the feldspar surface. However, their precipitation experiments were conducted in spiked solutions containing excessive amounts of dissolved Si and Al, which are not encountered in natural environment or conventional dissolution experiments.

Recently, Aagaard and Helgeson (1) and Helgeson et al. (14) have developed an overall mechanism consistent with the steady-state model of silicate hydrolysis where the reaction rate is surface controlled. They have tested their model with numerous experimental results published earlier. This mechanism is based on the transition state theory in which a critical activated complex is formed at the surface of the mineral. The rate limiting step corresponds to decomposition of the critical activated complex, which is irreversible. Far from equilibrium the rate is controlled solely by decomposition of this complex, but near equilibrium it is proportional to the chemical affinity of the overall reaction and depends thus also on the final products of the reaction. According to this theory, the activated complex is in equilibrium with its reactants.

The more recent tendency in the literature is thus to consider that the rate limiting step during silicate dissolution is related to a surface phenomenon, and the hypothesis of diffusion control is now commonly discarded because of the failure to identify a residual layer with the aid of modern spectroscopic methods. Our present work indicates that this tendency may not be entirely justified.

It is also obvious from the literature that more carefully defined experiments need to be conducted. As pointed out by Dibble and Tiller (9) surface reactions are usually very sensitive to the composition of the solution. In the batch reactor experiments, the composition of the solution changes continuously with time and it is difficult to identify and to quantify the influence of the individual dissolved compounds on the reaction rate. So far only the influence of pH on the rate has been investigated in some detail. Also, very often, not all the dissolved components of the weathering of feldspars were determined throughout the course of the reaction. Our main goal was to develop a continuous flow-through reactor based on the fluidized bed technique in order to study the dissolution of feldspar under better-controlled conditions.

EXPERIMENTAL METHODS

We will only briefly summarize here our experimental methods which have been described elsewhere in more detail (5). The fluidized bed reactor is similar to the packed column used by Correns and von Engelhardt (8) where water is continuously percolated through a bed of mineral grains. However, in our reactor, the solid is maintained in suspension by the liquid flow and continuously well mixed. There are thus no strong gradients of concentration in the aqueous and solid phases as exhibited in the packed column. The concentration of the dissolved species can be easily maintained at levels well below saturation with respect to any possible precipitates. Therefore, one can study exclusively the dissolution of the silicate minerals without interference from secondary precipitation. Furthermore, it is possible to evaluate the effect of various chemical conditions on the rate of dissolution of the same sample of solid, by changing abruptly the composition of the input solution without manipulating the solid phase. The rate of dissolution is obtained simply by multiplying the renewal rate by the difference in concentration of the elements between the input and the output solutions, and can then be normalized with respect to the total surface area of the solid.

Amelia albite of 50-100 μm is used in all dissolution experiments conducted with the fluidized bed reactor. Sample preparation was described in Chou and Wollast (5). The surface area of the starting material, obtained with B.E.T. measurement by nitrogen adsorption, is 750 ± 150 cm^2/g. All the experiments were conducted at room temperature and pressure. Most of the experiments were started with distilled-deionized water as the input solution. HCl solutions and various hydroxide solutions of different normalities were used as the acidic and alkaline media. At the output of the reactor, sample solutions were collected for pH, Si, Al and Na measurements. Si, Al and Na were determined by colorimetry, fluorimetry and flame atomic absorption respectively. The analytical methods were chosen for their high sentivities. The detection limits were found to be 1 ppb for Si, 0.1 ppb for Al and 1 ppb for Na respectively. The input solutions were also spiked with various amounts of Si, Al or Na in order to study the influence of their concentrations on the dissolution rate. Exchange experiments, however, were conducted using batch reactors.

THE INITIAL EXCHANGE STEP OF NA$^+$ BY H$^+$

The initial exchange of Na$^+$ by H$^+$ was studied using batch reactors where various amounts of albite were added to pure water (6). The exchange reaction can be represented by

$$NaAlSi_3O_8 + H^+ = HAlSi_3O_8 + Na^+ \qquad (1)$$

The suspension was then rapidly back-titrated with NaCl solutions. During the back-titration one observes a decrease in pH due to the displacement of hydrogen fixed at the feldspar surface by Na$^+$, confirming the reversibility of the above reaction (equation 1). The pH and pNa were continuously measured with electrodes and were recorded during the experiments. From the changes in pH and pNa, it was then possible to calculate, at each step of the back-titration, the amount of Na restored in the feldspar structure. The results of the back-titration experiments indicated that about 99% of the exchangeable Na sites were initially replaced by H$^+$ immediately after the addition of albite to pure CO$_2$-free water. It is thus possible to evaluate the total number of sites from the initial increase in pH. These results show that the total amount of Na exchanged during this rapid initial step corresponds to that initially present in a layer of the feldspar one unit cell thick, in agreement with the finding of Busenberg and Clemency (4).

If the back-titration is carried out rapidly (within 1-2 hours), the exchange reaction is perfectly reversible as observed by Tamm (28) and the equilibrium condition may be represented by

$$K_E = \frac{a_{Na^+} \cdot a_{HAlSi_3O_8}}{a_{H^+} \cdot a_{NaAlSi_3O_8}} \tag{2}$$

where a denotes the activity of the subscript species.

The mass balance condition allows us to calculate the mole fraction of the surface occupied by the species $HAlSi_3O_8$, here represented by N_{HX}. The mole fraction of $NaAlSi_3O_8$ is then given by $N_{NaX} = (1 - N_{HX})$. If we assume that these solid species act as regular solid solutions at the surface of the feldspar, then for two component systems the activity coefficient, γ_i, may be written as (27):

$$\gamma_i = \frac{a_i}{N_i} = e^{\frac{\omega}{RT}(1-N_i)^2} \tag{3}$$

where ω is the excess enthalpy of mixing of the solid solution, R is the gas constant and T is the absolute temperature. Introduction of this relation into equation (2) and expansion of the exponentials into series yield, as a first approximation (10),

$$K_E = \frac{a_{Na^+}}{a_{H^+}} \cdot \left(\frac{N_{HX}}{1-N_{HX}}\right)^n \tag{4}$$

where $n = 1 - (\omega/2RT)$.

Our experimental observations are in good agreement with the behavior predicted by equation (4) except in some cases where N_{HX} becomes small (< 0.25). Fig. 1 shows, as an example, that a straight line is obtained when $\log(a_{Na^+}/a_{H^+})$ is plotted as a function of $\log(N_{NaX}/N_{HX})$, as might be expected if equation (4) holds. The best fit is obtained for $n = 0.63$ ($\omega = 1.83$ Kj/mol) and $K_E = 10^{5.76}$. The large value for the exchange constant K_E indicates that the Hydrogen feldspar $HAlSi_3O_8$ is much more stable than albite in the presence of fresh water and that in almost all the experimental studies published in the literature the surface of the solid in contact with water is essentially depleted in alkali ions. It is interesting to note that in sea-water where $\log(a_{Na^+}/a_{H^+}) \simeq 7.8$, the reverse is true and most of the feldspar surface sites would be occupied by Na. Thus, these calculations confirm that during the weathering experiments of feldspars the aqueous solution is in contact with a sodium depleted layer of about one unit cell depth, after a few minutes of reaction. Another interesing conclusion provided by the exchange experiments is that the behavior of the solid species on the surface may be described in terms of a regular solid solution

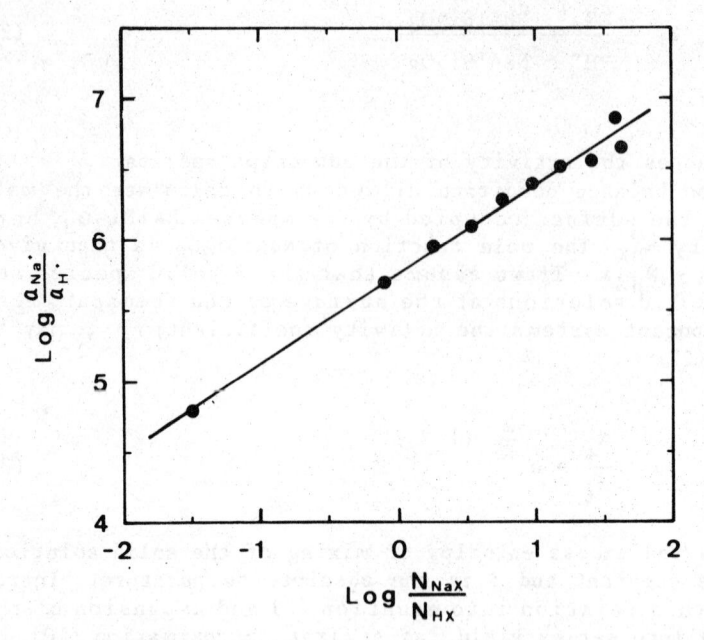

Figure 1. Exchange properties of albite in pure water.

model and approximated by equation (4). Our conclusion differs from that of Garrels and Howard (11) who suggested that the activity of $HAlSi_3O_8$ remained constant during their back-titration experiments conducted with orthoclase.

THE HIGH INITIAL RATE OF DISSOLUTION

As discussed above in the literature review, the high initial rate of dissolution of silicate minerals has been attributed to either the formation of a residual layer or the preferential dissolution of fine particles and of other active parts of the solid.

In order to distinguish between these two hypotheses, long-term dissolution experiments were carried out at room temperature under various pH conditions (pH 1-12). Fig. 2 shows as an example the results of an experiment which was started with distilled-deionized water open to the atmosphere and the reactor was then fed consecutively with HCl solutions of various concentrations. This figure gives the evolution of the concentrations of the feldspar components in the output solution of the reactor which are also directly related to the rate of release of the elements. The observed rates of release at the beginning of the experiment are very high for all three elementsand decrease

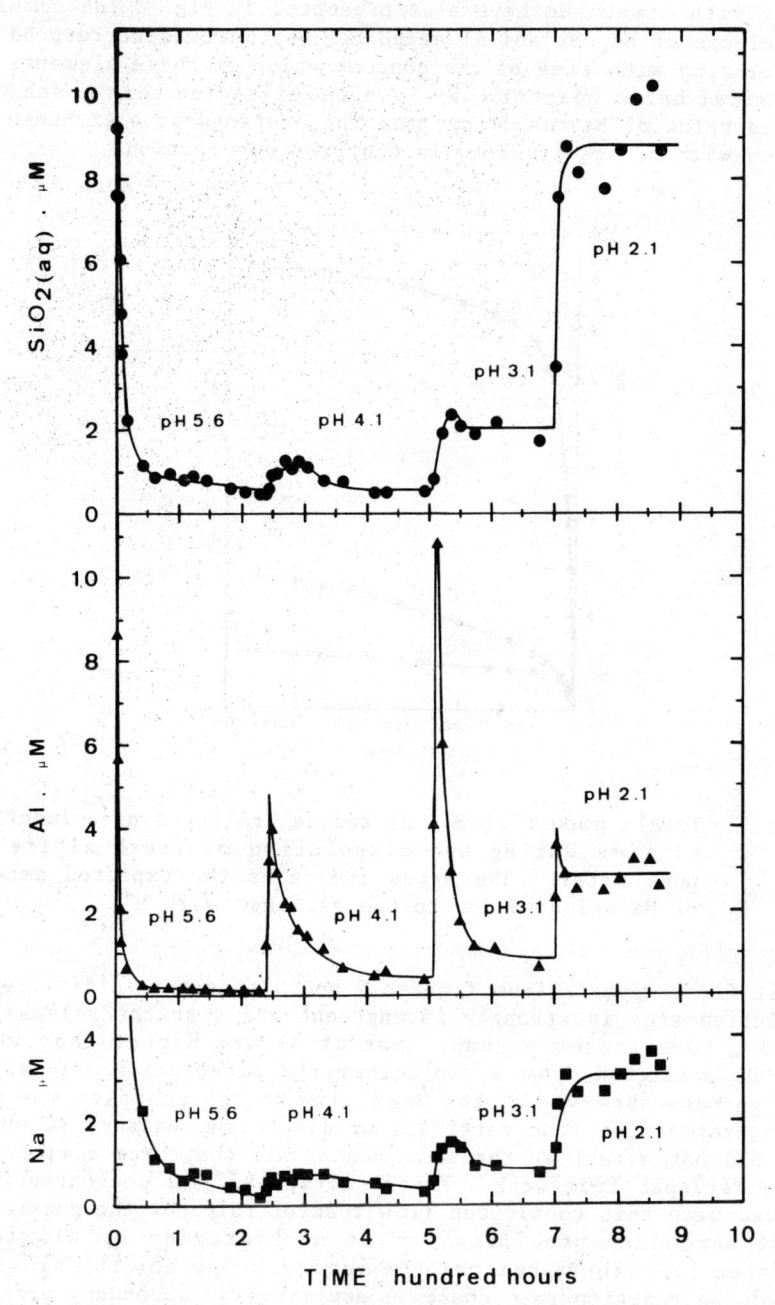

Figure 2. Evolution with time of the concentration of dissolved Si, Al and Na in the output solution of the reactor under various pH conditions.

rapidly with time. We have also presented in Fig. 3 the cumulative release of Na, Si and Al at pH 5.6 which would correspond to the evolution with time of the concentration of these elements in a classical batch reactor. We have indicated on this graph the expected value of Na resulting from the instantaneous exchange of this ion with H^+ as discussed in the previous section.

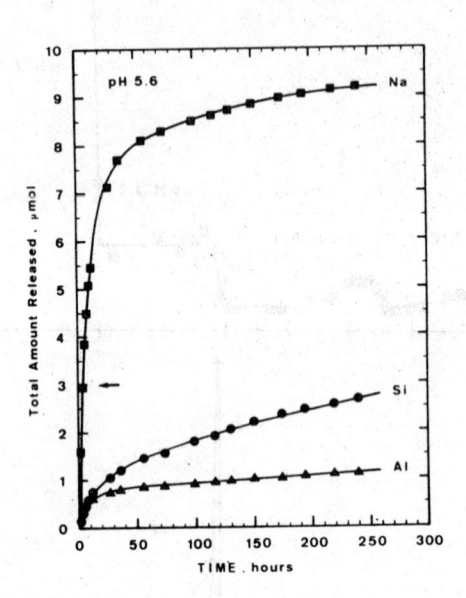

Figure 3. Total amount of Si, Al and Na released as a function of time during the dissolution of fresh albite in pure water. The arrow indicates the expected amount of Na released due to the exchange with H^+.

It first appears from figures 2 and 3 that the initial rapid dissolution step is strongly incongruent and that the release of Na and , to a lesser extent, that of Al are higher than what would be predicted from a stoichiometric dissolution. Thus, if the high rate observed at the beginning of the reaction was due to the presence of fine particles or disordered phases, it obviously did not affect in the same manner all the three components of the feldspar framework. Furthermore, the low concentrations observed with this continuous flow reactor rule out the possibility of secondary precipitation of Al-hydroxide or aluminosilicates (5). Thus, neither the incongruency nor the slowing down of the reaction rate could be explained by secondary precipitation. These observations thus favor very strongly a preferential leaching of Na and, to a smaller extent, of Al during this early stage of weathering. A consequence of the incongruent character of the initial dissolution step is the formation of a residual solid whose composition differs from the initial feldspar.

If the composition of the input solution to the reactor is changed towards more acidic conditions, it can be seen from Fig. 2 that the rate of dissolution of Al is strongly enhanced by this change. A similar enhanced rate of release of silica occurs when the input solution is switched from acidic to alkaline conditions (Fig. 4) and it can even exceed, as in this case, the initial rate obtained with fresh feldspar. Therefore, the presence of fine particles or disordered areas in the fresh material is not the only cause of the observed high initial rate.

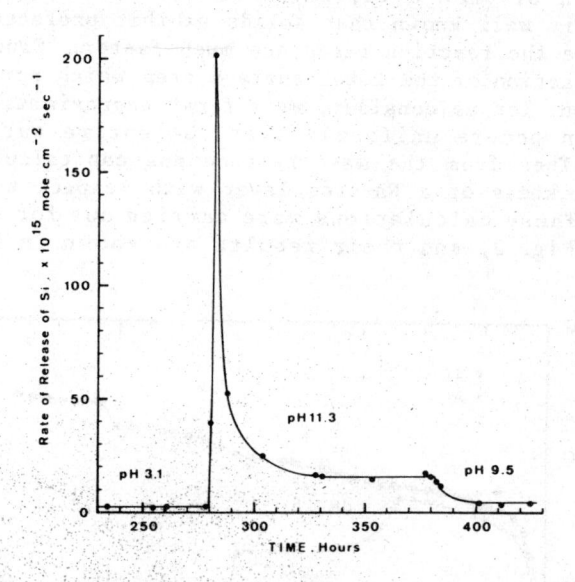

Figure 4. Evolution of the concentration of dissolved Si at the output of the reactor during the transition from acidic to alkaline solutions.

Our explanation is that when fresh feldspar is placed in contact with an aqueous solution, in addition to the instantaneous exchange of Na, the surface reacts rapidly until a residual layer builds up whose composition is dependent on the composition of the solution. Similarily, if weathered feldspar is placed into a solution of different composition, there is a rapid adjustment of the composition of this residual layer resulting in a preferential release of some components. It should be noted that the compositional change affects mainly the Si or Al content of the layer.

On the other hand, the experiments of Holdren and Berner (16) further confirmed by Schott et al. (25) on the suppression of the high initial rate of dissolution stage by pretreatment of the samples by HF, seems to demonstrate on the contrary the predominant influence of fine particles. However, Perry et al. (22) have shown that a pretreatment with highly concentrated HF solutions induces the formation of a fluorinated layer at the surface of the silicate minerals. We thus believe that pretreatment with HF as proposed by Holdren and Berner (16) may well result in the formation of a disturbed layer, which may affect strongly the kinetics of the early stages of dissolution.

It is tempting to try to evaluate the thickness and the composition of such a layer. This can be done only crudely because it is well known that solids exhibit preferential active sites, where the reaction rates are much faster. Since we do not know the fraction of the total surface area which participates in the reaction, let us consider, as a first approximation, that the dissolution occurs uniformly over the entire surface of the feldspar. Then from the mass balance one can calculate a theoretical thickness of a Na-free layer with respect to silica and alumina. These calculations were carried out for the run presented in Fig. 2, and their results are shown in Fig. 5. The

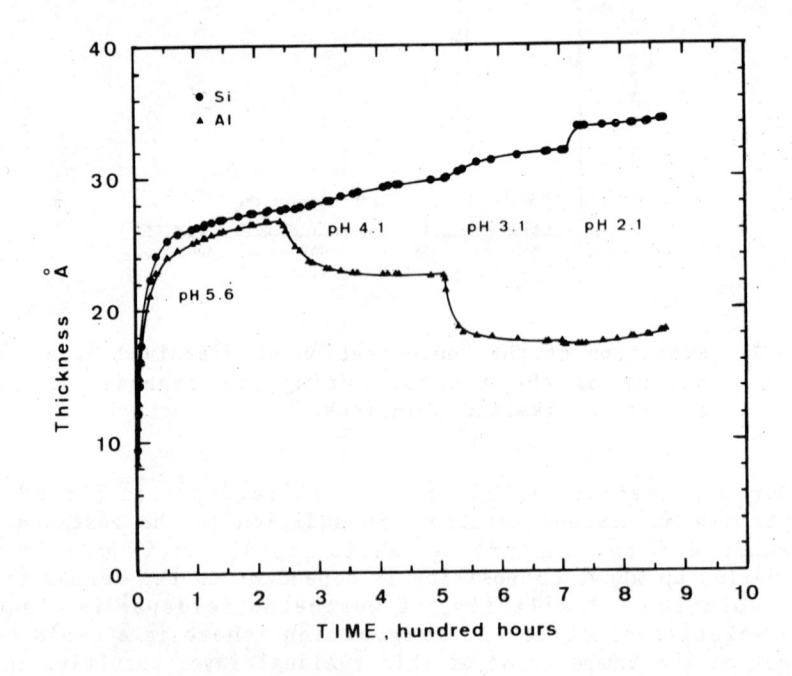

Figure 5. Calculated thickness of the residual Na-free layer with respect to Si and Al for the experimental run presented in Figure 2.

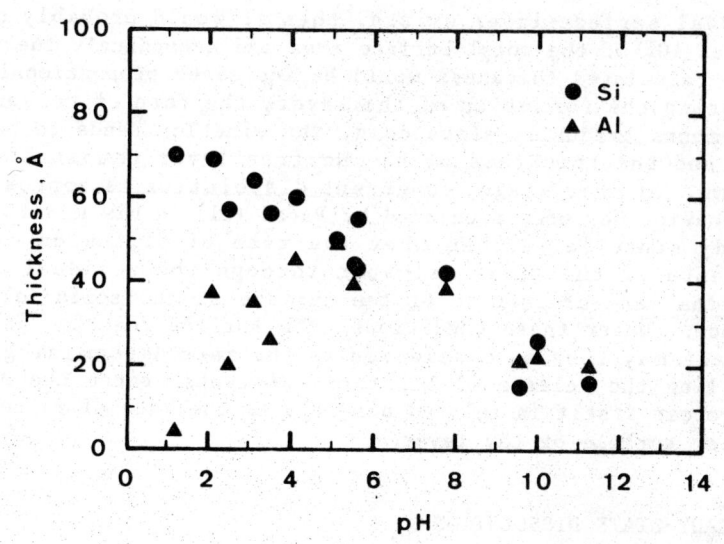

Figure 6. Influence of pH on the thickness of the remaining silica and alumina layer calculated from various dissolution experiments of albite.

thickness of this depleted layer is about 30 angstroms but differs for silica and alumina. When the pH is decreased the remaining silica layer becomes thicker than that for alumina or, in other words, the surface layer is progressively enriched in silica. Similar results for different runs have been presented in our previous paper (5) and are summarized in Fig. 6. In this figure we have presented the final thickness of the Na-free layer with respect to Si and Al, calculated for each pH condition. The thickness of the residual silica layer decreases almost linearly with increasing pH and approaches zero under extremely alkaline conditions. This can be easily explained if one considers that the replacement of Na in the feldspar framework results from the counter-diffusion of H^+ in the residual layer.

At pH above 5, the calculated thickness for Al is essentially the same as that for Si and thus we may consider that the Na-free layer has the same Al/Si ratio as the initial feldspar. The formation of this layer consists thus simply of the extraction of sodium from the feldspar framework. Below pH 5, Al is also extracted from this layer and around pH zero the remaining layer is composed of almost pure silica.

As pointed out earlier, not the entire surface is involved in the dissolution process. However, the XPS measurements of Holdren and Berner (16) seem to suggest that a large fraction of the total surface area is nonreactive, since they failed to identify the existence of a leached layer thicker than 20 Å. If a residual surface layer exists, then it would probably cover less than 10% of the total surface area and accordingly the value for the calculated thickness would be increased proportionally.

During the formation of the layer, the rate of release of the elements gradually slows down, the kinetics tends to become linear and the thickness of the Na-free layer remains fairly constant. At this stage, congruent dissolution is approached. This behavior has been predicted by Pačes (21) in his model where a steady state is reached when the rate of diffusion of the components of the fresh feldspar through the residual layer equals the rate of retreat of the surface at the solid-solution interface. Under these conditions, the surface reaction at this interface may thus be considered as the rate determining step controlling the release of all three elements, since the diffusion process itself is related directly to the rate of retreat of the outer surface of the layer.

THE STEADY-STATE DISSOLUTION

The steady-state stage of dissolution of feldspars corresponds to the linear increase of concentration observed in the case of batch reactors, in the absence of secondary precipitation. In the case of our continuous-flow reactor the steady state expresses itself by maintaining a constant concentration of all the dissolved components in the output solution. Our results indicate that the dissolution is then congruent within the limits of uncertainty of our experimental methods. Since we are following simultaneously the release of all three components of the feldspar we obtain thus three coherent evaluations of the rate of dissolution of feldspars during the steady state.

We will first consider (Fig. 7) the influence of pH on the steady state deduced from the experiment presented in Fig. 2 and from many others performed during this and previous studies (5). Fig. 7 indicates that the influence of pH on the rate r can be described empirically by a sum of three terms:

$$r = 10^{-13.69} \cdot a_{H^+}^{0.49} + 10^{-16.15} + 10^{-18.15} \cdot a_{H^+}^{-0.30} \tag{5}$$

where r is expressed in terms of mol cm^{-2} sec^{-1}.
Helgeson et al. (14) have also estimated the rate dependence on pH from the experimental results published in the literature:

$$r = 10^{-12.7} \cdot a_{H^+} + 10^{-15.4} + 10^{-18.6} \cdot a_{H^+}^{-0.4} \tag{6}$$

The two rate expressions (equations 5 and 6) are similar except at low pH where the slope that we have obtained (0.49) is significantly lower than that of 1 estimated by Helgeson et al. (14)

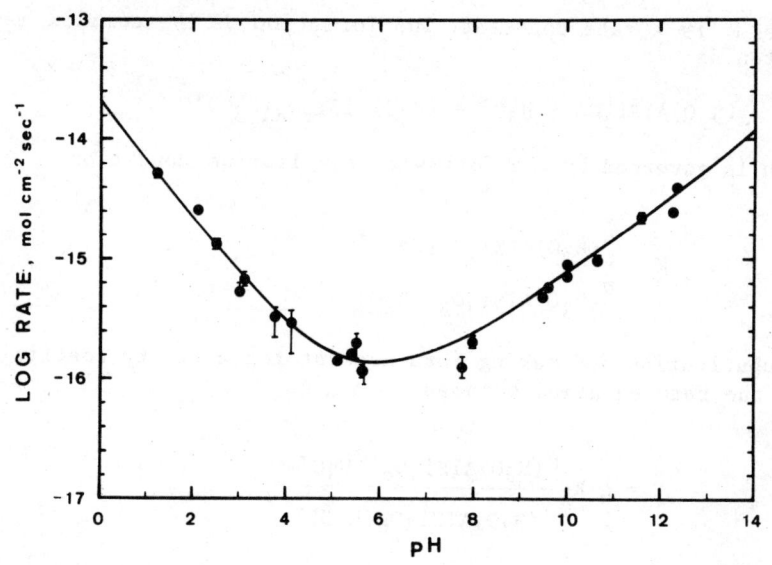

Figure 7. Influence of pH on the rate of congruent dissolution of albite during the steady-state stage. The bars represent the extreme values observed.

mainly based on the high temperature and pressure data of Lagache (17) extrapolated to 25°C and 1 atm. The more abundant data that we have now obtained in the acidic range make our evaluation more reliable.

Aagaard and Helgeson (1) and Helgeson et al. (14) have attributed the influence of pH on the reaction rate to the formation of various activated surface complexes, which are pH dependent. According to the transition state theory, the formation of these complexes is governed by an equilibrium condition involving the reactants and products, and the rate determining step is related to the decomposition of the activated complexes. According to these authors the most likely activated complexes responsible for limiting the rate of hydrolysis of feldspars are $(H_3O)AlSi_3O_8^+$, $(H_3O)AlSi_3O_8(H_2O)_n$ and an activated complex of stoichiometry $NaAl(OH)_{0.4}Si_3O_8^{-0.4}$, each one being predominant in the acidic, neutral and alkaline range respectively. If the rate is dependent only on the concentration of the activated complexes, then the influence of pH on the reaction rate is related to the equilibrium condition governing the formation of the activated complex. Considering more specifically the acidic conditions the rate is proportional to the concentration C of the prevailing activated complex:

$$r = k \cdot C_{(H_3O)AlSi_3O_8(H_3O)^+} \tag{7}$$

where k is a rate constant. The formation of the complex may be written as:

$$(H_3O)AlSi_3O_8 + H_3O^+ = (H_3O)AlSi_3O_8(H_3O)^+ \tag{8}$$

which is governed by the following equilibrium condition

$$K = \frac{a_{(H_3O)AlSi_3O_8(H_3O)^+}}{a_{(H_3O)AlSi_3O_8} \cdot a_{H_3O^+}} \tag{9}$$

By substitution and taking into account the activity coefficients γ_i, the rate equation becomes

$$r = k \cdot K \cdot \frac{a_{(H_3O)AlSi_3O_8} \cdot a_{H_3O^+}}{\gamma_{(H_3O)AlSi_3O_8(H_3O)^+}} \tag{10}$$

Helgeson et al. (14) assume furthermore that $a_{(H_3O)AlSi_3O_8}$ and $\gamma_{(H_3O)AlSi_3O_8(H_3O)^+}$ are constant, in which case the reaction rate is simply proportional to $a_{H_3O^+}$ in agreement with the previous assumption made by these authors (see Fig. 7). Similarly, in the neutral range the concentration of the activated complex $(H_3O)AlSi_3O_8(H_2O)_n$ depends only on the activity of water which may be considered as constant and the resulting rate of hydrolysis of the silicate is thus constant and independent of pH in this range. In fact, the assumption that $a_{(H_3O)AlSi_3O_8}$ and the activity coefficients of the activated complexes are constant is valid if one assumes that the concentration of the activated complex remains always very small compared to the total number of $(H_3O)AlSi_3O_8$ sites.

This assumption may not be justified if one considers the broad range of pH covered. In the case of the formation of charged complexes at the surface of oxides and silicates, Stumm et al. (26, this volume) have shown that it is easy to titrate completely with a strong acid or base the total number of active sites. Also, the results of the exchange experiments discussed in a previous section indicate that it is possible to cover the whole range of occupancy of the $NaAlSi_3O_8$ sites by H^+ at the surface of the feldspar if the pH is adequately changed. Thus if the mole fraction of the activated sites does not remain small with respect to the remaining H-feldspar sites the assumption of Helgeson et al. (14) is no longer applicable and an approximated equilibrium condition such as equation (4) valid for regular solutions is preferable. The equilibrium condition for the formation of the activated complex under acidic conditions represented by equation (9) can then be rewritten as:

$$K = \frac{1}{a_{H_3O^+}} \cdot \left(\frac{X}{1-X}\right)^n \tag{11}$$

where X denotes the mole fraction of the activated complex $(H_3O)AlSi_3O_8(H_3O)^+$ and $(1-X)$ is that of the remaining $(H_3O)AlSi_3O_8$ sites. From equation (11), one obtains:

$$X = \frac{(a_{H_3O^+} \cdot K)^{1/n}}{1 + (a_{H_3O^+} \cdot K)^{1/n}} \tag{12}$$

If C_T is the total number of sites ($C_{(H_3O)AlSi_3O_8(H_3O)^+} + C_{(H_3O)AlSi_3O_8}$), then

$$C_{(H_3O)AlSi_3O_8(H_3O)^+} = X \cdot C_T \tag{13}$$

and the rate equation (7) becomes

$$r = k \cdot C_T \cdot \frac{(a_{H_3O^+} \cdot K)^{1/n}}{1 + (a_{H_3O^+} \cdot K)^{1/n}} \tag{14}$$

The rate dependence on pH becomes then more complex and the reaction order is no longer an integer. If $(a_{(H_3O)^+} \cdot K)^{1/n}$ is negligeable with respect to 1, then equation (14) reduces to

$$r = k \cdot C_T \cdot K^{1/n} \cdot (a_{H_3O^+})^{1/n} \tag{15}$$

The more extensive measurements that we have collected concerning the effect of pH on the rate of dissolution of albite indicates that n is close to 2 in this case. A fractional order of dependence on pH of the rate of dissolution of minerals has often been observed and can be explained by our approach.

On the other hand, it is also possible to explain fractional orders by the existence of activated complexes of a more complicated stoichiometry, such as the one proposed by Helgeson et al. (14) for the alkaline range. We believe that it is premature at this stage to speculate about the nature of these complexes as long as we ignore the influence of the concentration of other species on the reaction rate.

To illustrate this point, we will consider the influence of the concentration of Na, Al and Si on the rate of dissolution of albite in the acidic range. HCl solutions at pH 3, spiked with various amounts of these elements were used as input solutions to our fluidized bed reactor and the rate of dissolution during the steady-state stage were measured. The results are presented in Fig. 8. Dissolved Si and Na affect only slightly the rate of dissolution at this pH since the apparent order with respect to these elements is only around -0.1. In the case of dissolved Al,

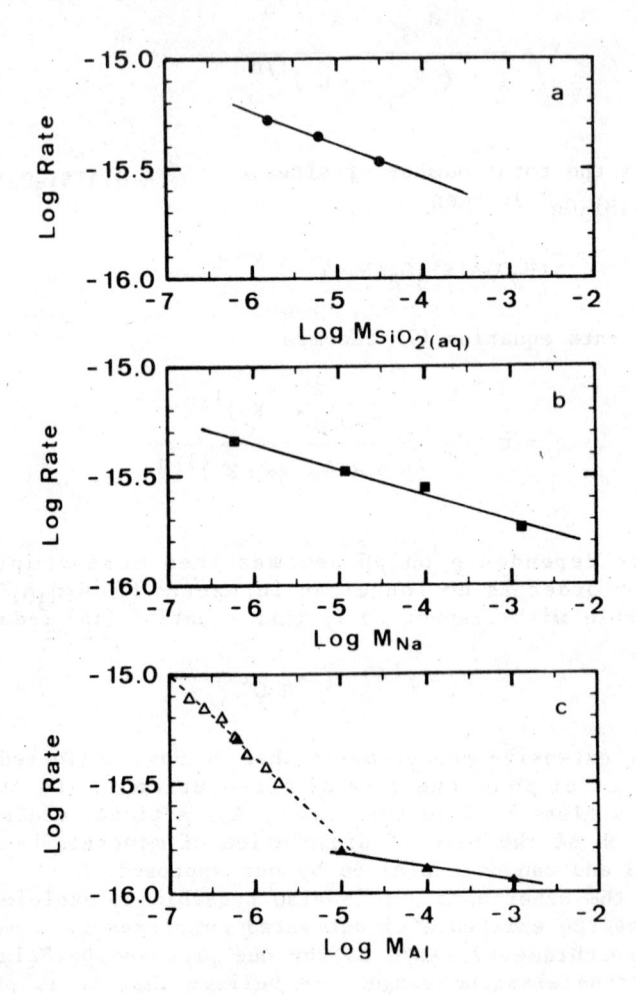

Figure 8. Influence of the concentration of dissolved (a) Si, (b) Na, and (c) Al on the steady-state rate of dissolution of albite at pH 3. The open triangles correpond to a separate set of experiments where the concentration change was obtained by changing the pumping rate. The rate and concentration are expressed in terms of mol cm^{-2}sec^{-1} and mol/l respectively.

the concentration of this element affects significantly the rate of hydrolysis of albite at concentrations below 10^{-5} mol/l. It should be noticed that this range of concentration covers the usual conditions encountered during our other experiments as well as those observed in batch reactors. This assertion and the fact that dissolved Al plays a major role in influencing the rate of dissolution are demonstrated by the experimental points represented in Fig. 8c (open triangles) and corresponding to the measured rates of weathering of albite observed when the pumping rate of the input solution is changed. In this case, the contact time between the solid and the solution in the reactor is modified and accordingly the concentration of the dissolved species are also changed. Fig. 8c shows that it is possible to explain entirely the observed rate by considering only the influence of the concentration of dissolved Al. If we want to interpret these results in terms of the surface activated complex theory, the stoichiometry of this complex should reflect the influence of dissolved Al. In other words, the equilibrium constraints of the formation of the complexes should include the concentration of Al with a appropriate exponent. Our main efforts are presently oriented toward this direction.

CONCLUSIONS

From our experimental study of the dissolution of albite with a continuous flow reactor, three successive steps could be identified for the weathering of fresh feldspars. The first one is the almost instantaneous and reversible exchange of the surface alkali ions with protons, which may be described in terms of a chemical equilibrium between an ion exchanger and the aqueous solution. The ion exchanger can be regarded as a regular solid solution.

The dissolution reaction itself starts with a rapid build-up of a residual layer whose composition depends on the composition of the aqueous phase. This layer is always depleted in alkali ions and its thickness increases with decreasing pH. In the case of albite, our results indicate that above pH 5 the Al/Si ratio in this layer remains the same as that in the initial feldspar but below this pH, Al is also progressively removed from the layer, leaving an almost pure silica residue at pH close to zero. During this stage the elements depleted in the residual layer diffuse from the fresh feldspar interface through the layer and the remaining elements dissolve at the outer surface, i.e. the solid-solution interface.

As the reaction progresses, the rate of release of all three components of the feldspar slows down until they reach a steady value. At this stage the thickness of the layer remains constant and the dissolution becomes congruent. This behavior corresponds to the steady state predicted by the theoretical model of Paĉes (21) where the rate of diffusion hindered by the build-up of the layer becomes equal to that of the retreat of the layer. Under these conditions, one may consider that the observed congruent rate of dissolution is controlled by the surface reaction occurring between the residual layer and the aqueous solution.

A study of the influence of pH, and of the concentration of dissolved Na, Al and Si indicates that this surface reaction depends not only on pH. At pH 3 under which condition the dissolution of albite was studied more extensively, the rate is also affected by the concentration of dissolved Al. Application of the theory of the activated complex to the study of the kinetics of feldspar dissolution, as demonstrated by Helgeson et al. (14), seems to be promising. However, considerably more experimental data are needed in order to identify the nature of the activated complexes and to quantify the rate equations.

It must be emphasized that in the natural environment, the steady state dissolution step is the main process controlling the weathering of feldspars. If the residence time of the solution in contact with the rocks is sufficiently long, the concentration of the dissolved elements may eventually reach oversaturation values high enough to induce secondary precipitation. The overall weathering reaction will then exhibit an apparent incongruent character.

ACKNOWLEDGEMENT

We would like to thank W. M. Murphy who kindly reviewed our manuscript and provided critical comments which improved the clarity of our paper. J. I. Drever made helpful editorial suggestions. The interesting discussions with many of the participants of this NATO conference are also acknowledged. This work was partially supported by a grant from the Solvay Company.

REFERENCES

1. Aagaard P. and Helgeson H. C. 1982, Thermodynamic and kinetic constraints on reaction rates among minerals and aqueous solutions. I. Theoretical considerations. Amer. J. Sci. 282, pp. 237-285.
2. Berner R. A. and Holdren G. R. Jr. 1979, Mechanism of feldspar weathering: II. Observations of feldspars from soils. Geochim. Comochim. Acta 43, pp. 1173-1186.
3. Busenberg E. 1978, The products of the interaction of feldspars with aqueous solutions at 25°C. Geochim. Cosmochim. Acta 42, pp. 1679-1686.
4. Busenberg E. and Clemency C. V. 1976, The dissolution kinetics of feldspars at 25°C and 1 atm CO_2 partial pressure. Geochim.

Cosmochim. Acta 40, pp. 41-49.
5. Chou L. and Wollast R. 1984, Study of the weathering of albite at room temperature and pressure with a fluidized bed reactor. Geochim. Cosmochim. Acta (in press).
6. Chou L.and Wollast R. (in preparation), Study of the initial exchange reaction of albite in aqueous solutions.
7. Correns C. W. 1963, Experiments on the decomposition of silicates and discussion of chemical weathering. Clays and Clay Min. 10, pp. 443-459.
8. Correns C. W. and von Engelhardt W. 1938, Neue Untersuchungen über die Verwitterung des Kalifeldspates. Chemie der Erde 12, pp. 1-22.
9. Dibble W.E. Jr. and Tiller W. A. 1981, Non-equilibrium water/rock interactions. I. Model for interface-controlled reactions. Geochim. Cosmochim. Acta 45, pp. 79-92.
10. Garrels R. M. and Christ C. L. 1965, Solutions, Minersls and Equilibria. Freeman, Cooper & Company, San Francisco, 450 pp.
11. Garrels R. M. and Howard P. 1959, Reactions of feldspar and mica with water at low temperature and pressure. In Proc. Sixth National Conference on Clays and Clay Minerals. pp. 68-88. Pergamon Pres.
12. Grandstaff D. E. 1978, Changes in surface area and morphology and the mechanism of forsterite dissolution. Geochim. Cosmochim. Acta 42, 1899-1901.
13. Helgeson H. C. 1971, Kinetics of mass transfer among silicates and aqueous solutions. Geochim. Cosmochim. Acta 35, pp. 421-469.
14. Helgeson H. C. and Murphy W. M. and Aagaard P. 1984, Thermodynamic and kinetic constraints on reaction rates among minerals and aqueous solution. II. Rate constants, effective surface area, and the hydrolysis of feldspar. Geochim. Cosmochim. Acta (in press).
15. Holdren G. R. Jr. and Adams J. E. 1982, Parabolic dissolution kinetics of silicate minerals: An artifact of nonequilibrium precipitation processes? Geology 10, pp. 186-190.
16. Holdren G. R. Jr. and Berner R. A. 1979, Mechanism of feldspar weathering. I. Experimental studies. Geochim. Comochim. Acta 43, pp. 1161-1171.
17. Lagache M. 1965, Contribution à l'étude de l'altération des feldspaths, dans l'eau, entre 100 et 200°C sous diverses pressions de CO_2, et application à la synthèse des minéraux argileux. Bull. Soc. Fr. Miner. Crist. 88, pp. 223-253.
18. Lagache M. 1976, New data on the kinetics of the dissolution of alkali feldspars at 200°C in CO_2 charged water. Geochim. Cosmochim. Acta 40, pp. 157-161.
19. Luce R. W., Barlett R. W. and Parks G. A. 1972, Dissolution kinetics of magnesium silicates. Geochim. Cosmochim. Acta 36, pp. 35-50.
20. Nash V. E.and Marshall C. E.1956, The surface reactions of silicate minerals. Pt. I. The reactions of feldspar surfaces

with acidic solutions. Univ. Missouri Agr. Expet. Sta. Res. Bull. 613, 36 pp.
21. Pačes T. 1973, Steady-state kinetics and equilibrium between ground water and granitic rock. Geochim. Cosmochim. Acta 37, pp. 2641-2663.
22. Perry D. L., Tsao L. and Gaugler K. A. 1983, Surface study of HF and HF/H_2SO_4-treated feldspar using Auger electron spectroscopy. Geochim. Cosmochim. Acta 47, pp. 1289-1291.
23. Petrović R. 1976, Rate control in feldspar dissolution. II. The protective effect of precipitates. Geochim. Cosmochim. Acta 45, pp. 1675-1686.
24. Petrović R., Berner R. A. and Goldhaber M. B. 1976, Rate control in dissolution of alkali feldspars. I. Study of residual feldspar grains by X-ray photoelectron spectroscopy. Geochim. Cosmochim. Acta 40, pp. 537-548.
25. Schott J., Berner R. A. and Sjöberg E. L. 1981, Mechanism of pyroxene and amphibole weathering. I. Experimental studies of iron-free minerals. Geochim. Cosmochim. Acta 45, pp. 2123-2135.
26. Stumm W. 1984, The effects of complex-forming ligands on dissolution of oxides and silicates (this volume).
27. Swalin R. 1972, Thermodynamics of Solids. 2nd ed. John Wiley & Sons, New York, 387 pp.
28. Tamm O. 1930, Experimentelle Studien über die Verwitterung und Tonbildung von Feldspaten. Chemie de Erde 4, pp. 420-430.
29. Wollast R. 1967, Kinetics of the alteration of K-feldspar in buffered solutions at low temperature. Geochim. Cosmochim. Acta 31, pp. 635-648.

INTERSTRATIFIED CLAY MINERALS AND WEATHERING PROCESSES

M.J. Wilson and P.H. Nadeau,

Department of Mineral Soils, The Macaulay Institute for Soil Research, Craigiebuckler, Aberdeen AB9 2QJ, Scotland, U.K.

ABSTRACT

Although interstratified clay minerals have been intensively studied, their actual physical character remains uncertain and, from a thermodynamic point of view, it is still not clear whether they should be treated as single phases or as polyphase aggregates. In this paper the major types of interstratified clay minerals are reviewed with particular emphasis on those that are characteristic of a weathering environment. It is concluded that these minerals form mainly by means of transformation reactions involving relatively large crystals of mica or chlorite. On the other hand, those interstratified clay minerals that are generally associated with diagenesis - which include regularly interstratified, partially ordered and randomly interstratified illite-smectite - consist of extremely fine particles that are typically some tens of Angstroms in thickness. For these clays it might be considered that interstratification is more apparent than real as it can be accounted for by an interparticle diffraction effect, whereby the interfaces between thin particles adsorb ethylene glycol and are, therefore, perceived by XRD as a smectite component. This concept is supported by experimental evidence using both X-ray diffraction and transmission electron microscopy. It is concluded that regularly or partially ordered interstratified clays can be regarded as single phase for thermodynamic purposes but that randomly interstratified clays must be generally treated as mixtures of two or more phases.

1. INTRODUCTION

Interstratified clay minerals occur widely in soils, sediments and rocks and have been studied intensively by X-ray diffraction and other methods. Nevertheless, the actual physical character of these minerals remains uncertain. At present, interstratified clays are viewed as consisting of physically separable crystallites that are themselves made up of a mosaic of discrete domains with variable compositions, but which collectively describe the average composition of the sample. The coherent domains within the crystallites are comprised of fixed sequences of silicate layers involving repetition of unit cells of different thicknesses, arranged in a random, ordered, or partially ordered way (42). Not surprisingly, interstratified clays have given rise to difficulties to those geochemists who seek to explain the conditions of formation and the occurrence of clay minerals from a thermodynamic point of view. Zen (65) first drew attention to this problem and pointed out that interstratified clays may behave thermodynamically as single phases or as polyphase aggregates. If a given interstratified mineral really is a single phase, as would indeed be indicated by the current concept described above, then it should be plotted as a single point on a phase diagram according to its bulk composition. On the other hand, it may be that an interstratified clay, particularly if randomly arranged, is no more than an intimate mechanical mixture of different components (66) and in these circumstances must be plotted as two or more phases. More recent treatments of the stability relationships of clay minerals have still had to contend with this problem. For example, Aagaard and Helgeson (1) analyzed the thermodynamic consequences of treating interstratified clays as solid solutions, assuming ideal mixing of atoms on equivalent structural sites, even though they recognized that there was considerable field and laboratory evidence for regarding these minerals as mixtures. It was emphasized that the activity/composition relations between soil/sediment waters and clay minerals did not allow definite conclusions to be drawn with respect to the phase status of interstratified clays in geochemical processes. It is clearly desirable, therefore, to clarify the phase status of these clay minerals in as direct a manner as possible.

In this paper the major types of interstratifications are reviewed with particular emphasis on those that form in weathering environments. X-ray diffraction (XRD) and transmission electron microscope (TEM) evidence is then presented for a new conceptual model of interstratified clays which has general implications for a better understanding of the origin of these minerals in the various environments in which they are found.

2. NATURE OF THE MAIN INTERSTRATIFIED MINERALS

The major types of interstratified clay minerals have been

TABLE 1

The major types of interstratified clay minerals classified according to the nature of the non-expansible component and the ordering of the structure

Non-expansible component / Ordering	Mica Di	Mica Tri	Chlorite Di	Chlorite Tri	Kaolin
Regular	Rectorite Mica-vermiculite	Hydrobiotite	Tosudite	Corrensite	
Partial	Illite-smectite	Mica-vermiculite	?	?	?
Random	Illite-smectite	Mica-vermiculite	Chlorite-smectite	Chlorite-smectite	Kaolin-smectite

extensively discussed by Sudo and Shimoda (54), a review highlighting the fact that interstratified minerals nearly always contain an expansible phase, either smectite or vermiculite. Interstratified clays where neither component is expansible are known, for example interstratified mica-chlorite (48), but such occurrences are very much the exception to the general rule. The non-expansible component usually consists of a mica or chlorite-like phase and may be dioctahedral or trioctahedral, corresponding approximately to aluminous and ferromagnesian compositions. In the present paper the various types of interstratified clay minerals are subdivided and discussed in terms of the nature of the non-expansible phase and the type of ordering, as indicated in Table 1.

2.1. Interstratifications with a mica-like phase

Rectorite (Allevardite). This dioctahedral mineral is one of the most completely regularly interstratified clay minerals known, being made up of alternating mica-like and smectite-like layers (4). The high contents of fixed sodium and aluminium suggest that the mica- and smectite-like layers are paragonitic and beidellitic respectively (5, 22). Typically, the XRD pattern of a highly-dispersed, well-oriented sample of rectorite shows an intense basal reflection at ~26Å (Fig. 1i) in the ethylene-glycol-solvated state with a large number of higher orders. This pattern corresponds to a regular 50:50 interstratification between a 9.6Å mica phase and a 16.9Å swelling phase. In hand specimen rectorite resembles soft, matted paper and it is, therefore, possible to examine thin flakes of undispersed material by single crystal techniques. The "single crystal" patterns so obtained reveal that the mineral consists of exceedingly fine platy crystals whose a and b axes are randomly orientated around the c^* axis (22). Under the electron microscope rectorite is seen to be composed of plates and ribbons which may be folded upon themselves in a highly characteristic manner. The thickness of individual particles is ~20Å which corresponds to the thickness of two 2:1 layers, the basic structural unit (60).

Partially ordered illite-smectite. Rectorites like those described above are highly restricted in occurrence but rectorite-like clays, where the interlayer cation in the non-expanding unit is potassium rather than sodium, occur widely in deeply buried pelitic sediments. Although a high-spacing reflection may be observed in these clays it is weaker and more poorly defined than in true rectorite (Fig. 1ii), as is the whole diffraction pattern. Furthermore the sequence of 00ℓ reflections may be non-integral. Reynolds and Hower (43) concluded that these types of clays were made up of rectorite-like units randomly interstratified with illite. In general, the proportion of illite layers exceeds 60%. Similar clays are found in hydrothermally altered material. They frequently show high spacings between 25 and 30Å (53) and the

probability parameters for the layer sequence, as deduced by
the transform method of MacEwan (29), typically indicate that one
expansible layer is never followed by another expansible layer (46,
53, 58). Where the proportion of illite approaches 80% then
non-nearest neighbour effects become evident in these types of
clays (43). The XRD trace typically shows a strong 9.9Å peak with
a subsidiary peak at 11.5Å after glycol treatment (Fig. 1iii).
Reynolds and Hower (43) concluded that this effect arises from
the long-range ordered interstratification of an IMII superlattice
with illite and that virtually all illite-montmorillonites with
expandibilities of <35% are partially ordered.

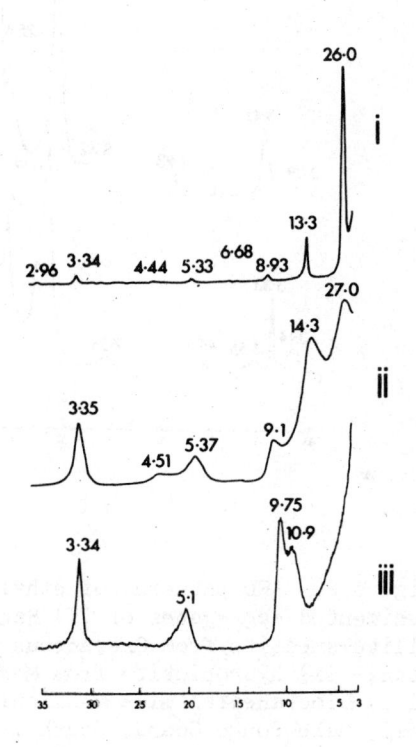

Figure 1. XRD patterns of ethylene glycol solvated
sedimented aggregates of <0.1 µm fractions of (i)
Rectorite from Baluchistan, (ii) Rectorite-like clay
from Cretaceous bentonite, Canon City, Colorado and
(iii) Long-range IMII interstratification from Devonian
bentonite, Tioga, New York. Spacing in Å and scale in
°2θ Coα radiation.

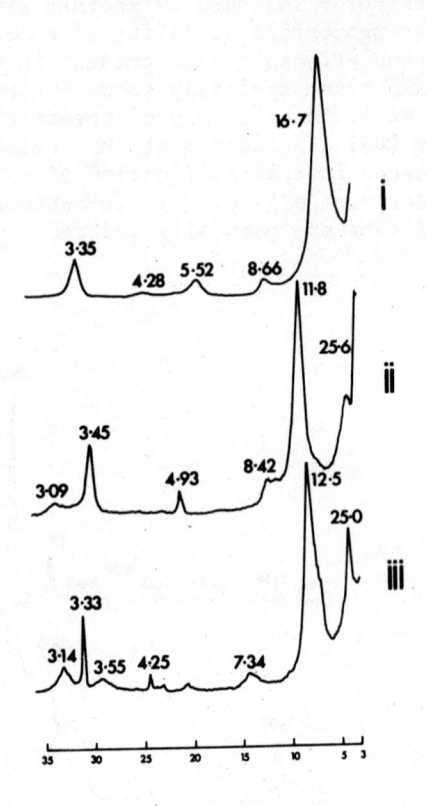

Figure 2. XRD patterns of ethylene glycol solvated sedimented aggregates of (i) Randomly interstratified illite-smectite from Cretaceous bentonite, Westwater, Utah, (ii) Hydrobiotite from Morogora, Tanzania (iii) Dioctahedral mica-vermiculite from a podzolized soil, Malborough Sound, South Island, New Zealand; this sample also contains a small amount of kaolin mineral and quartz.

Randomly interstratified illite-smectite. This type of interstratification is extremely widespread in sediments and in the air-dried state yields a strong broad basal reflection between 10 and 15Å. With smectite contents of between 40 and 60% and after ethylene glycol treatment, the XRD pattern typically shows a broad maximum at 17Å with a high background towards the low angle side and a non integral series of 00ℓ reflections (Fig. 2i). Both the partially ordered and randomly interstratified illite-smectites are aluminous and dioctahedral in nature.

Interstratified mica-vermiculite. This type of interstratification occurs both in trioctahedral and dioctahedral forms. The trioctahedral form is often in large crystals consisting of regularly alternating layers of biotite and vermiculite and yields an XRD pattern characterized by a high spacing reflection at ~24Å (Fig. 2ii) and a rational or nearly rational series of higher orders. This material is known as hydrobiotite. When the biotite:vermiculite ratio is other than 1:1 the XRD patterns often show rather broad peaks indicating some stacking disorder. There are, however, few detailed studies of the layer sequence in hydrobiotite. Dioctahedral mica-vermiculite yields an XRD pattern similar to that of hydrobiotite although the reflections may be sharper and better defined (Fig. 2iii).

2.2. Interstratifications with a chlorite-like phase

Corrensite. This mineral has a ferromagnesian composition and is now defined as a 1:1 regular interstratification of trioctahedral chlorite and trioctahedral smectite (2). Typically, the XRD pattern shows a high spacing basal reflection at 29Å in the air-dry state, which expands to 31Å after ethylene glycol treatment (Fig. 3i) and contracts to ~24Å after heating at 550°C. A large number of higher orders deriving from the high spacing reflection are observed, although the odd order reflections are very weak. A number of clay minerals similar to corrensite have been described where the swelling component is not truly smectitic, but is better described as swelling chlorite or vermiculite. In fact, corrensite was first described in terms of a regular interstratification of trioctahedral chlorite and trioctahedral swelling chlorite (26). In the broad sense corrensite occurs in many different environments, almost always as clay-sized material. Sutherland and MacEwan (55) have described single crystals of a chloritic material that was very close to corrensite (as originally defined) although the characteristic 28Å spacing was weak and diffuse and only observed occasionally.

Tosudite. This highly aluminous mineral is now defined as a 1:1 regularly interstratified dioctahedral chlorite-dioctahedral smectite (2). The sequence of (00ℓ) reflections and their responses to ethylene glycol and heating are very similar to corrensite but the d (060) reflection occurs at 1.50Å. The mineral occurs in clay-size particles in the form of thin, irregular flakes (6).

Randomly interstratified chlorite-smectite. These types of clay minerals, based on both tri- and di-octahedral structures, occur extensively. Their XRD characteristics are incomplete swelling with ethylene glycol, yielding a peak between 14 and 17Å, and incomplete contraction on heating, yielding a reflection

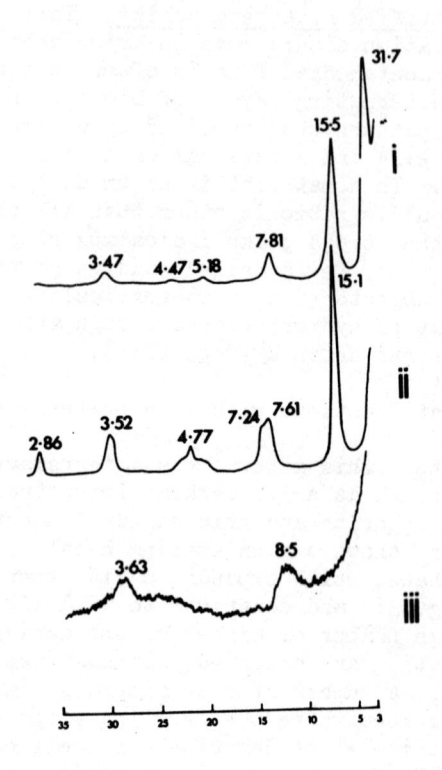

Figure 3. XRD patterns of ethylene glycol-solvated sedimented aggregates of (i) Corrensite from a vein filling in a Carboniferous dolerite from Hillhouse Quarry, Ayrshire, Scotland, (ii) Randomly interstratified chlorite-smectite from altered Devonian basalt, Angus, Scotland: the sample is impure and (iii) Randomly interstratified kaolinite-smectite from Tepakan, Yucatan, Mexico.

between 10 and 14Å. No high-spacing peak is observed, nor is a regular series of orders found (Fig. 3ii). The layer sequence of these clay minerals has not been studied in detail. In principle, there seems to be no reason why partially ordered chlorite-smectite should not occur in the same way as partially ordered illite-smectite but so far this type of interstratification has not been described.

2.3. Interstratifications of kaolin and smectite

This type of interstratification was first reported in the so-called acid clays of Japan (52). The XRD patterns of these clays show a very diffuse maximum in the low angle region superimposed on which are several broad peaks at about 15, 10, 8 and 7Å. Schultz et al. (47) showed that the most diagnostic XRD feature of these types of clays is the fact that the broad reflection in the 7 to 8Å region (Fig. 3iii) appears to increase in spacing and to intensify after heating at 300°C. TEM shows that the clays consist predominantly of extremely fine platy particles about 0.05-0.10 μm in diameter. Interstratified kaolin-smectite nearly always occurs in a random arrangement, as assessed by comparison with the calculated profiles of Cradwick and Wilson (10), although Schultz et al. (47) report some examples which seem to show a higher degree of order.

3. INTERSTRATIFIED MINERALS IN THE WEATHERING ENVIRONMENT

The extensive occurrence of interstratified clays in rocks and sediments means that in the right circumstances all varieties of these minerals could be inherited by soils and saprolites developed on these materials. It is more meaningful to enquire, therefore, into the kinds of interstratified minerals that are actually formed - as well as those that are not formed - in the weathering environment.

3.1. Interstratifications with a mica-like phase

There is no doubt that hydrobiotite, which consists of large crystals of regularly alternating biotite and vermiculite layers, does result to a large extent from the weathering of biotite. Hydrobiotite has been observed many times in sedentary soils developed upon biotite-rich material (9, 20, 63) and seems to form rather rapidly. This point was well-illustrated by a recent study of some youthful (<3000 years old) skeletal soils developed from quartz-mica-schist on the recently de-glaciated Signy Island in maritime Antarctica, where it was found that the fine sand fractions (63-125 μm) contained abundant hydrobiotite (Wilson, unpublished results). Moreover, hydrobiotite can be readily synthesized using dilute solutions to exchange interlayer potassium (41) or by oxidation of structural iron during cation exchange (12). Norrish (38) proposed a convincing mechanism for the formation of hydrobiotite, which involved sympathetic movement of the orientations of the hydroxyl on either side of an octahedral sheet. Removal of potassium from one side of the silicate sheet increases the angle of the (OH) bond direction on that side of the octahedral sheet, but causes a decrease in the angle of the (OH) bond on the other side. This results in the interlayer potassium on this side being placed in a more negatively charged and hence more stable environment. Thus, alternating interlayer regions are built up within single crystals where potassium is either more or less removed or held very strongly.

Regularly interstratified dioctahedral mica-vermiculite is occasionally found in soil clay fractions where it clearly derives from weathering of mica (7, 8). Such an origin was also demonstrated where this mineral was found in large amounts in some New Zealand podzols (25). The dioctahedral mica separated from the soils' parent rock - a fine-grained indurated greywacke - could easily be transformed into a regularly interstratified mineral, yielding high spacing reflections similar to those found in the soil clays, merely by gentle acid treatment (Wilson, unpublished results). A similar transformation of dehydroxylated sericite has been studied in detail by Tomita (57). In this case the regular interstratification was related to the 2 M mica polytype, as similar experiments with the 1 M form resulted in a random arrangement.

Further weathering of vermiculitized mica may result in a lower negative charge on the silicate sheet leading to the formation of smectite-like products which are capable of expanding with glycerol or ethylene glycol. This transformation can occur with both trioctahedral (17, 28) and dioctahedral (8) micas and in the latter instance leads to the formation of regular interstratifications with XRD characteristics similar to those of rectorite. These materials differ from true rectorite when examined under the TEM, however, in that they consist of thick, well-defined, platy crystals yielding a single crystal-type electron diffraction pattern. True rectorite typically consists of highly dispersible, fine-grained material characterized by a turbostratic-type electron diffraction pattern and is usually considered to be of diagenetic or hydrothermal origin. The partially ordered IIS and IIIS types of illite-smectites described by Reynolds and Hower (43) are also of diagenetic origin and, as far as the authors are aware, no evidence has been presented to show that such material forms in weathering environments. The same may be true of the randomly interstratified illite-smectites that are so characteristic of vast thicknesses of the sedimentary column (33, 34, 39, 59). Where this type of interstratification has been described in soils it is often the case that the clays have been inherited from the underlying argillaceous parent material (23, 61) and formed originally in a diagenetic environment.

3.2. Interstratification with a chlorite-like phase

Interstratified chloritic minerals occur extensively in soils, but usually in the form of the so-called "intergrade" or "intergradient" minerals (18). These minerals form frequently in acid soils following the introduction of non-exchangeable hydroxy-aluminium into the interlamellar spaces of expansible minerals.

Both di- and tri-octahedral vermiculites may be interlayed in this way (21, 36, 44, 62) as well as smectites (18). Magnesium interlayering of expansible minerals can also occur in Mg-rich environments. Interstratification is nearly always random unless there was a pre-existing, regularly interstratified silicate structure. Regularly interstratified chloritic minerals can, however, occur in soils and weathered material following the vermiculitization of chlorite, although the mechanism by which this process occurs is far from clear. A regularly interstratified chlorite-vermiculite yielding a high spacing of 28Å with many lower orders was described by Johnson (19) from the C horizon of a soil developed upon metamorphosed basalt in Pennsylvania and a similar mineral was reported by Herbillon and Makumbi (16) in a recent tropical soil derived from chlorite-schist in Zaire. In the latter instance at least, it was evident that the chlorite from which the interstratified product developed was already partly vermiculitized (30) but the experiments of Ross and Kodama (45) show conclusively that some true chlorites will break down to a regularly interstratified product. Although no definite mechanism for the transformation was suggested, the results indicated that oxidation reactions were implicated in the same way as found by Farmer and Wilson (12) for the biotite to hydrobiotite conversion. It is concluded, therefore, that regularly interstratified corrensite-like material can form by the weathering of chlorite although it seems that such occurrences are rather unusual. In general, both corrensite and tosudite must be regarded as being more characteristic of diagenetic and/or hydrothermal environments. Corrensite is usually associated with evaporitic sequences (13, 26, 51) carbonates (3, 40), hydrothermally altered basic and ultrabasic rocks (14, 50) and deep diagenesis (11). Tosudite nearly always seems to result from hydrothermal activity (54) although there is some evidence that it may form during diagenesis (24).

3.3. Kaolinite-smectite interstratifications

Although interstratified kaolinite-smectite has not been widely reported from soils and weathered materials, it may be more common than is appreciated at present. Thus Norrish and Pickering (38) concluded that is was possibly a very common major component of Australian soils. It was identified as the only clay mineral in about 40 profiles and because of its weak diffraction effects could have gone undetected in the presence of other well-crystallized clay minerals in many other profiles. Kaolinite-smectite has also been reported in soils from Scotland (64) and as a major component in some soils from Cameroun (15). It is also widespread in altered volcanic material in Japan, where it is thought to result from the alteration of montmorillonite under acid conditions (54). Indeed, this interstratification has recently been synthesized by Srodon (49) from acidified suspensions of montmorillonite.

3.4. Conclusions

The previous discussion suggests that, in general, those interstratified minerals that form in weathering environments do so by means of transformation reactions involving extensive layer sequences inherited from pre-existing layer silicate structures. Thus, the removal of interlayer K from mica and the decomposition of the brucitic sheet in chlorite leads to regular and irregular interstratified mica-vermiculite and chlorite-vermiculite respectively and, with further weathering, these minerals may become more smectitic. On the other hand, it seems that many of the most frequently studied interstratified species, such as rectorite, partially ordered illite-smectite, randomly interstratified illite-smectite, corrensite and tosudite, may not be really characteristic of weathering environments and the question arises as to the physical nature of these minerals. Is it the same as that of the interstratified minerals that form during weathering?

4. INTERPARTICLE DIFFRACTION AND A NEW CONCEPT FOR INTERSTRATIFIED MINERALS

In this section XRD and TEM evidence is presented to show that the minerals that are not associated with a weathering environment generally consist of extremely fine particles and exhibit the phenomenon of interparticle diffraction (34). This phenomenon, which leads to a new concept of interstratification (35) will be illustrated firstly with reference to pure smectite and then to regularly and randomly interstratified clays themselves.

4.1. Smectite

Interparticle diffraction reconciles the apparently contradictory nature of TEM and XRD observations made on the same smectite sample. Thus, TEM observations of dried dilute suspensions of Na-saturated montmorillonite indicate that the bulk of the clay has dispersed to 10Å thick elementary particles (Fig. 4), which consist, therefore, of single silicate sheets. It would be predicted from XRD theory that samples consisting largely of such very thin particles would show virtually no Bragg reflection from their basal planes, but intense basal maxima are, in fact, recorded (Fig. 4). When the breadth of these reflections is related to the number of coherent diffracting layers (N) by using the Scherrer equation it is found that N=9, a result clearly at odds with the TEM data. The incompatible nature of the TEM and XRD data of similar clay materials has been noted previously (31, 56). Nadeau et al. (34) reconciled these observations by postulating that a sedimented aggregate of such elementary particles, dried on to a flat surface such as a glass slide, would show an interparticle diffraction effect, whereby the

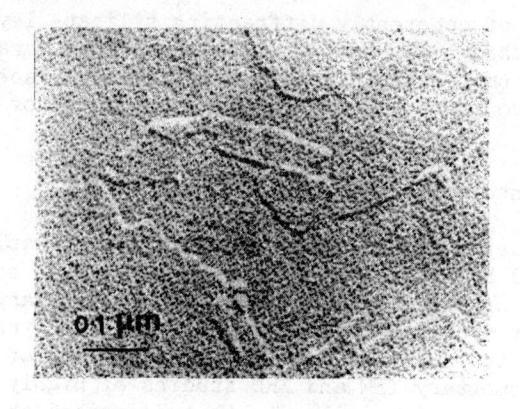

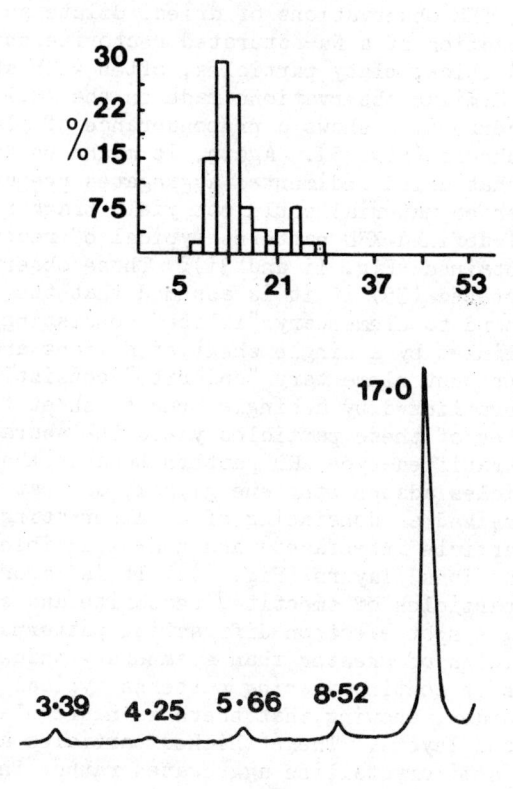

Figure 4. Top. TEM of Pt-shadowed Na-dispersed <0.1 μm fraction of Wyoming bentonite. Middle. Particle thickness distribution data. Bottom. XRD pattern of ethylene glycol solvated sedimented aggregate of the same <0.1 μm sample.

effective number of coherently diffracting silicate layers would be greatly increased because of the well-developed parallelism between adjacent particles. Furthermore, the interfaces between these particles would be capable of adsorbing water or organic molecules.

4.2. Regularly interstratified clays

The XRD patterns of completely regular interstratified clays are characterized by a high spacing basal reflection equal to the sum of the basal spacings from the different elementary layers and a large series of orders related to the high spacing reflection. However, despite the apparently well-defined nature of these materials, complementary TEM and XRD studies of highly dispersed samples reveal an anomaly similar to that described above for smectites. Thus, TEM observations of dried, dilute suspensions of the <0.1 μm fraction of a Na-saturated rectorite show that it is dominated by ∼20Å thick, platy particles, often with straight edges (Fig. 5). Similar observations made on the <0.1 μm fraction of Li-dispersed corrensite shows a preponderance of platy particles about 24Å in thickness (Fig. 5). Again, it might be anticipated from XRD theory that dried sedimented aggregates prepared from such highly dispersed material would not yield Bragg reflections, but in fact, well-defined XRD patterns typical of rectorite and corrensite are obtained (Fig. 1i and 3i). These observations can be readily interpreted (35) if it is assumed that the 20Å thick particles correspond to elementary "illite" consisting of two 2:1 silicate sheets linked by a single sheet of K^+ ions and that the 24Å particles represent elementary "chlorite" consisting of two 2:1 silicate layers linked by a single brucite sheet (Fig. 6). Oriented aggregates of these particles yield the characteristic regularly interstratified-type XRD pattern because the interfaces between the particles adsorb ethylene glycol, so that the overall structure is perceived as consisting of an alternating sequence of expansible (particle interfaces) and non-expansible (sheets with brucite or K^+ ions) layers (Fig. 6). It is important to note that elementary particles of smectite, rectorite and corrensite always yield single spot electron diffraction patterns. On the other hand, particles of greater than elementary thickness yield the discontinuous or continuous ring patterns typical of rotational turbostratic disorder, showing that there is no regular arrangement between the various layers. These thicker particles are, therefore, better viewed as semi-crystalline aggregates rather than crystals *per se*, as they possess periodicity, only in the c direction by virtue of the uniform thickness of their elementary particles, but show no such periodicity in the a, b directions.

The concept of interparticle diffraction can also be used to interpret the XRD patterns of illite-smectites that are regarded as partially ordered. Thus Nadeau *et al.* (35) presented electron microscope evidence to show that samples with IIIS-type ordering

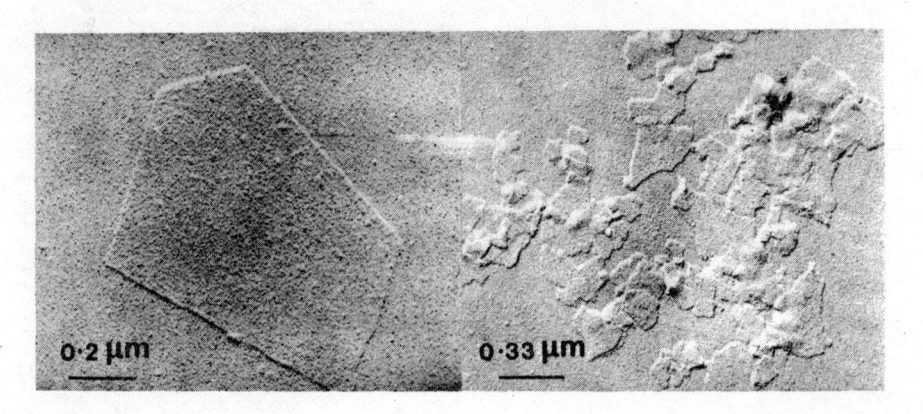

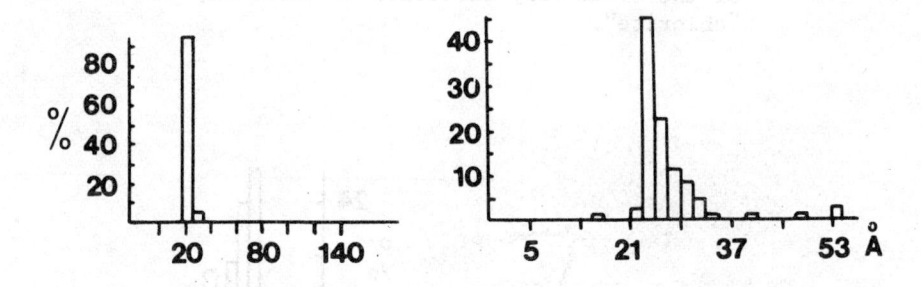

Figure 5. TEM and particle data of the <0.1 μm fraction of (left) Na-dispersed Baluchistan rectorite and (right) Li-dispersed Hillhouse Quarry corrensite.

consist largely (but not entirely) of fundamental 30-40Å thick "illite" particles respectively (Fig. 7). The smectite interlayers detected by XRD would again correspond to ethylene glycol sorption in the interfacial region between the fundamental particles. When these particles exceed about 50Å in thickness, (Fig. 8) diffraction is essentially intraparticle and a normal 10Å illite spacing is observed (Fig. 8). It is probable that similar interparticle diffraction effects would be applicable to partially ordered thin "chlorites", which would appear to be chlorite-smectite by XRD. So far however, such clays have not been described.

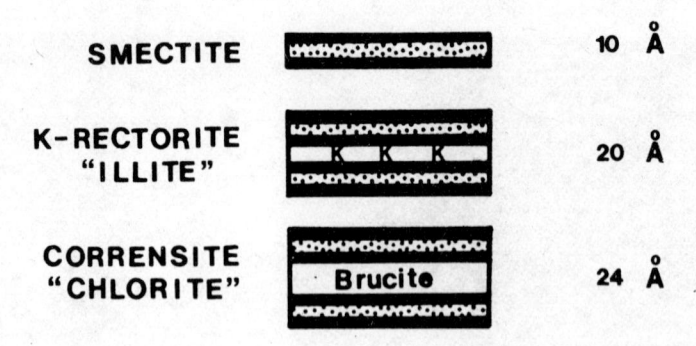

Figure 6. Diagrammatic representation (not to scale) of the elementary particles of smectite, "illite" and "chlorite".

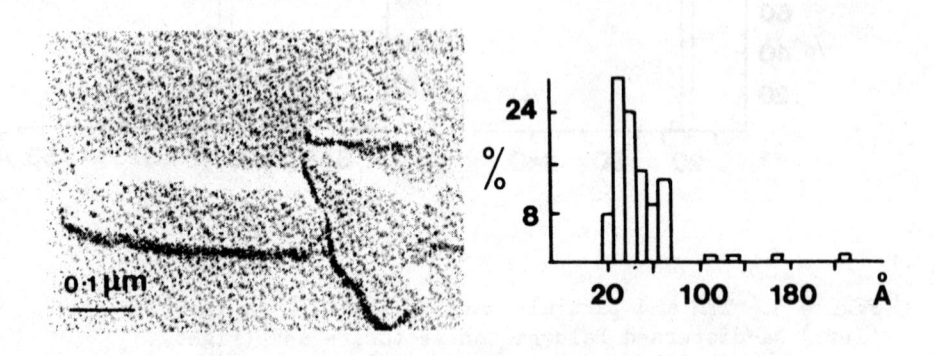

Figure 7. TEM and particle thickness distribution data of Na-dispersed <0.1 μm fraction of Tioga bentonite, a long range ordered IMII interstratification. For XRD pattern see Fig. liii.

4.3. Randomly interstratified clays

Nadeau *et al.* (34) showed that preparations made by mixing suspensions of smectite and rectorite-like regularly interstratified illite-smectite, both dispersed to their elementary particles of 10 and 20Å thickness respectively, exhibited all the XRD characteristics of randomly interstratified illite-smectite. It was concluded that the random layer sequence deduced from XRD examination of the sedimented aggregates merely reflects the random, but strongly oriented, association of the component elementary particles. It was further shown that the proportions

INTERSTRATIFIED CLAY MINERALS AND WEATHERING PROCESSES 113

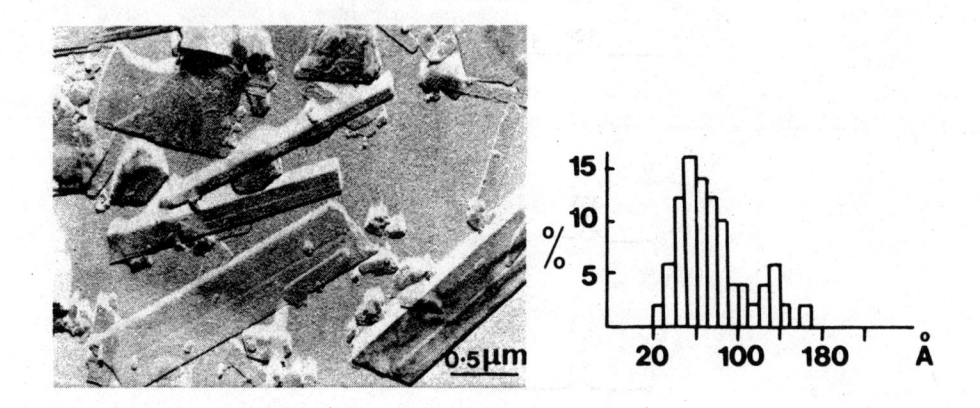

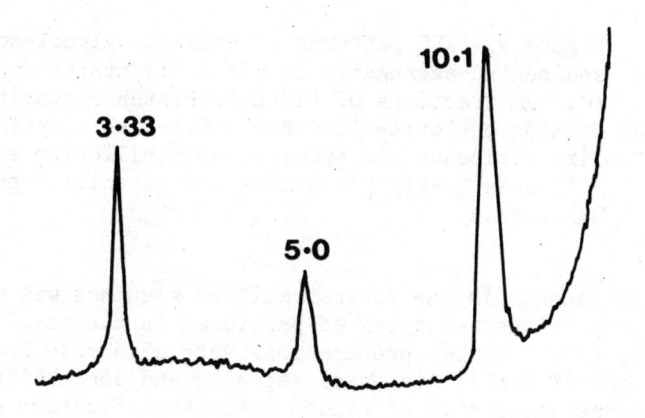

Figure 8. Top. TEM and particle thickness distribution
data for Na-dispersed <0.1 μm fraction of illite from
Permian Rotliegend Sandstone, UK, North Sea.
Bottom. XRD pattern of ethylene glycol-solvated
sedimented aggregate of the same <0.1 μm fraction.

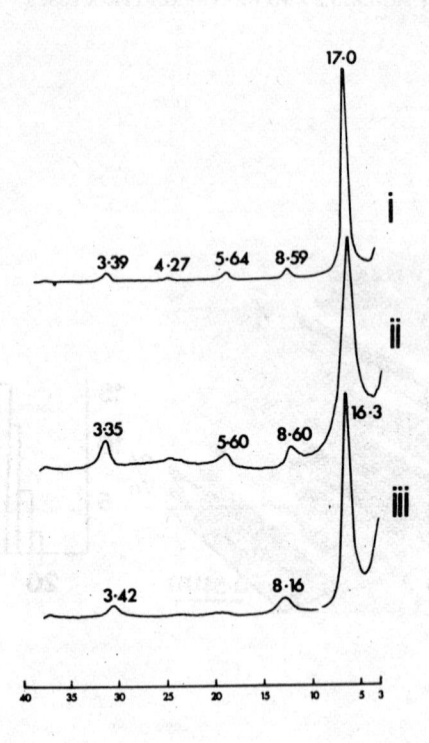

Figure 9. XRD patterns of ethylene glycol-solvated sedimented aggregates of mixed 1:1 suspensions of the <0.1 μm fractions of (i) Baluchistan rectorite and Wyoming bentonite (ii) Rectorite-like clay from Canon City, Colorado and Wyoming montmorillonite and (iii) Hillhouse Quarry Corrensite and saponite from Ballarat, California.

of the components in the interstratified sequence was dependent on the proportions and types of particles in the mixed suspensions (Fig. 9i, ii). Similar preparations were also made from mixed suspensions of highly dispersed saponite and corrensite and XRD patterns characteristic of random interstratification again obtained (Fig. 9iii). On the other hand, preparations made from mixed suspensions of elementary smectite and a diagenetic illite with a particle thickness ≥50Å yielded XRD patterns of discrete mixtures only (34). These experiments demonstrate the phenomenon of interparticle diffraction and strongly suggest that many naturally occurring randomly interstratified clays are really mixtures of very fine particles. Direct electron microscope evidence of this was obtained by determining the particle thickness distribution of a highly dispersed bentonitic clay which yielded a pattern identical to that of randomly interstratified illite-smectite with 70% smectite layers (Fig. 10). The results show that the clay consists primarily of 10Å (smectite) and 20Å (illite) thick particles in the appropriate proportions (Fig. 10).

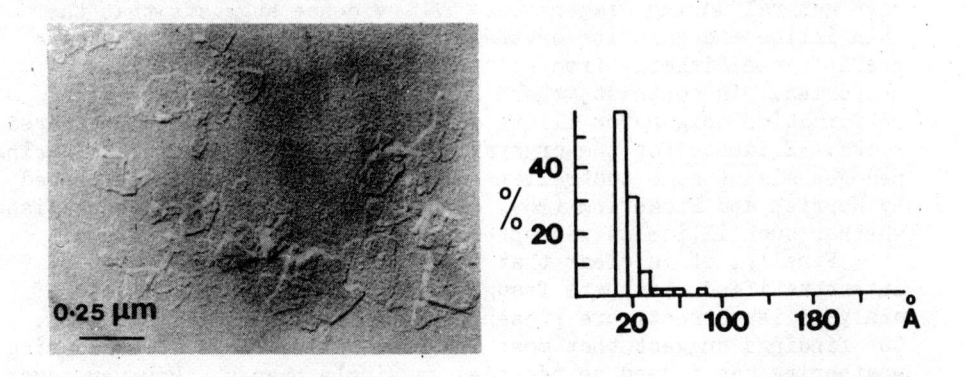

Figure 10. TEM's and particle thickness distribution data for the <0.1 μm fraction of a randomly interstratified illite-smectite from Westwater, Utah. Note the predominance of 10 and 20Å thick particles representing elementary smectite and "illite" particles respectively. For the XRD pattern of the sample see Fig. 2i.

5. CONCLUSIONS

The results and discussion presented above suggest that fundamentally different types of interstratified minerals are formed during weathering and during diagenesis. During weathering, interstratified minerals tend to form by transformation reactions of relatively coarse-grained layer silicate minerals as exemplified by hydrobiotite, dioctahedral mica-vermiculite, chlorite-vermiculite and the weathering products derived therefrom. It is possible that interstratified kaolin-smectite may be an exception to this general rule and its physical nature remains to be evaluated. The interstratified minerals formed during diagenesis, on the other hand, consist of extremely fine-grained particles, typically some tens of Angstroms in thickness when examined in their primary form. For these clays it might be considered that interstratification is more apparent than real as it can be accounted for by an interparticle diffraction phenomenon, whereby the interfaces between thin particles adsorb ethylene glycol and are, therefore, perceived by XRD as a smectite component. The clay minerals which exhibit this effect include rectorite, corrensite, partially ordered illite-smectite and randomly interstratified illite-smectite. It is possible, however, that randomly interstratified illite-smectite may also form in a weathering environment and further evidence is required on this point.

The different types of interstratified clays are a direct reflection of their mode of origin in so far as transformation dominates in the weathering environment whereas neoformation is more general during diagenesis. TEM evidence suggests that the thin illite and chlorite crystals found in diagenetic clays have precipitated directly from solution and must be regarded as neoformed. In contrast, there is little evidence to support a neoformation origin for illite and chlorite in soils and weathered rocks. Evidence for the crystallization of iron-rich illite during pedogenesis of some Australian soils has, however, been presented by Norrish and Pickering (38), although it remains to be established whether such illite is widespread in the soil of arid regions.

Finally, it is clear that thermodynamic studies of interstratified clays are fraught with difficulty and, if our analysis is correct, are presently in a state of some confusion. Our findings suggest that most interstratified clays formed during weathering can indeed be regarded as single phases. However, such clays have not been well-characterized compositionally and although there is more abundant data on diagenetic interstratified clays, it seems unwise to use this data to characterize weathering reactions. In conclusion, our findings indicate that for thermodynamic purposes regularly or partially ordered interstratified clays, whether formed by weathering or diagenesis, can be regarded as single phases, whereas, random interstratifications are often mixtures of two or more phases. The consequences of such distinctions. which have never previously been made, remain to be explored.

REFERENCES

1. Aagaard, P. and Helgeson, H.C. 1983, Clays Clay Miner. 31, pp. 207-217.
2. Bailey, S.W. 1981, Clay Science 5, pp. 305-311.
3. Bradley, W.F. and Weaver, C.E. 1956, Am. Miner. 41, pp. 497-504.
4. Brown, G. and Weir, A.H. 1963a, Proceedings International Clay Conference Stockholm 1, pp. 27-34.
5. Brown, G. and Weir, A.H. 1963b, Proceedings International Clay Conference Stockholm 2, pp. 87-90.
6. Brown, G., Bourguignon, P. and Thorez, J. 1974, Clay Miner. 10, pp. 135-144.
7. Churchman, G.J. 1978, Mineralogy N.Z. Journ. Sci. 21, pp. 467-480.
8. Churchman, G.J. 1980, Clay Miner. 15, pp. 59-76.
9. Coleman, N.T., Leroux, F.H. and Cady, J.G. 1963, Nature Lond. 198, pp. 209-210.
10. Cradwick, P.G. and Wilson, M.J. 1972, Clay Miner. 9, pp. 395-405.

11. Dunoyer de Segonzac, G. 1969, Mem. Serv. Carte Geol. d'Alsace Lorraine, No. 29, 317 pp.
12. Farmer, V.C. and Wilson, M.J. 1970, Nature Lond. 226, pp. 841-842.
13. Grim, R.E., Droste, J.B. and Bradley, W.F. 1960, Clays Clay Miner. 8, pp. 228-236.
14. Hayashi, H., Inaba, A. and Sudo, T. 1961, Clay Sci. 1, pp. 12-18.
15. Herbillon, A.J., Frankart, R. and Vielvoye, L. 1981, Clay Miner. 16, pp. 195-201.
16. Herbillon, A.J. and Makumbi, M.N. 1975, Geoderma 13, pp. 89-104.
17. Ismail, G.T. 1969, Am. Miner. 54, pp. 1460-1466.
18. Jackson, M.L. 1963, Clays Clay Miner. 11, pp. 29-46.
19. Johnson, L.J. 1964, Am. Miner. 49, pp. 556-572.
20. Kapoor, B.S. 1972, Clay Miner. 9, pp. 383-394.
21. Kato, Y. 1965, Soil Sci. Plant Nutrition, Tokyo, 11, pp. 114-122.
22. Kodama, H. 1966, Am. Miner. 51, pp. 1035-1055.
23. Kodama, H. and Brydon, J.E. 1966, Clays Clay Miner. 13, pp. 151-173.
24. Kulke, H. 1969, Contr. Miner. and Petrol. 20, pp. 135-163.
25. Lee, L., Bache, B.W., Wilson, M.J. and Sharp, G.S. 1984, Journ. Soil Sci. (in press).
26. Lippmann, F. 1954, Heidelb. Beitr. Mineralog. Petrog. 4, pp. 130-134.
27. Lippmann, F. 1956, J. Sedim. Petrol. 27, pp. 125-139.
28. MacEwan, D.M.C. 1954, Clay Min. Bull. 2, pp. 120-126.
29. MacEwan, D.M.C. 1956, Kolloidzeitschrift 149, pp. 96-108.
30. Makumbi, M.N. and Herbillon, A.J. 1972, Bull. Group Franc. Argiles 24, pp. 153-164.
31. Mering, J. and Oberlin, A. 1971, The Electron-Optical Investigation of Clays (edited by J.A. Gard), Min. Soc.
32. Nadeau, P.H. and Reynolds, R.C. 1981a, Clays Clay Miner. 29, pp. 249-259.
33. Nadeau, P.H. and Reynolds, R.C. 1981b, Nature, Lond. 294, pp. 72-74.
34. Nadeau, P.H., Tait, J.M., McHardy, W.J. and Wilson, M.J. 1984a, Clay Miner. 19, pp. 67-76.
35. Nadeau, P.H., Wilson, M.J., McHardy, W.J. and Tait, J.M. 1984b, Clay Miner. (in press).
36. Nagasawa, K., Brown, G. and Newman, A.C.D. 1974, Clays Clay Miner. 22, pp. 241-252.
37. Norrish, K. 1972, Proceedings International Clay Conference, Madrid, pp. 417-432.
38. Norrish, K. and Pickering, J.G. 1983, "Soils: an Australian Viewpoint" Division of Soils, CSIRO, pp. 281-308. C.S.I.R.O. Melbourne Academic Press: London.
39. Perry, E. and Hower, J. 1970, Clays Clay Miner. 18, pp. 165-177.
40. Peterson, M.N.A. 1961, Am. Miner. 46, pp. 1245-1269.

41. Rausell-Colom, J., Sweatman, T.R., Wells, C.B. and Norrish, K. 1965, Proc. 11th Easter School Agr. Sci. University of Nottingham. Experimental Pedology. Butterworth, London, pp. 40-72.
42. Reynolds, R.C. 1980, "Crystal Structures of Clay Minerals and their X-ray Identification", Mineralogical Society, Lond. pp. 249-303.
43. Reynolds, R.C. and Hower, J. 1970, Clays Clay Miner. 18, pp. 25-36.
44. Rich, C.I. 1968, Clays Clay Miner. 16, pp. 15-30.
45. Ross, G.J. and Kodama, H. 1976, Clays Clay Miner. 24, pp. 183-190.
46. Sato, M., Oinuma, K. and Kobayashi, K. 1965, Nature, Lond. pp. 179-180.
47. Schultz, L.G., Shepard, A.O., Blackmon, P.D. and Starkey, H.C. 1971, Clays Clay Miner. 19, pp. 137-150.
48. Shirozu, H., Ozaki, M. and Hagashi, S. 1972, Clay Sci. 4, pp. 45-52.
49. Srodon, J. 1980, Clays Clay Miner. 28, pp. 419-424.
50. Steiner, A. 1968, Clays Clay Miner. 16, pp. 193-213.
51. Stephen, I. and MacEwan, D.M.C. 1951, Clay Min. Bull. 1, pp. 157-162.
52. Sudo, T. and Hayashi, H. 1956, Clays Clay Miner. 4, pp. 389-412.
53. Sudo, T., Hayashi, H. and Shimoda, S. 1962, Clays Clay Miner. 9, pp. 378-392.
54. Sudo, T. and Shimoda, S. 1977, Minerals Sci. Engng. 9, pp. 3-24.
55. Sutherland, H.H. and MacEwan, D.M.C. 1962, Clays Clay Miner. 9, pp. 451-458.
56. Tettenhorst, R. and Roberson, H.E. 1973, Am. Miner. 58, pp. 73-80.
57. Tomita, K. 1978, Clays Clay Miner. 26, pp. 209-216.
58. Tomita, K. and Dozono, M. 1973, Clays Clay Miner. 21, pp. 185-190.
59. Weaver, C.E. 1956, Am. Miner. 41, pp. 202-221.
60. Weir, A.H., Nixon, H.L. and Woods, R.D. 1962, Clays Clay Miner. 9, pp. 419-423.
61. Weir, A.H. and Rayner, J.H. 1974, Clay Miner. 10, pp. 173-187.
62. Wilson, M.J. 1966, Miner. Mag. 35, pp. 1080-1093.
63. Wilson, M.J. 1970, Clay Miner. 8, pp. 291-303.
64. Wilson, M.J. and Cradwick, P.D. 1972, Clay Miner. 9, pp. 435-437.
65. Zen, E. 1962, Geochim. Cosmochim. Acta. 26, pp. 1055-1067.
66. Zen, E. 1959, Amer. J. Sci. 257, pp. 29-43.

FORMATION OF SECONDARY IRON OXIDES IN VARIOUS ENVIRONMENTS*

U. Schwertmann

Institut für Bodenkunde, T.U. München-Weihenstephan,
8050 Freising, F.R.G.

ABSTRACT*

Iron oxides are among the most common minerals formed during rock weathering. They vary in mineral species (goethite, hematite, lepidocrocite, maghemite, ferrihydrite) and, additionally for any mineral, in crystallinity and Al- for Fe-substitution in the structure. All three parameters may reflect the weathering environment. For example, the ratio of goethite to hematite, which is the most important pair of Fe oxides, varies widely with climatic conditions (as caused by latitude and altitude), with topographic position in a landscape, and with profile depth. This can be partly explained by such factors as soil temperature, moisture, organic matter and pH. A kinetic model based on synthesis experiments is proposed, according to which goethite may be formed from various Fe sources via a solution-nucleation-crystallization process, whereas hematite is formed from ferrihydrite via a dehydration-rearrangement process within the defect hematite-like structure of ferrihydrite. These two processes are competitive. Their relative rate depends on the environmental conditions and determines the final goethite/hematite ratio. This kinetic concept is preferred over a thermodynamic stability concept with which, because of kinetic hindrance, experimental results are often in disagreement.

The precursor of hematite - ferrihydrite - often occurs in weathering environments and can usually be considered a young metastable Fe oxide of low structural order and high surface area. Its transformation to more stable forms may be considerably retarded by adsorbed silicate and organics.

Environmental effects can also influence the formation of lepidocrocite and maghemite.

In routine X-ray diffraction, line widths are often related to the crystallinity of fine grained minerals. Line broadening was shown to originate mainly from small crystal size rather than from structural disorder, and may vary considerably for any one Fe oxide, e.g. goethite. It may therefore reflect the conditions of crystallization in the weathering environment.

This is also the case for Al- for Fe-substitution in the Fe oxide structure, which again varies with the environment depending on such factors as Al activity in solution, and also on the type of the Al species.

Following these results, an intensivated use of Fe oxides in the weathering zone as indicators of the weathering environment is advocated.

* A detailed report on this subject will be published in "Advances in Soil Science", Vol. 1, 1984.

PHYSICAL CONDITIONS IN ALUNITE PRECIPITATION AS A SECONDARY
MINERAL

Rodriguez-Clemente R., Hidalgo-Lopez A.

Institute of Geology. C.S.I.C.
c/ Jose Gutierrez Abascal, 2. Madrid-28006. SPAIN

ABSTRACT

The problem of alunite formation is analyzed in the light of equilibrium with the species from which it forms. The results of several synthesis experiments show the yield of precipitate as a function of the initial physical and chemical parameters. Accurate pH-controlled precipitation experiments are presented, and from their results we propose a mechanism of precipitation controlled by the $Al(H_2O)_4(OH)_2^+$ concentration. Nucleation is presented as the basic mechanism for alunite condensation as a solid. Crystal growth and ripening are presented as low rate processes which do not significantly affect the precipitation of alunite.

1. INTRODUCTION

Alunite is a basic sulphate common in a wide range of geological environments, especially in hydrothermal alteration (6,10,15) and metallic mineralization. The genesis of alunite is attributed to the alteration of Al and K-rich rocks interacting with $SO_4^=$-rich solutions, in an acid medium. Usually there is little doubt as to the origin of the Al and K, but the source of the $SO_4^=$ raises at least two basic hypotheses: supergene origin by oxidation of sulphide deposits, and the hypogene rise of sulphatic solutions, the dissociation of which increases with falling temperatures.

The alunite group of minerals has a composition indicated by the formula:

$$AB_3(SO_4)_2(OH)_6$$

where A = K^+, Na^+, NH_4^+, Ag^+, and B = Al^{3+} or Fe^{3+}. The mineral alunite has A = K^+ and B = Al^{3+}.

In deposits formed by alteration of silicate rocks, K-rich alunite is the most common form, but substitution by Na^+ and H_3O^+ are frequent. Usually alunite is found in association with jarosite (the Fe^{3+} member), but this paragenesis is strongly related to the pH of the parent solution (17); only at pH$\sim$3 can they be formed simultaneously, above this value iron hydroxide precipitates. Usually alunite does not incorporate Fe^{3+}, but jarosite does incorporate Al^{3+}. In nature, the size of alunite crystals varies between 5 and 25 μm.

The crystalization of alunite and jarosite has received much attention (7,12) due to their importance in geological environments, metalurgical processes and other processes. However, little work has been done on the morphological properties of the precipitates. The work of Aslanian et al. (1) regarding the dependence of habit on temperature is important for its implications in the definition of the conditions of the rock alteration processes.

Synthetic alunites normally have a smaller unit cell than natural ones in the K^+ end member of the K-Na series, and they have higher content of water and lower percentages of alkalis and alumina. These differences are due to substitution of K^+ by H_3O^+ in synthetic alunites; by heating them to 300° C, water is expelled and the lattice parameter becomes very similar to that of natural specimens.

2. AIM OF THE WORK

Alunite is not only an important mineral for the understanding of the alteration processes, but is also a raw material for several industrial processes for instance: charge in papermaking, source of potash, alums and alumina, and in general as a reserve ore for Al. Recently a new application has been found for the basic sulphates, mainly alunite and jarosite: their use as a charge in the plastic used in greenhouses for intensive agriculture. For this purpose, large crystals of pure alunite and low cost production are required. We have tried to define the working conditions of the classical method (1,7) of alunite synthesis to improve the grain size. We have also tried to interpret the genetic conditions for natural alunites, especially those from Spanish localities.

3. EXPERIMENTAL

The synthesis of alunite was done by spontaneous and induced hydrolisis, for reasons that will be explained later. The method was similar to the one used in previous works (1,7). We always used the same volume of initial solution (200 ml.) with

TABLE I

Exp.	Initial Conditions K2SO4 Molality	Al2(SO4)3 Molality	pH at 25°C	Adjusted pH	T of reaction	Time of experiment	Agitation	Relation K/Al	Final Condition Final pH	$(Al) mol/kg	Yielding %
1	8,54364.10⁻³	0.02563	3.088	2.5	98.5°C	7 days	not controlled	1:3	1,713	0,01553	7,94
2	8,54364.10⁻³	0.02563	3.088	1.5	98.5°C	7 days	"	1:3	1,475	0,01702	0,34
3	0.02437	0,07312	2.776	2.5	98.7°C	7 days	"	1:3	1.275	0,03189	29,70
4	0.02437	0.07312	2.788	1.5	98.8°C	7 days	"	1:3	1.265	0,04242	11,15
5	0.04746	0.14240	2.584	2.5	98.3°C	7 days	"	1:3	1.015	0,05763	32,18
6	0.04762	0.14287	2.589	1.5	98.4°C	7 days	"	1:3	1.070	0,07350	19,15
7	0.09067	0.27201	2.345	2.5	98.9°C	7 days	"	1:3	0.798	0,19432	33,91
8	0.09055	0.27165	2.349	1.5	98.4°C	7 days	"	1:3	0.949	0.14776	14.68
9	0.04761	0.14285	2.593	2.5	98.5°C	30 days	"	1:3	0.811	0,04221	46,69
10	0.04772	0.14316	2.583	1.5	98.5°C	30 days	"	1:3	0.821	0,05028	39,77
11	0.04757	0.14272	2.607	2.5	98.5°C	3 days	"	1:3	1.296	0,07682	16,14
12	0.04767	0.14303	2.620	1.5	08.5°C	3 days	NO	1:3	1.361	0,09212	2.85
13	0.04766	0.14299	2.602	2.5	98.5°C	3 days	NO	1:3	1.471	0,08767	6.94
14	0.04762	0.14287	2.598	2.5	90°C	7 days	not controlled	1:3	1.712	0,08326	10.56
15	0.04764	0.14293	2.600	1.5	90°C	7 days	"	1:3	1.367	0,09425	0.92
16	0.04766	0.14299	2.608	2.5	75°C	7 days	"	1:3	2.340	0,09432	0.89
17	0.04765	0.14296	2.524	1.5	75°C	7 days	"	1:3	1.454	0,09489	0.38
18	0.04767	0.14302	2.587	2.5	98.5°C	3 days	200 rpm	1:3	1.562	0,08921	5.40
19	0.04771	0.14315	2.606	2.5	98.5°C	3 days	400 rpm	1:3	1.402	0,09411	4.17
20	0.04752	0.14261	2.587	2.5	50°C	7 days	not controlled	1:3	2.5	0,09190	2.80
21	0.04755	0.14266	2.588	1.5	50°C	3 days	"	1:3	1.5	0,09510	0
22	0.04756	0.14292	2.58	Alcaline	98°C	3 days	"	1:3	1.118	0,06778	22.35
23	0.04764	0.14292	2.58	"	98°C	3 days	"	1:3	1.191	0,06351	27.97
24	0.04765	0.14295	2.58	"	98°C	3 days	"	1:3	1.202	0,06353	27.98
25	0.04761	0.14285	2.58	"	40°C	3 days	"	1:3	2.763	0,08889	5.60
26	0.04761	0.14285	2.58	"	40°C	3 days	"	1:3	2.790	0,09322	1.77
27	0.04783	0.14351	2.58	"	25°C	3 days	"	1:3	2.764	0,09567	0.01
28	0.04575	0.13726	2.58	"	25°C	3 days	"	1:3	2.774	0,08897	5.61
29	0.04575	0.13726	2.266	2	98.4°C	3 days	"	1:6	1.397		7.58
30	0.03575	0.10726	2.203	2	98.2°C	3 days	"	1:9	1.440		21.21
31	0.03475	0.10426	1.986	2	98.9°C	3 days	"	1:12	1.417		20.45

different proportions of K_2SO_4 and $Al_2(SO_4)_3$, different initial pH values, different times of reaction and different temperatures. The results are shown in Table I. In most of the experiments, a stoichiometric molal ratio $Al/K = 3$ was used, but we also varied this ratio in order to investigate its influence, as well as the effect of agitation.

The initial and final pH were measured with a CRISON 517 digital pH-Meter, which gives an accuracy of 0.001. Experiments at boiling temperature were done in thermostatic vessels. Solid precipitates were separated by filtration and washed with 2M HCl, distilled water, alcohol and acetone; they were then dried at 100°C for 24 hours. All the materials obtained were analyzed by X-ray powder diffraction with an internal standard. The crystal size was analyzed by optical and electron microscopy, (Figs. 1 and 2). The Al and K content and the pH of the final solutions were measured.

In order to study the mechanism of precipitation, a series of experiments were done by carefully and continuous controlling the temperature and pH throughout the reaction (see Fig. 6). The initial conditions are given in Table II. Conductivity measurements were not possible due to the boiling character of most of the solutions. As a rule, during heating the pH steadily decreases to a fixed value corresponding to that at the boiling temperature. If the solution is concentrated enough, the precipitation starts before boiling is reached, normally the system starts to boil and the pH remains stable for an induction period before precipitation starts.

4. ALUNITE PRECIPITATION

There are many references on hydrolysis processes, especially for Al (2,3,8,21). Alunite stability diagrams and syntheses have also attracted a lot of attention (1,7,12,15,17,20). In spite of this, little is known about the basic mechanism of alunite precipitation and growth.

Al^{3+} exists in aqueous solution as a complex of the type: $Al(H_2O)_6^{3+}$, or even forms dimers and other polymeric forms if the concentration is high enough. As the enthalpy of its first hydrolysis step is positive and close to the dissociation enthalpy of water (13.3 Kcal/mol), $Al(H_2O)_6^{3+}$ hydrolyzes easily as temperature

TABLE II

Exp. Fig. 6	K_2SO_4 (molal)	$Al_2(SO_4)_3$ (molal)	pH at 25° C	pH before precip.	T of precip.	Induction Time
32 (·)	$8.543 \cdot 10^{-3}$	0.0256	2.788	2.030	98.5°C	20 min.
33 (x)	0.0474	0.1424	2.553	1.960	98.5°C	2 min.
34 (o)	0.0906	0.2720	2.345	1.921	75 °C	0
35 (+)	0.4746	0.1424	2.550	2.754	98.5°C	0

(a)   (b)

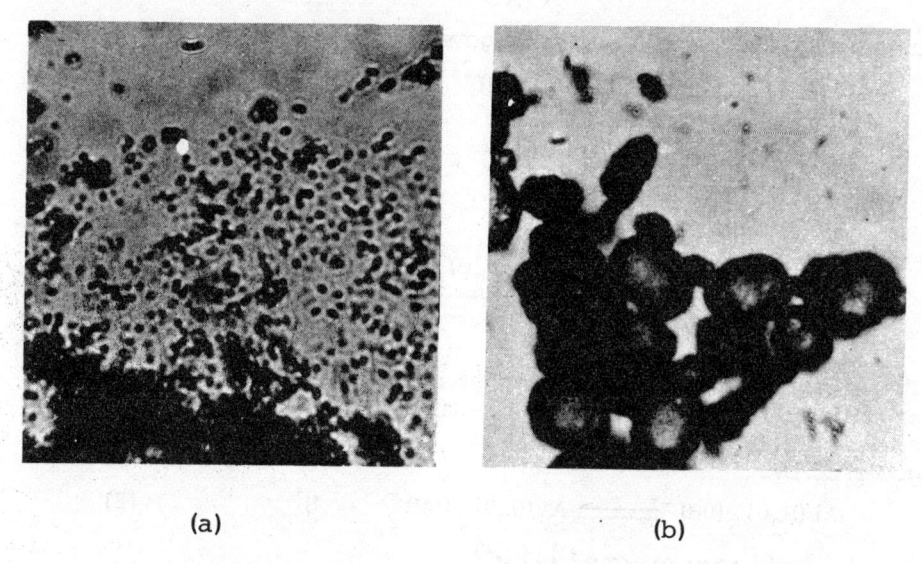

Figure 1, Alunite precipitates (1000x). (a) average size $\bar{r}$ = 1-2 μm, (b) $\bar{r}$ = 10-15 μm

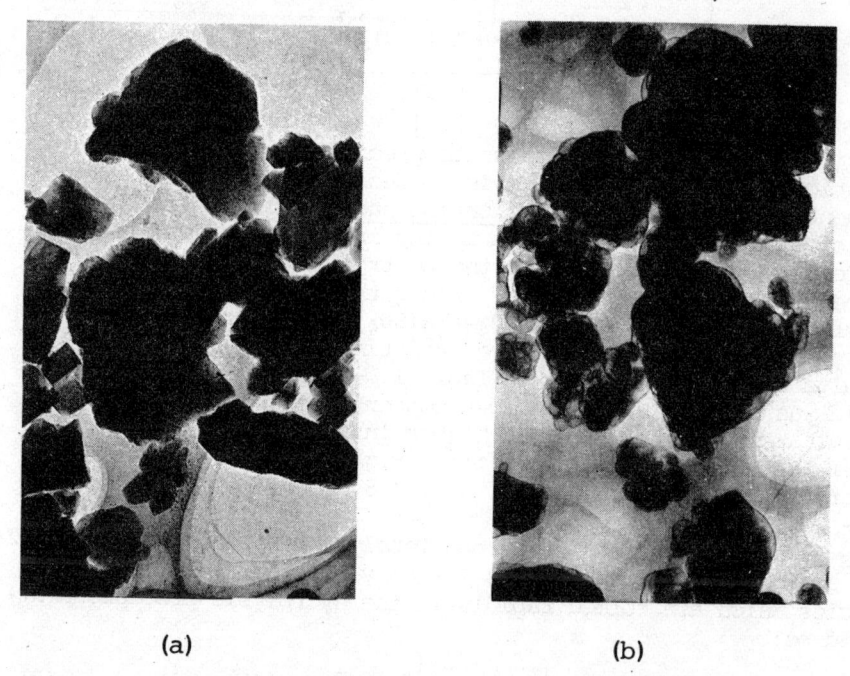

(a)   (b)

Figure 2. Electron microscope photographs of alunite (50000x). (a) Laminar crystals. (b) Rounded crystals.

increases, following a sequence governed by:

$$Al(H_2O)_6^{3+} \rightleftharpoons Al(H_2O)_5(OH)^{2+} + H^+ \qquad (1)$$

$$K_1 = \frac{\left[Al(H_2O)_5(OH)^{2+}\right]\left[H^+\right]}{\left[Al(H_2O)_6^{3+}\right]}$$

$$\left[Al(H_2O)_5(OH)^{2+}\right] = \frac{K_1\left[Al(H_2O)_6^{3+}\right]}{\left[H^+\right]}$$

The enthalpy of the second step of mononuclear hydrolysis is not known, but it can be considered similar to that of the polynuclear species $Al_2(OH)_4^{4+} = 18.7$ Kcal/mol. Its formation would be governed by:

$$Al(H_2O)_5(OH)^{2+} \rightleftharpoons Al(H_2O)_4(OH)_2^+ + H^+ \qquad (2)$$

$$K_2 = \frac{\left[Al(H_2O)_4(OH)^+\right]\cdot\left[H^+\right]}{\left[Al(H_2O)_5(OH)^{2+}\right]}$$

$$\left[Al(H_2O)_4(OH)_2^+\right] = \frac{K_2\cdot K_1\left[Al(H_2O)_6^{3+}\right]}{\left[H^+\right]^2}$$

The interpretation of alunite precipitation raises several questions, the first of which is to decide which are the constituent species acting in nucleation and growth, and the second is to calculate the potentials acting in the precipitation process and their relation to the controlling variables of pH, concentration and temperature. Along this line we made several experiments in the system K_2SO_4-$Al_2(SO_4)_3$-H_2O, at varying initial pH, concentrations, temperatures and times of reaction, the results of which are shown in Table I. As expected, the higher yield was obtained with the most concentrated solutions (Fig. 3), in agitated vessels, with the higher initial pH, (Fig.4), higher initial Al/K ratio and longer time (Fig.5). At 50º C there is no precipitation at an initial pH = 1.5, but it occurs with pH = 2.5.

As both salts dissociate completely and K^+ does not hydrolyze, the species which form alunite (if only consider the mononuclear species which are formed rapidly in the hydrolysis processes (2)) would be:

$$6H^+ + KAl_3(SO_4)_2(OH)_6 \rightleftharpoons 3Al(H_2O)_6^{3+} + K^+ + 2SO_4^= \qquad (3)$$

$$5H^+ + KAl_3(SO_4)_2(OH)_6 \rightleftharpoons 3Al(H_2O)_5(OH)^{2+} + K^+ + 2SO_4^= \qquad (4)$$

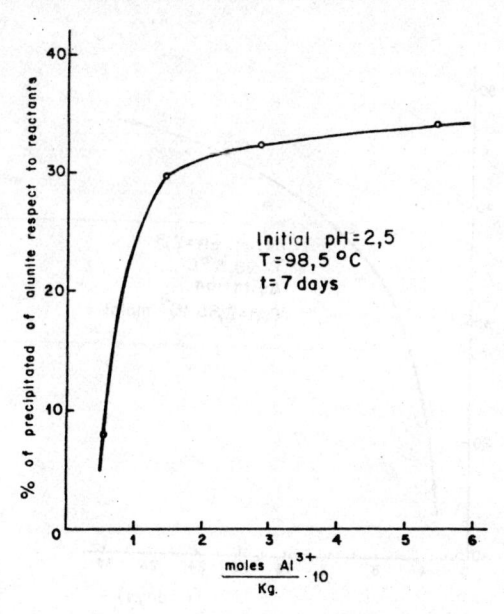

Figure 3. Yielding of Alunite respect to initial concentrations.

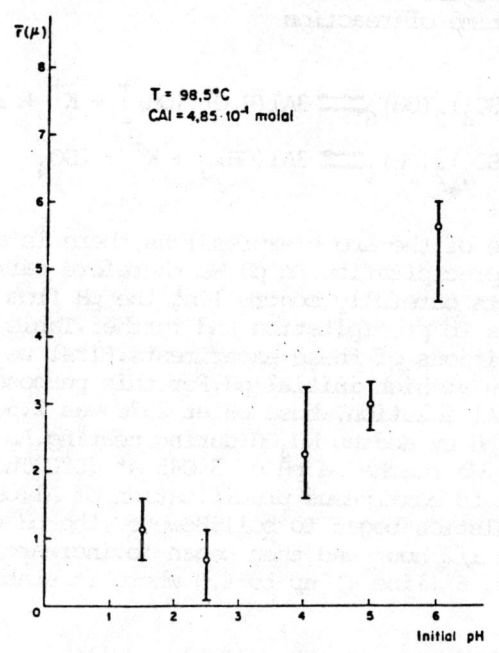

Figure 4. Crystal size as a function of the initial pH.

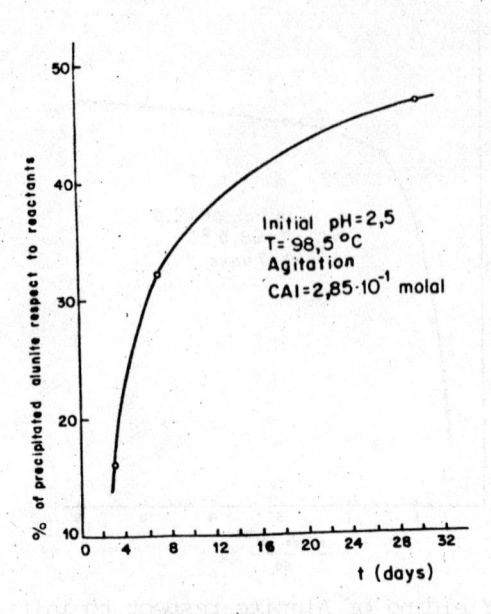

Figure 5. Yield of alunite as a function of the time of reaction

$$4H_2O + KAl_3(SO_4)_2(OH)_6 \rightleftharpoons 3Al(H_2O)_4(OH)_2^+ + K^+ + 2SO_4^= \qquad (5)$$

$$3OH^- + KAl_3(SO_4)_2(OH)_6 \rightleftharpoons 3Al(OH)_3 + K^+ + 2SO_4^= \qquad (6)$$

In three of the above expressions, there is a dependency of the alunite precipitation on pH. We therefore carried out a series of experiments carefully controlling the pH from the early stages of hydrolysis to precipitation and further. Table II describes the initial conditions of these experiments. First we studied alunite precipitation at high initial pH. For this purpose we prepared a standard K/3Al solution, whose pH at 25°C was 2.505, and we tried to raise the pH by adding NH_4OH during heating. As expected, $Al(OH)_3$ precipitates. We reached a pH of 3.045 at 70°C. The pH decreases to 2.750 due to continuous precipitation of $Al(OH)_3$ and at that point the solution began to boil. However, the pH continued to decrease for 1/2 hour and then began to increase continuously as shown in Fig. 6 (line +) up to 4.9 where it stabilized.

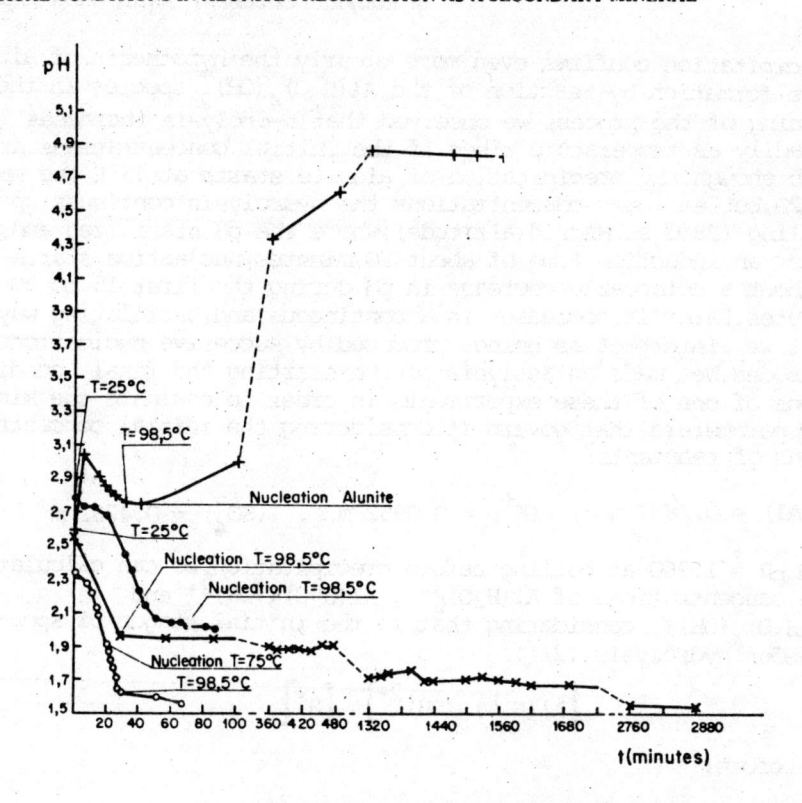

Figura 6. pH evolution from the initial conditions given in Table II.

Spontaneous precipitation was carried out without any kind of induced hydrolysis. K/3Al solutions of different initial concentrations (Table II) were heated and the process was followes by pH measurements. The results are also shown in Fig. 6 (lines., o,x). The above experiments clarify, in our opinion,the precipitation process. The experiment at high initial pH (line +) started with Al^{3+} hydrolysis until it was completely neutralized by NH_4OH in the form of amorphous $Al(OH)_3$, then alunite began to precipitate consuming $Al(OH)_2^+$ and inducing a displacement of the equilibrium to the right in the reaction:

$$Al(OH)_{3\,(amorphous)} \rightleftharpoons Al(OH)_2^+ + OH^-$$

which led to precipitation in accordance with Eq. (5) and an increase in pH. This reaction causes a more rapid variation in pH than does the spontaneous precipitation because of the absence of intermediate hydrolysis states. Spontaneous precipitation confirms even more clearly the hypothesis of alunite formation by reaction of the $Al(H_2O)_4(OH)_2^+$ species. In the beginning of the process we observed that hydrolysis increases steadily as temperature rises. If the initial concentrations are high enough, the precipitation of alunite starts at 75°C and pH= 1.920, but at lower concentrations the hydrolysis continues up to boiling (98°C at Madrid altitude), where the pH stabilizes and, after an induction time of about 20 minutes, nucleation starts without a noticeable decrease in pH during the first 15 to 30 minutes. Later it decreases in a continuous and oscillating way that we interpret as being produced by succesive nucleation episodes. We made an analysis of the starting and final conditions of one of these experiments in order to evaluate the kinetic parameters that govern it. Considering the initial concentrations of reactants:

$$\Sigma(Al) = 0.2857 \text{ m} , \quad (K^+) = 0.0952 \text{ m} , \quad (SO_4^=) = 0.4762 \text{ m}$$

and pH = 1.960 at boiling before precipitation, we can calculate the concentrations of $Al(H_2O)_6^{3+}$, $Al(H_2O)_5(OH)^{2+}$ and $Al(H_2O)_4(OH)_2^+$ considering that in the initial stages of spontaneous hydrolysis (21):

$$\left[Al(H_2O)_5(OH)^{2+}\right] = \left[H^+\right]$$

Therefore:

$$pK_1 = 2pH - pAl(H_2O)_6^{3+}$$

we discounted the contribution of the second hydrolysis step because its equilibrium constant is 100 times lower than for the first step (8). Applying the same reasoning:

$$\left[Al(H_2O)_6^{3+}\right] + \left[Al(H_2O)_5(OH)^{2+}\right] = 0.2857 \text{ m}$$

from these expressions we obtain: $pK_1 = 3.36$, which is of the same order of magnitude than the value obtained by using the temperature coefficient of the ionic product of water, because their ionization mechanisms are similar (21).

$$pK_1 (98°C) = pK_1 (25°C) - 73 \times 0.02 \text{ units/°C} = 3.08$$

pK_2 is difficult to estimate and, in fact, has been evaluated as 10.5 (2) to 6 (8), in any case we can make an approach by giving it a value 100 times lower than K_1, and therefore our result would be $pK_2 = 5.36$.

With these constants, we obtained the following concentrations of hydrolysis species: $\left[Al(H_2O)_6^{3+}\right] = 0.2747$ molal

$$[Al(H_2O)_5(OH)^{2+}] = 0.0109 \text{ molal}$$
$$[Al(H_2O)_4(OH)_2^+] = 10^{-5.361} \text{ molal}$$

The activity coefficients of the constituents were calculated approximately by means of:

$$\log f = \frac{-Z^2 A \sqrt{I}}{1 + B_1 \sqrt{I}} \quad , \quad B_1 = B \cdot a_o \tag{9}$$

where $A = 0.366069$ and $B = 2.9441684 \times 10^7$ and $a_o = 4.10^{-8}$

$f(SO_4^=) = 0.577$

$f(Al^{3+}) = 0.363$

$f(AlOH^{2+}) = 0.770$

$f(Al(OH)_2^+) = 0.994$

$f(K^+) = 0.919$

Using Eq. (5) we can calculate the ionic product (I.P.) of the reactants at the beginning of precipitation:

$$I.P. = a^3_{(Al(H_2O)_4(OH)_2^+)} \cdot a_{K^+} \cdot a^2_{(SO_4^=)} = 10^{-18.271}$$

which compared with the solubility Product (S.P.) of alunite at 100° given in the literature (12,17) allows us to know the β supersaturation acting before precipitation:

$$\beta \text{ ref. (17)} = \frac{I.P.}{S.P.} = 10^{4.638} \quad , \quad \beta \text{ ref. (12)} = 10^{12.98}$$

We have followed the process for 1 month and at the end of this period the parameters were:

$\Sigma (Al) = 0.1266$ molal, pH $= 0.870$, then:

$[Al(H_2O)_6^{3+}] = 0.12509$

$[Al(H_2O)_5(OH)^{2+}] = 0.00151$

$[Al(H_2O)_4(OH)_2^+] = 10^{-7.882}$

$[K^+] = 0.0422$

$[SO_4^=] = 0.2013$

which gives an I.P. $= 10^{-27.80}$. This value is lower than K_5 at 100° given by Lopez et al. (17) $= 10^{-22.91}$ but higher than the value given by Hladky et al. (12) $= 10^{-31.25}$. We can then suppose that the value of K_5 given by Hladky et al (12) better describes alunite precipitation in the system K_2SO_4-$Al_2(SO_4)_3$-H_2O.

4.1 Nucleus Size

The solubilities of sparingly soluble salts such as alunite are always difficult to measure, but considering that in our experiments log ($SO_4^=$) and log (K^+) are -2, the minimum solubility (S) of alunite, as deduced from the $Al(H_2O)_4(OH)_2^+$ content of the solution, can be estimated at 100° as log $S_{alunite}$ = -8.25. According to (19) such a solubility can be correlated with a specific surface energy: $\sigma \sim 200$ erg./cm^2. At this high supersaturation, we can imagine that nucleation would be homogeneous, and the size of the critical nucleus can be calculated through the Gibbs-Thompson equation:

$$r^+ = \frac{2 \cdot M \cdot \sigma}{R \cdot T \cdot \rho \ln \beta} \tag{10}$$

where M is the molecular weight and ρ the crystal density.

At the supersaturation $\beta = 10^{12.98}$, the radius of the critical nucleus would be 6.3Å, which corresponds to a volume of 1047×10^{-24} cm^3, or 1.43 unit cells. In other words, the calculated critical nucleus would be formed by 4.29 alunite molecules. This result brings us to the interesting question of the nature of nucleation. It is well known that Eq. (10) holds only for a sufficiently large nucleus and lower degrees of supersaturation (23), therefore we can imagine that at this extremely high supersaturation, non-classical nucleation occurs (9,23), with nuclei of such size that the surface energy component in the formation energy is not relevant.

4.2 Crystal Growth

The preceding paragraphs led us to the major problem of alunite crystal growth. This mineral is found in nature as minute crystals of 1 to 25 microns, this range of size is similar to that obtained in the Laboratory. We have only indirect references (M. Maleev, personal communication) to the finding of several-mm.-sized crystals of alunite from a unique Ukranian deposit. Of course alunite grows, and most probably by means of one of the two classical layer growth mechanisms (two-dimensional nucleation or spiral growth) corresponding to crystals with high surface energy, because the surfaces are flat, with sharp edges and even with regularly shaped growth hillocks (24) as seen by electron microscope (26), (Fig.2). The fact is, that in spite of the evident growths from the critical nucleus, we have not succeeded in obtaining big crystals of at least microscopic size.

The reasons for this unsuccessful growth could again be found in the analysis of Eqs. (1) (2) and (5). When the system arrives at the equilibrium of the first hydrolysis step the solution pH stabilizes, but as the enthalpy associated with the second step is higher than that of the first, the rate of formation of the growth units $Al(H_2O)_4(OH)_2^+$, is lower and the system needs a certain time to reach equilibrium for this second process. At this moment the supersaturation of the alunite phase is very high and nucleation begins, producing a consumption of $Al(H_2O)_4(OH)_2^+$, which does not alter significantly the equilibirum of Eq. (1); therefore, pH remains constant for a certain time. The slowness of the rate of growth, typical of the high energy surface crystals, together with the huge production of new $Al(H_2O)(OH)_2^+$ through the reestablishment of equilibrium in Eq. (2), makes the solution continously supersaturated with respect to alunite and makes nucleation a quicker means of condensing matter than growth. In fact, the oscillatory decrease of pH in the spontaneous hydrolysis can be attributed to sequences of nucleation followed by redissolution of subcritical nuclei, which makes the pH decrease in a general step-like manner. Also, the increase in alunite solubility with decreasing pH contributes to this step-like configuration.

We think that this step by step production of the growth units, together with the high surface energy and the difficulties for ripening in supersaturated solutions, is responsible for the periodical non-classical nucleation episodes, the small average size (but good crystal perfection (22,24)), and the lack of evident ripening in comparison with other substances of similar surface energy and solubility such as $BaSO_4$.

5. MORPHOLOGICAL PROPERTIES

The relationship between growth conditions and morphological properties is not easy to establish. Previous work (1) indicates a predominance of rhombohedral and pinacoidal habits for alunite formed at about 200°C, with an increasing dominance of pinacoids as the temperature increases. The same work indicates a rhombohedral habit for low temperature (100°C) alunites. Our experiments were done in agitated vessels and, possibly, the collision between particles deforms the crystal morphology. Nevertheless, the electron microscope pictures (Fig. 2) show, besides crystals with ill defined and rounded shapes (Fig. 2b), a certain homogeneity of laminar rectangular shapes (Fig. 2a).

The size of the crystal is another problem. In our experiments, we obtained tiny crystals ranging in size from 0.5 to 15 microns, which is roughly the average size of natural crystals. We have analysed the possibilities of Ostwald Ripening (4) for these crystals, but as the time to reach equilibrium, i.e. total hydrolysis of Al^{3+}, is very long, we have not observed an increase of size with time for experiments lasting 1 month (Fig. 7).

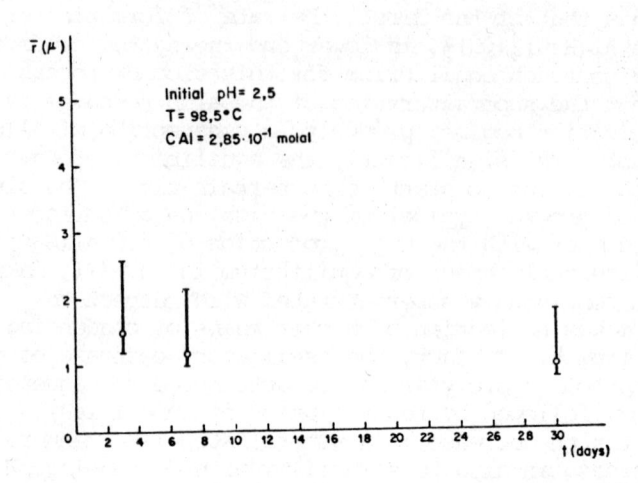

Figure 7. Variation of grain size $\bar{r}$, (μm) with the time of reaction.

We have only observed an increase in the average size of the crystals (which may be agglomerates) with an increase in the Al/K ratio , probably due to the increase in concentration of the Al(H$_2$O)$_4$(OH)$_2^+$ growth units relative to the other two species in the parent solution. This growth unit, in our opinion, governs the entire formation process.

Because alunite precipitates by a nucleation controlled process, we have studied the effect of temperature on the growth process. Eq. (10) indicates an inverse relation between r^+ and T. We performed experiments at several temperatures varying from 50° to 98.5° C (Fig. 8) with the same initial composition, and we observed that no precipitation occurs at 50° during one month of observation for an initial pH = 1.5; at 75° we obtained crystals with an average size larger than at 98.5°. These results could indicate an improvement of the growth processes relative to nucleation at low temperature.

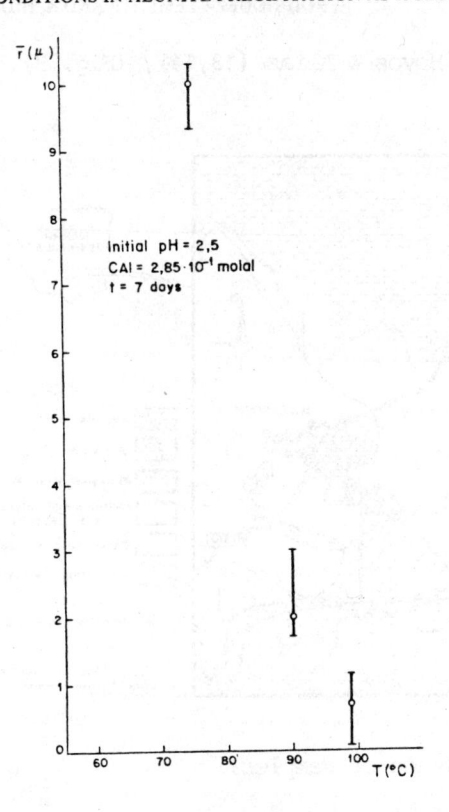

Figure 8. Relation between grain size $\bar{r}$ (mm) and reaction temperature.

6. THE ALTERATION OF THE MAZARRON AREA

The above study represents a basis for the new industrial application of alunite. It is therefore, interesting to re-evaluate the known deposits and to improve the actual knowledge of the natural mechanism of formation.

Most of the theoretical and field studies, have established the major importance of hydrothermal alteration of silicate rocks in the formation of the main alunite deposits (5,6,11, 16,20,25,27). We therefore, consider it interesting to investigate the natural processes and contrast them with the laboratory results. One of the main regions of alunite deposits in Spain is the volcanic Mazarron area, and one of the most important individual deposits is "Cerro San Cristobal", which has been previously described by Hoyos & Alias (13,14), (Fig. 9).

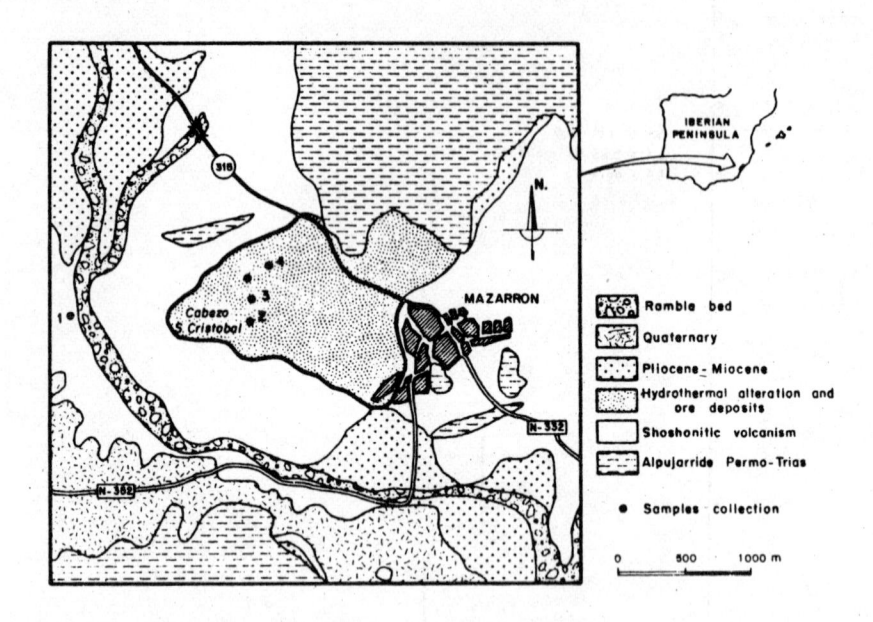

Figure 9. Location of samples.

It is interesting to consider the alteration process of this area in light of the model proposed by Hemley et al. (11) for the alteration of silicate rocks in hydrothermal environments. Here three grades of increased alteration indicated by different mineral assamblages are distinguished. In this model alunite is associated with advanced hydrolytic attack.

The geological studies in this area (13,14) suggest that alunite has been formed from the attack of dacites, rich in sanidine, (and therefore in K and Al), by a rising acidic hydrothermal solution rich in $SO_4^=$, which could be of postmagmatic origin or a product of alteration of underlying sulphides by acidic fluids. This mechanism is similar to that proposed by Lombardi (16) for the Italian deposits.

In order to study the alteration sequence, we have taken samples from the locations indicated in Fig. 9. The unaltered rock can be classified as a toscanite (18) and its mineralogy and composition are given in Table III. To verify the effectiveness of the hydrothermal alteration process, we have treated samples with different degrees of alteration with an acidic boiling solution of initial pH=2.5 for 3 days. The results are shown in Fig. 10, in which the evolution of the mineralogy of the alteration products with respect to the unaltered rock can be observed by X-ray diffractograms.

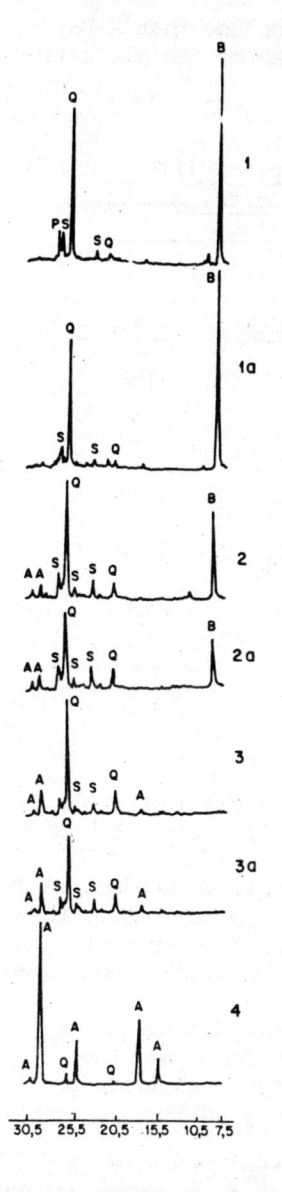

Sample 1 is the unaltered rock, formed of biotite and quartz as main constituents, and Na-Ca plagioclase, sanidine and cordierite as minor constituents. After treatment, the height of the biotite and plagioclase peaks decreases notably, while those of quartz and sanidine remain more or less constant. We can also observe the appearance of a weak alunite main peak.

Sample 2 corresponds to an altered rock inside the San Cristobal Hill. Its mineralogy consists of biotite, quartz, sanidine and an appreciable amount of alunite. After treatment, the biotite peak decreases appreciably.

Sample 3 is a strongly altered rock composed almost exclusively of quartz, sanidine and alunite. It was collected near a sulphide outcrop. With the treatment we observed a slight increase in the height of the alunite peaks.

Sample 4 is almost pure alunite with a little quartz, and represents the extreme alteration product of the original volcanic rock.

Several conclusions can be drawn from these results, the first is the sensitivity of biotite towards hydrothermal attack, followed by plagioclase and sanidine. We can also observe that the relative abundance of alunite with respect to the original constituents of the rock increases steadily from the unaltered wall-rock, and can represent an indicator for the extent

Figure 10. Evolution of the alteration profiles, see text. The -a diagram corresponds to samples after treatment. A-alunite, B-biotite, P-plagioclase, Q-quartz, S-sanidine.

of alteration (5). Therefore, we can conclude that X-Ray diffraction provides a good pictorial view of the alteration sequence.

TABLE III: Composition of Sample 1 (18)
Mineralogical composition: biotite, quartz, plagoiclase, sanidine, cordierite.

Bulk Chemical Composition:

Oxide	%	Oxide	%
SiO_2	64.19	MgO	1.58
Al_2O_3	16.22	Fe_2O_3	0.81
K_2O	4.43	TiO_2	0.63
FeO	3.26	P_2O_4	0.42
CaO	2.57	MnO	0.05
Na_2O	2.25		

7. CONCLUSIONS

Alunite precipitation is governed by the hydrolysis of the Al^{3+} cation and therefore, is strongly related to temperature and initial pH.

The rate of production for the hydrolysis product is much higher than the rate of alunite growth. The system is thus continuously supersaturated while free Al^{3+} is present. The mechanism of precipitation is a three dimensional non-classical nucleation.

The advance in spontaneous hydrolysis decreases the pH of the parent solution, and therefore decreases the production of hydrolysis products. However, equilibrium is only reached a pH near 1, which is very unlikely to occur in nature.

In induced hydrolysis experiments the situation is reversed, $Al(H_2O)_4(OH)_2^+$ is produced at the expense of amorphous $Al(OH)_3$, and therefore alunite precipitation increases the pH.

Ripening is not observed in long experiments, and the grain size of the crystals ranges from 0.5 to 15 microns. We deduce that crystal growth is extremely slow by a step displacement mechanism, due to the high surface energy, and therefore does not contribute significantly to the solute precipitation.

Hydrothermal alteration of shoshonitic rocks proceeds by successive attacks on plagioclase, biotite and sanidine in order to produce alunite. X-Ray diffraction provides a good tool to follow the alteration sequence.

Acknowledgments:

This work was done within the framework of C.S.I.C. Program "Materials and Solid Surfaces", and had been financed by a CAICYT grant: Project 1147/1983. Dr. E. Rodríguez Badiola kindly helped us in the sample collection. Mr. Luis Puebla from the Institute "Elhuyar" of the C.S.I.C., kindly made the electron microscope picture. Miss R. Torres kindly revised and typed the manuscript.

REFERENCES

1) Aslanian, S., Velinov, I. and Velinova, M. 1976. "Genetic types of alunites and experimental conditions for their synthesis". Bull. Acad. Sc. Geochem. Min. Petrol., 5, pp. 144-155.
2) Baes, Ch. Jr. and Mesmer, R.E. 1976. "The hydrolysis of cations". J. Wiley & Sons, N.Y.
3) Baes, Ch., and Mesmer, R.E. 1981. "The thermodynamics of cation hydrolysis". Am. J. Sc., 281, pp. 936-962.
4) Baronet, A. 1982. "Ostwald ripening in solution, the case of calcite and mica". Estud. Geologicos, 38, pp. 185-198.
5) Bladth, K.W. 1982. "The formation of goethite, jarosite and alunite during the weathering of sulfide-bearing felsic rocks". Econ. Geol., 77, pp. 176-184.
6) Caballero, A., Menendez del Valle, F. and Martin Vivaldi, J. L. 1974. "Yacimientos españoles de las bauxitas y alunitas". Bol. Geol. Min. España, 85-I, pp. 32-42.
7) Dutrizac, J.E. and Kaiman, S. 1976. "Synthesis and properties of jarosite-type compound". Can. Min., 14, pp. 151-158.
8) Frink, C.R. and Peech, M. 1963. "Hydrolysis of the aluminum ion in dilute aqueous solutions". Inor. Chem., 2, pp. 473--478.
9) Garcia Ruiz, J.M. and Amoros, J.L. 1980. "Sobre la precipitación polimorfica del carbonato calcico". Estud. Geologicos, 36, pp. 193-200.
10) Grabau, A.W. 1920. "Geology of the non-metallic mineral deposits other than silicates". Vol. I. McGraw Hill Inc., N.Y.
11) Hemley, J.J. and Jones, W.R. 1964. "Chemical aspects of hydrothermal alteration with emphasis on hydrogen metasomatism". Econ. Geol., 59, pp. 538-569.

12) Hladky, G. and Slansky, E. 1981. "Stability of alunite minerals in aqueous solutions at normal temperature and pressure". Bull. Miner., 104, pp. 468-477.
13) Hoyos, A. and Alias, L.J. 1962. "Mineralogia y genesis del yacimiento de alunita del Cerro de San Cristobal (Mazarrón, Murcia) . I. Difracción de rayos X, ATD y estudio microscopico de los materiales más alunitizados". Notas y Com. Inst. Geol. Min. España, 67, pp. 109-120.
14) Hoyos, A. and Alias, L.J. 1965. "Mineralogia y genesis del yacimiento de alunita del Cerro de San Cristobal, Mazarron (Murcia). III. Estudio mineralogenetico". Notas y Com. Inst. Geol. Min. España, 80, pp. 109-120.
15) Hemley, J.J., Hostetler, P.B., Gude, A.J. and Mountjoy,W.T. 1969. "Some stability relations of alunite". Econ. Geol.,64, pp. 599-612.
16) Lombardi, G. 1977. " Alunite in Italy occurrence and genesis". 8th Intern. Kaolin Symp. and Meeting on alunite Madrid-Roma, nº A-3.
17) Lopez Aguayo, F., La Iglesia, A., Doval, M. and Menendez,F. 1977. "New data on stability of alunite and jarosite". 8th Intern. Kaolin Symp. and Meeting on Alunite. Madrid-Roma, nº A-4.
18) Lopez Ruiz, J. and Rodriguez Badiola, E. 1980. " La region volcanica neogena del Sureste de España". Estud. Geol., 36, pp. 5-63.
19) Nielsen, A.E. and Söhnel, O. 1971. "Interfacial tensions electrolyte crystal-aqueous solution, from nucleation data". J. Crystal Growth, 11, pp. 233-242.
20) Nordstrom, D.K. 1982. "The effect of sulfate on aluminum concentrations in natural waters: some stability relations in the system $Al_2O_3-SO_3-H_2O$ at 298K". Geochim. Cosmochim. Acta, 46, pp. 681-692.
21) Schofield, R.K. and Taylor, A.W. 1954. "The hydrolysis of aluminum salt solutions". J. Chem. Soc., 4445-4448.
22) Sheftal, N.N. 1959. "Trends in real crystal formation and some principles for single crystal growth". In "Growth of Crystals". Vol. 1, ed. Shubnikov & Sheftal. Consultant Bureau, N.Y., 185-210.
23) Stoyanov, S. 1979. "Nucleation theory for high and low supersaturations". In "Current Topics in Materials Science". Ed. Kaldis, Vol. 3, pp. 423-462. North-Holland Pub. Amsterdam.
24) Sudo, T. and Takahashi, H. 1971. "Other minerals in clays". In "The Electron Optical Investigation of Clays". Ed. Gard, J.A. Miner. Soc. Monograph, 3, London.
25) Ubanell, A.G., Garzon, G., de la Peña, J.A., Bustillo, A. and Marfil, R. 1978. "Estudio de procesos de alteracion hidrotermal en rocas graniticas y sedimentarias (provincia de Avila)". Estud. Geologicos, 34, pp. 151-160.

26) Velinov, I.A. and Aslanian, S. 1981. "Vein-type alunite from Bulgaria". Compt. Rend. Acad. Bul. Sci., 34, pp. 1417-1419.
27) Wan, H.M. 1977. "Mineralogy of alunites in northern Taiwan, Chinkuashin copper-gold and tatun sulpher-pyrite deposits". 8th Intern. Kaolin Symp. and Alunite Meeting. Madrid-Roma, nº A-6.

THIS PLANET IS ALIVE

--WEATHERING AND BIOLOGY, A MULTI-FACETTED PROBLEM--

W. E. Krumbein and B. D. Dyer*

Geomicrobiology Division, University of Oldenburg,
P. O. Box 2503, D-2900 Oldenburg, West Germany
*Department of Biology, Boston University,
Boston, Mass. 02215 USA

ABSTRACT

The age of planet Earth is about 4.5 to 5.0 ga. Life established itself above, at and below the planet surface about 3.8 ga ago. Weathering processes, although they may take place also on planets without life, are controlled at and under the surface of Earth by biological processes. These include direct mechanical and chemical, and indirect action of the metabolic products of organisms. The organisation and reorganisation of the material of the crust and portions of the mantle are ruled by kinetics of a biological nature. These include the amount and kind of minerals and rocks destroyed and formed, and the accumulation of minerals at specific places. Furthermore the speed of chemical reactions and physical activities is controlled by biological parameters and processes. Some examples and principles of biological weathering and transport are described with emphasis on the interaction of the biotic and abiotic world. Life can be regarded as a chemical process and almost no chemical process on Earth takes place without being a part of Life. Life catalyses the rates of physical-chemical processes and in some cases runs chemical reactions which are thermodynamically improbable. Life and chemistry combine into material transfer activities, which create the geological history of the planet. These transfer reactions have been in co-existance for at least 3 ga. The chemical state of the Earth and its buffering margins are set and controlled by Life.

INTRODUCTION AND HISTORY

" The Lord spoke to Moses and Aaron and said: When you have entered the land of Canaan which I give you to occupy, if I inflict a fungus infection upon the house in the land you have occupied, its owner shall come and report to the priest, that there appears to be a patch of infection on his house. The priest shall order the house to be cleared before he goes in to examine the infection, or everything in it will become unclean. After this the priest shall go in to inspect the house. If on inspection he finds the patch on the walls consists of greenish and reddish depressions, apparently going deeper than the surface, he shall go out of the house and put it under quarantine for seven days. On the seventh day he shall come back to inspect the house, and if the patch has spread on the walls, he shall order the infected stones to be pulled out and thrown away outside the city in an unclean place." (Leviticus 13, 14).

The knowledge about microbial rock-deteriorating processes is certainly older than written records of scientific endeavour. The nomads of the Sinai Desert chose the surfaces on which they engraved their Petroglyphs, avoiding rocks covered by epilithic lichens, selecting rock surfaces carrying a biogenic rock varnish (32, 14) not susceptible to the activity of endolithic lichen pitting. During the process of hammering on the rock surface they must have produced the same green sap that is produced also today, when we "scientifically" study rock surfaces in dry areas. The gren spots are produced by chlorophyll of endolithic photosynthetic bacteria and algae.

Also the ancient Egypt, Greek and Roman scientists speculated about the biological cycling of materials and the influences of human populations on a large scale. Pliny the Elder in his "Natural History" summarizes biological erosion in bitter words as he ascribes a lot of it to the wildest animal, Homo sapiens, and warns against environmental destruction and pollution in the same passage.

Albertus Magnus (ca. 1250), the Saint of Science, wrote the following words about biogeochemical cycles:

"Quocirca propter hoc quod unius generatio est alterius corruptio, et e converso unius corruptio est alterius generatio, est generatio transmutatio quae nunquam secundum naturam quiscit."

In approximate translation: Since the generation of one natural object at the same time represents the destruction of another and in turn the destruction of one is the generation of the other, the process of generation is a change (or exchange), that never comes to stillstand.

Albert of Lauingen, Duke of Bollstedt (Albertus Magnus) discussed the effect of gaseous compounds transferred through trees with Hildegard of Bingen and was fully aware of the intimate relationship of living and mineral matter. At one point he was amazed about the decay and preservation potential of roman rock sculptures he happened to find at the occasion of the foundation of the Cologne cathedral.

Theophilus Bombastus von Hohenheim named Paracelsus (about 1535) described another process in rock cycling this way:

"Then I saw the pebbles in the river and they were covered with a mucus. And that mucus transferred into stone and thus glued the pebbles into hard rock. And when I moved the material into my laboratory and put it into a cucurbite (cucumber shaped glass vessel, later named Erlenmeyer) I saw the same slime growing and thus produced the rock within my laboratory."

Bombastus also claimed, that one can find rich accumulations of metals underneath plant growth, which is yellowish or grey and unhealthy. Thus he was probably the first who applied biogeochemical prospecting methods.

Giordano Bruno (1584) argued about astrophysics with the Roman Catholic Church, he also wrote long books on the intimate relationship between life on Earth, life in Space and the biogeochemical cycles. He followed the classical Greek pattern of discussion symposia and submitted his biogeological arguments via a man named Theophilus in the following way:

"Don't you see, that which was seed will get green herb and herb will turn into ear and ear into bread. Bread will turn into nutrient liquid, which produces blood, from blood semen, embryo, men, corpse, Earth, rock and mineral and thus matter will change its form ever and ever and is capable of taking any natural form."

Giordano Bruno was certainly referring to the biogeochemical cycles and made little distinction between biotic and abiotic matter. He was one of the founders of the views, propagated also today, claiming that life rules the whole Universe and certainly his planet.

The ideas expressed by Bruno, which he himself based on Cusanus (Nikolaus von Kues), were later transferred by Leibniz and in part by Ehrenberg in the 17th, 18th and 19th Century. A continuation of these ideas and models about biogeochemical cycles and the control of life on these cycles was presented in Russia by Vernadsky (45) and later in a modified form by Lovelock and Margulis with their Gaia-concept and Daisy-World models (34, 35, 36).

Many modern textbooks on weathering and erosion, however, handle the subject with a few lines or omit it altogether (47).

For example Amoroso and Fassina (1) showed scanning electron micrographs with detailed microbial weathering features and refer only to thenardite formation as a weathering agent in the figure legend. Although the tremendous impact of life on the cycle of weathering, erosion and rock formation is often not acknowledged, the effect of human pollution is usually exaggerated. The biological factor is often treated exclusively in physical and chemical terms.

THE FLYING MOUNTAIN

"We can say nothing, but what has been said" (Burton, 1605)
Or:
"Indeed, nothing is said that hasn't been said before" (Publius Terenz Afer, 190-159 B. C., see also 31).

Once upon a time there was a king of Mesopotamia who developed a method to encourage his people to come to his court to pay their taxes. He constructed huge dams in the mountains. He would not open the gates and let the minerals and waters down to the fields to wet and fertilize them so that corn could grow unless the people delivered the taxes. One day the gods inflicted a plague of locusts on the king's fields. The locusts devoured all that was growing from the fertile soils and when they found no more food they started to fly. For two days and one night the sky was dark above the Red Sea. A little later a scientist (10) calculated, that the weight of the swarm was 4.4×10^7 tons, which equals an amount of material closer to a mountain than a hill. It was calculated, that all of the mineral matter, that was biologically motivated to move from Mesopotamia to Ethiopia across the Red Sea equalled in mass, for example, all zinc and copper mined throughout the 19th Century.

Unfortunately all the locusts were killed by famine in the highlands of Ethiopia but the dead bodies served as a perfect source of nutrients to rock-dwelling lichens, cyanobacteria, and

THIS PLANET IS ALIVE - WEATHERING AND BIOLOGY

ammonia and nitrite oxidizing bacteria waiting in the dust and rock crevices. These were activated and started new transformations and biological acid production, which accelerated mineral transfers considerably.

Thus the "leprous disease of rock" quoted before from the Bible took the appearance of "galloping consumption" by this "injection of nutrients". Cavities were forming on the rocky desert slopes, in the centers of which large amounts of nitrates were collecting until a sudden rain would wash them down into the valleys. Pitting and exfoliation were speeding up (13, 26, 32). As a consequence much more rock was eroded, dissolved and washed downhill.

Unknowingly the poor farmers of the Nile Delta ultimately took profit from the plague in Mesopotamia as rich and fertile waters brought down dissolved nitrate, phosphate and undissolved but finely dispersed potassium with the grains of rock detached by the "biological rock erosion front" activated by the dying locusts. The productivity of the waters of the nile in those years was extremely high due to the locust transfer. Later a sage man named Thienemann (Schröder, 42) started to call fertile waters eutrophic and the process of fertilizing water was named eutrophication. But soon after a few other wise people called eutrophication a bad process and fertilizatin was overfertilization (42).

The pharaoh and his kinsmen liked to swim in clear oligotrophic water and to walk along blue, and crystal clear rivers in leisure. But the locust swarm that initiated the outburst of bioerosion in Ethiopia had made the Nile waters too fertile. The good growth of crops made the pharaoh and his people rich and attracted many people from poorer regions. One of the pharaohs thought that the ugly looking waters around his palace might have been due to overpopulation, and cut down on the food supply. He was pressing especially hard on the people of Abraham. It is well known, that the Gods inflicted another plague upon Egypt. All waters turned into blood, the air smelled of fever and the stink of hell and all silver turned black (11, 23).

The plague enabled the people of Israel to move across the Red Sea back into Asia. They left behind the pyramids, which were biologically piled up from biogenic rocks that had precipitated long ago in the delta (28, 29). At the same time however, when these carbonate rocks were exposed to weathering (as stated by Herodotus 450 B. C.) new carbonates were formed by the activities of tiny animalcula, which precipitated limestone from slime and made the rocks grow. Thus it was expressed by Ehrenberg.

TRACES OF BYGONE BIOSPHERES

The word biosphere was reintroduced into 20th Century science by Vernadsky (45) and others (27, 30, 31, 36). In the introduction to "La Biosphere" Vernadsky has stated, that "a book was lacking on the biosphere, this huge integrated block, which represents the manifestation of the regular and organized mechanism of the Earth's Crust. Vernadsky continues that generations of scientists, geologists, chemists and physicists have studied the geological and planetary phenomena without realizing that the phenomena they sutdy are nothing else but manifestations of life."

Vernadsky (45) regarded the biosphere as the only actual transforming force of extraterrestrial energy input. He was inspired himself by the materialistic biological view of R. Mayer, the founder of modern thermodynamics, and by Moleschott and Büchner, Mayer in 1845 coined the beautiful picture of the sunrays being caught in flight and transformed into static energy and chemically bound energy by the green carpet covering the Earth. Vernadsky states (45, p. 27): "There is no phenomenon on the surface of the Earth, which is independent of life." He continues (p. 29): "If biogenic oxygen were to vanish from the atmosphere ... all agents of weathering would vanish as well. There would be no action of rain, no carbonic acid. Under the thermodynamic conditions of the Earth, water is a relatively strong chemical agent; but this 'natural' water is rich in chemically agressive substances only by biological processes (grace a l'existance de la vie) and especially by the activities of microorganisms" (48). Vernadsky also introduced the term "geochemical energy of life in the biosphere". Vernadsky himself only repeated what had been said before by Hooke and Ehrenberg (20, 14) when they talked about the force of the invisible tiny life, that creates and destroys rocks and materials. Winogradski (48), Bachmann (2), Bassalik (3, 4) and Nadson (38) have precisely described the microbial processes, which activate and balance the mineral transfer cycles, between 1880 and 1913. All of them have seen the traces of the biosphere in old sediments and rocks and have speculated about the precision of the biological cycling. This was expressed already by the Ancient scientists (e. g. Lucrez in his De Rerum Natura, written about 2020 B. P.).

"NEW LOOKS AT THE BIOSPHERE"

Vernadsky reintroduced the concept of biotic influence on the reactive parameters of Earth to Russian scientists, but this was ignored by most Western scientists until Lovelock reintroduced the idea as the Gaia hypothesis (34). The hypothesis may be summarized: the reactive compounds of the atmosphere and hydrosphere of Earth and the temperature of Earth are maintained in rheostasis in a range suitable for the biota, by means of a biotic feedback mechanism. The reactive parameters of Earth are in fact maintained in a chemical disequilibrium compared to Earth's moon, Venus, and Mars. For example very reactive gases, oxygen and methane are in disequilibrium in Earth's atmosphere, while on Mars, all reactive gases have long ago reacted and the atmosphere is in equilibrium, with mostly carbon dioxide. Water is short in supply and no reactive biogenic radicals are available. This is one of the reasons, why the 26 km high volcano located on Mars looks like a sculpture of a volcano freshly washed and cleaned with distilled water.

An atmosphere in chemical disequilibrium and large scale biological transfer of the lithosphere are characteristics of a planet with a biota.

The hypothesis of the coexistance of biological with geological forces as equivalent and powerful forces has been further extended to include the influence of the biota on the lithospere (46). The weakest link in the Gaia hypothesis, the mechanism of the biotic feedback system has recently been strengthened (Margulis, personal comm.) The mechanism is as follows:

Prokaryotes are the most important organisms in the frame of the biological shaping of Earth. In contrast, the impact of human pollution is orders of magnitude less in most cases than the output of prokaryotes (34). The kingdom of prokaryotes is the most diverse and prokaryotes are present in almost every environment on Earth, often in complex communities such as microbial mats. They represent the most persistent and stabilized genetic pools on Earth. There is literally no example of a genus of bacteria that has become extinct. Probably very few biochemical innovations have occurred in the last two billion years. And certainly very few have been developed by eukaryotes. The diversity of a prokaryotic community at any given point is the "standing diversity" and it has specific interactions with the environment in the form of inputs and outputs of compounds. However the "potential diversity" of a community is much larger, consisting of metabolically inactive organisms or genes, organisms in very low numbers, and organisms in resting stages such as cysts and spores. Any change in an environmental parameter affects the standing diversity of a community and thus a new standing diversity will be selected from the potential diversity, and a new set of imputs and outputs will affect the environment. The mechanism is analogous to an environmental change which selects a specific phenotype from the gene pool of a population of organisms. Thus the environment itself has about the equivalent power and influence as the biota and both communicate with each other in equilibrated and successful ways to keep the total system going.

The Darwinian paradigm is oversimple and may be soon outlived completely. Among the many modifications, that had to be added with the years the following may be the most important ones:

Organisms as living natural bodies are not passive pieces in a game between chance mutation and chance environmental change. The variability of a genome has been shown to be due in part to many built-in mechanisms such as jumping genes which may alter the genotype of an organism. The environment of an organism is made up of other organisms as well as physical parameters; environmental information, however comes back to the genome through several filters and is well perceived and transformed into new genetic and sexual transformations of the genome. There are no monocultures in nature, and of course the organisms themselves have a very powerful control over the physical parameters of their environment. Thus they create their environment and in response to their environment they create themselves.

DEFINITIONS CONCERNING THE PROCESSES BY WHICH COMPONENTS OF THE LITHOSPHERE ARE TRANSFERRED.

Both the antiquity and the multidisciplinary nature of this field of study have generated many problems of terminology which obscure the basic concepts and which isolate the field from scientific approaches (chemical, physical and by these naturally biological ones). In this section some of the terminology is reviewed and then presented in a new synthesis, based mostly on chemical terminology. In this synthesis it is stressed that life is a chemical process and no chemical processes on the surface of this planet and close to its surface can be conceived without the participation of life.

"Weathering" with its meteorological connotations is the general heading under which the folowing terms in alphabetical order, are usually grouped.

<u>Abrasion</u> is a special case of erosion. It is defined as the friction or mechanical effect of any kind of solid or hard

surface on another solid or hard surface leading to erosion. Thus abrasion is the destruction of a river bed or coastal rocky shore by stones, pebbles or sand; it also is wind-polishing of desert rocks. <u>Bioabrasion</u> would be the action of organisms on rocks with their hoofs, beaks and tongues, roots, branches, leaves. The chiton, for example, abrades rock with its magnetite rasp or <u>Patella</u> drills a biogenic abrasive hole with its shell on the rocky shore. There are myriad examples of effective bioabrasion.

<u>Alteration</u> is the change of the mineralogical composition of any rock or mineral by chemical means. The term is mainly used for the attack of hydrothermal solutions and alterations during metamorphism. In this case biological processes play a minor role. Bioalteration thus is not used in the geological and geomorphological literature. Unfortunately "altération" and "bioaltération" are common terms in the French language and may produce some confusion in translation. In French they have the meaning of weathering and bioweathering.

<u>Decomposition</u> is used in geological nomenclature as a synonym for chemical weathering by e. g. humic substances. In biology the term decomposition is used to describe the process of formation of minerals from organic compounds via respiratory and fermentative pathways.

<u>Degradation</u> is usually used as a term for the combined effects of erosion and weathering, i. e. the wearing down of the Earth's surface by physical, chemical and biological action in the case of biodegradation. Degradation of a soil describes the processes leading to the structuring of new lattices via base exchange and bioleaching in order to reduce (degrade) the exchangeability of cations.

<u>Deterioration</u> and biodeterioration do not exist in the Glossary of Geology nor in the Oxford English Dictionary as geological terms, although international biodeterioration societies and journals exist. Deterioration means "to make things worse; an act of decline to the negative". In this respect the terms can hardly be ascribed to natural phenomena since humans do not understand what is better or good for nature. Perhaps the term should be restricted to objects constructed by humans when exposed to the natural environment e. g. the chemical deterioration of the pyramids, the chemical and biological deterioration of the Angkor Vat Temples, the stained glass of mediaeval cathedrals or the bronze Quadriga at Venice.

<u>Erosion</u> is usually defined as the removal and transport of substances by mechanical and physical action. Some authors also include solubilisation and transport of solutes while others

assign this to weathering at large followed by transport of the dissolved material by water or air.

Leaching is a technical term derived from mining and metal processing. It can be effected by acids and bases or by electrochemical methods. Bioleaching or biological leaching is effected by biogenic acids as e. g. sulfuric acid or by the direct leaching activity of chelating agents and direct surface contacts of microorganisms as e. g. fungi, cyanobacteria and chemoorganotrophic bacteria.

We suggest and we intend to use in our own work, the following terminology:

Parts of the lithospere may be either "chemically" or "physically" transferred. Under the heading of physical transfer are all of biotic and abiotic processes which mechanically remove particles of all sizes from rocks and minerals and transfer them into the hydrosphere or atmosphere as particulates, colloids and aerosols. Chemical transfer of parts of the lithosphere includes chemical reactions, both biotic and abiotic, which remove parts of the lithosphere into the hydrosphere or atmosphere in the form of gases, solutes, colloids, aerosols, and particulates.

A transferred part of the lithosphere may then be transported across distances ranging from microns to kilometers, and perhaps secondarily returned to the lithosphere or otherwise immobilized. Biotic and abiotic mechanisms include precipitation, flocculatin, mechanical sorting, chelation, and integration into the biota.

EXAMPLES

(1) A biogenic salt dome of Permian salt and carbonates pushed an island ridge up into the North Sea. By the end of the ice age the Island of Helgoland produced by salt tectonics, emerged into the North Sea. The tidal currents produce a mechanical transfer pattern, with rates of 4 to 10 cm year^{-1} on vertically exposed cliffs, a little less on coasts with a smaller angle. At the same time the rocky littoral and sublittoral developed in a more or less humid cold climate a considerably active but certainly less impressive microbe, plant and animal ecosystem than e.g., in tropical areas. The Echinus esculentus-Polydora ciliata community developed. Within a short while the combined boring activity of Polydora ciliata and the grazing activity of Echinus esculentus produced a biological transfer front directed downwards which approximates 50 cm 100 years^{-1}, i. e. 10 times faster, than the mighty waves of storm surf and

tides (31). But still probably slowlier than at several other coasts.

(2) Worldwide and especially in the tropical areas we see tremendous cliffs of sedimentary rocks with undercuttings at the tidal surf level. They seem to be closely correlated to the tidal movement and storm tides and exposure. Looking closer by using a microscope, we realise that the whole undercutting represents the niche of the (endolithic) Hyella ssp. Chiton spp. community (25). The cyanobacteria bore the rock and the chiton and other littoral forms graze on the cyanobacterium. Since the best way to reach their food (as it is the case with the Echinus esculentus-Polydora ciliata community) is to rasp away the rock itself and eat the rock with the rock-boring cyanobacteria, there is a tremendous speed of transfer. Natural genetic engineering and changes on large scale are necessary to establish this rock-eating machinery. The chiton is equipped with a mechanism rather similar to the crown of an oil drill. Its tongue is equipped with thousands of teeth made of pure magnetite, a mineral ranking high on the Mohs scale of hardness.

(3) Throughout a period of 100 years Humans fertilized an area, such as the Great Plains, USA. The area became covered with much more green then ever before. The former cover (climax community) of sparse green was replaced by a dense growth of crops. The organic matter produced is immediately taken away with the crop and transferred in millions of tons of corn, which in turn are biotransported to e. g. the Soviet Union (Another biological mass transfer of the locust swarm dimension). Several years without rain, overproduction of corn and a shift of labor from farming to car production leaves the fields devastated, depleted of humus, enriched in nutrients such as nitrate and without their natural over of resistent dryland brushes and grasses. Wind and water floods wash the soil away rapidly.

(4) In a very speculative manner it is even possible to ascribe a considerable part of the differences between elevated and deep areas of the globe to biological activities. The filling of oceanic basins as well as the transfer of rock material from the mountains and continents into the basins is so largely influenced by the biologically controlled processes described in 1-3 that one may conceive, that even the machinery of plate tectonics, which is responsable for orogenesis is largely modified by the biota. The primitive Earth is always described as different in its mantle and crustal behaviour (41). Perhaps it was only long after the firm establishment and takeover of life, that the firm crustal plates formed, which today govern the morphological basis of erosion. When some people object to Lyell's philosophy ((1), p. 318) that one better patiently unties the gordian knot instead proposing drastic cuts

such as bolid impacts and other catastrophies, then we suggest firmly, that the smoothest and most drastical impact, that ever may have happened to Earth, was the establishment of Life on this planetary body and we will not argue about its origin. We may, however, try to state, that even the firmest (iridium data evidenced) bolid impact theory still remains a theory and that ice ages can be promoted both ways biologically and (perhaps) abiologically. These notes have been added in order to mollify the somewhat exaggerated statement of biota interactions with plate tectonics. One must, however, seriously consider the large amount of organics such as hydrocarbons and the well known effects of organic polymers on the plasticity of clays as well as the fact, that internal pressure may build up at many places of the Earth crust by many different means and that even continental plates undergo the rules of buoyancy not much unlike the buoyancy of cyanobacteria via production and distruction, increase and decrease of vacuoles. The equilibrium of plate tectonics may be more fragile, and more subdued to minute modifications than we conceive it today. At least the Veda mythology of India tells about Earth quakes being the "thunder" of rocks scraping the metal carcass of a giant turtle.

(5) Transfer of rock material is observed in all altitudes, latitudes and longitudes. In all cases the rock surface and the first millimeters, sometimes centimeters and more of the rock are inhabited by a multitude of different genera of microorganisms. These do not only live there, they also make their living on the water and light and other energy sources available. The whole rock to them is a huge reservoir of nutrients as well as a holdfast and protective covering. If we place a culture of bacteria in a dialysis bag together with several particles of granite and add glucose outside the bag but no mineral nutrients such as nitrate, phosphate, potassium or magnesium, the suspension of bacteria will settle on granite surfaces and in granite crevices. Those bacteria that have been able to settle on a surface will divide faster then those floating or actively swimming in the liquid. A little later, a plate count of the liquid will deliver very low numbers ml^{-1}, while a plate count of bacteria g^{-1} rock will deliver figures usually exceeding those in the liquid by 3-4 orders of magnitude. A little later, potassium, iron, sodium and magnesium, which have not been added to the liquid medium will occur outside the dialysis bag by molecular diffusion. Powdering the material will increase the speed of the process by a factor of 10 to 100. A sterile control experiment will also deliver some nutrients by the action of glucose-enriched distilled water, but it will take months with vigorous shaking until significant values are reached.

(6) This next scenario is exactly the same as in example 5. Only this time the microorganisms are placed with the glucose outside the dialysis bag, and the rock with the nutrients in the mineral lattices and interfaces of clays or feldspars is not directly available for the organisms. In this case the rate of biological transfer is much slower but still higher by 2-3 orders of magnitude than in the case of distilled water alone under sterile conditions. The shortage of nutrients initiates the bacterial production of acids, bases or chelating compounds, which are excreted into the surroundings. The bacteria will be found in larger numbers close to and settling on the outer membrane of the dialysis bag.

MICROBIAL TRANSFER PROCESSES

Carbon Metabolism

Microbes deliver huge amounts of carbonic acid into the lithosphere, hydrosphere and atmosphere. Up to 50 % of oceanic productivity is caused by nannoplanktic cyanobacteria. The carbonic acid excreted and the oxygen both react with water and make it reactive. The agressive water dissolves minerals at rates exceeding by far those of the solvent water alone.

Another impact of the carbon metabolism on transfer rates is the enormous variability and number of biogenic acids, many of which are highly agressive and in addition act as chelating factors. The combination of the abilities to dissolve and to chelate is sometimes advantageous. For example the organisms may dissolve some minerals that are nutritionally important, but they may also dissolve poisonous minerals both of which may be immobilised by a chelate formation. The nutrients may then be further processes at a later date, while the toxic substances are detoxified (e. g. manganese in rock varnishes (22, 28). Polysaccharides and peptide bonded saccharides play an important role in the availability of nutrients from feldspars and clay minerals. These interact not only with inorganic soil absorbed nutrients, but also with energy donators absorbed to soil (16).

Sulfur Metabolism

Elemental sulfur and sulfate are not reactive under oxidising conditions and they are very slowly reactive under reducing conditions. The reductive sulfur cycle is practically exclusively run by microbial activity on this planet. Indirectly even the oxidation of sulfides is a biological process, for the oxygen which serves as oxidant is exclusively of biological origin, and is kept at a steady state level by biological means. The microbial sulfur cycle hooks in into the geochemical large-

scale cycles in the distribution of sulfides as compared to
sulfates and elemental sulfur on global and historical scales.
Practically every valence state of sulfur and the associated
compounds is microbially metabolised. Sulfuric acid production
is important for the dissolution of minerals and increases
transfer tremendously (24, 26, 29, 41).

Hydrogen Metabolism

Hydrogen is the most important electron donor for many
microorganisms under anaerobic and reducing conditions.
Methane, carbon dioxide, anoxygenic photosynthesis, petroleum
transformations and sulfate reduction are some of the biological
reactions dependent on hydrogen production and consumption by
bacteria. One of the most significant examples for transfer is
the anaerobic transfer of electrons and cations of steel, iron
and other metals, which is caused by hydrogen consumption by
anaerobic bacteria in soil and water. The scale is global
without doubt (29, 41).

Nitrogen Metabolism

Dinitrogen is completely inert and unreactive under abiotic
conditions. The nitrogenase system has been mentioned often
enough in newspapers to make its importance for the biogeochemical cycles clear. Many studies exist on free-living cyanobacteria involved in nitrogen reduction under very specific conditions (19, 43, 9). Microbial nitrogen reduction and oxidation
of any reduced nitrogen compound to nitrate controls and speeds
up many transfer processes. Local microenvironmental concentrations of nitric acid have been often described. They may dissolve other elements and minerals and the processes involved
have large-scale back-coupling with the total productivity of
the biosphere (24, 27, 34).

Metal Metabolism and cation transfer

Iron, manganese and other metals in their reduced and oxidized valence states are biologically transferred. The biological importance of these is much more obscure than their effects
on the acceleration of transfer processes. It is still not
known precisely whether iron and manganese oxidation really
serves as a considerable energy source or electron donor. Reduced iron may have served as electron donor for photosynthesis
in early periods of Earth history prior to water, and some
banded iron formations may have formed after intensive transfer
of iron via this process. There is better evidence that oxidized iron and manganese serve as electron acceptors, replacing
oxygen or nitrate in anaerobic respiration processes (e. g. 18,
37). The relative equations with the energetic efficiency are

quoted by Krumbein and Swart (29). Biological iron and manganese transformations either slow down or accelerate physicochemical kinetics within the field of redox-reactions outlined e. g. by Garrels and Christ (17). Nealson (39, 40) has summarized much of the information on the transfers of these elements. In the context of material transfer it is important that laminated epi- and endolithic communities (28) largely influence the transfer of iron, manganese and other heavy metals from the interior of rocks and minerals to the outer layers. Iron and manganese are depleted in the interior layers and enriched in the outer layers of rocks. These transfers are not destructive alone. Iron and manganese crusts and rock varnishes temporarily may serve as protective crusts according to the "Schutzkrustenmodel" of Linck (33).

Gold, uranium and silver have a biological transfer cycle. The biological activities in the solubilisation, stabilisation and re-solubilisation of several other elements as e. g. magnesium, alumina and even silica has been studied in detail by Berthelin (5, 6, 7). Biological transfers of chert (subcrystalline quartz) has been demonstrated by Krumbein and Jens 28) although the exact mechanisms involved are not yet understood in detail. Specific silicate dissolving bacteria have been isolated and described by Tesic and Tedorovic (44) although other investigations seem to indicate, that normal chemoorganotrophic bacteria and fungi of the soil microflora and rock microflora are active in the complete transfer of various kinds of silicates including quartz (28, 5, 8). In general the biological transfer of rocks and minerals is a faster process than chemical solution in the natural environment and, even more important, biological transfer is a selective process, which often acts against the thermodynamic conditions at the respective transfer site of weathering.

CONCLUSIONS

"Sans les éléments minéreaux dont certains seulement en très petite quantité il ny aurait pas de vie et en particulier pas de vie microbienne."

As a contrast to this statement of Bertrand (8) one might say:

"Without the multitude of living reactions as channels for energy and electrons and without the enormous capability of life to transfer with and against chemical gradients, there would be practically no or at least a very reduced geochemical rock cycle on this planet."

1) Transfer of the rocks and minerals of the lithosphere through time and space is considerably speeded up and directed by biological processes on all scales and through all the five kingdoms of the living natural bodies (biota).

2) Biological transfer of parts of objects of art and technology made of minerals, glasses and metals is considerable and by far exceeds the speed of inorganic transfer processes.

3) The transfer activity of the biosphere includes enrichments for practically all elements, purposeful arrangement according to the requirements of the biota, and rate limiting controls.

4) The abiotic transfer of rocks and minerals in a geomorphological sense is negligible under the thermodynamic conditions of Earth.

5) Biotic and abiotic states of matter are in homeorhesis. This is achieved by life as a "regularity of nature" (15) The transfer of rocks and minerals of the Earth's lithosphere are in the firm grip of the powerful force of life, because after all

this planet is alive.

BIBLIOGRAPHY

(1) Amoroso, G. G. and Fassina, V. 1983, Stone Decay and Conservation. Elsevier, Amsterdam.
(2) Bachmann, E. 1890, Ber. dtsch. Bot. 8, pp. 141-145.
(3) Bassalik, K. (1912) Z. Gärungs-Physiol. 2, pp. 1-32.
(4) Bassalik, K. (1913) Z. Gärungs-Physiol. 3, pp. 15-42.
(5) Berthelin, J. (1983) In: Microbial Geochemistry (Krumbein W. E., Ed.), pp. 223-262, Blackwell, Oxford.
(6) Berthelin, J. and Y. Dommergues (1972) Rev. Ecol. Biol. Sol 9, pp. 397-406.
(7) Berthelin, J. and A. Kogblevi (1972) Rev. Ecol. Biol. Sol 9, pp. 407-419.
(8) Bertrand, D. (1972 Rev. Ecol. Biol. Sol 9, pp. 349-396.
(9) Carpenter, E. J. (1973) Deep Sea Res. 20, pp. 285-288.
(10) Carruthers, G. T. (1890) Nature 41, pp. 153-156.
(11) Caumette, P. (1978) Dissertation, Montpellier.
(12) Danin, A., R. Gerson, K. Marton and J. Garty (1982) Palaeogeogr. Palaeoclimatol. 37, pp. 221-233.
(13) Dorn, R. I. and Th. M. Oberlander (1982) Progress in physical Geography 6, pp. 317-367.
(14) Ehrenberg, C. G. (1854) Mikrogeologie.
(15) Eigen, M. and P. Schuster (1979) The Hypercycle--a

Principle of Natural Self-Organisation. Springer, Berlin.
(16) Filip. Z. (1978) In: Microbial Ecology (Loutit M. W. and Miles J. A. R., Ed.).
(17) Garrels, R. M. and C. L. Christ (1965) Solutions, Minerals and Equilibria. Harper & Row, New York.
(18) Ghiorse, W. C. and P. Hirsch (1979) Arch. Microbiol. 123, pp. 213-226.
(19) Giani, D. (1983) N2-Fixierung bei Plectonema boryanum und anderen Cyanobakterien ohne Heterocysten. Ph. D. thesis, Oldbg.
(20) Hooke, J. (1665) Micrographia, or some physiological descriptions of minute bodies made by magnifying glasses, with observations and inquiries thereupon. Royal Soc. London.
(21) Imhoff, J. F., H. G. Sahl, G. S. H. Soliman and H. G. Trüper (1979) Geomicrobiol. J. 1, pp. 219-234.
(22) Jens, K. (1984) Dissertation, Oldenburg.
(23) Krumbein, W. E. (1969) Geol. Rdsch. 58, pp. 333-363.
(24) Krumbein, W. E. (1972) Rev. Ecol. Biol. Sol 9, pp. 283-319.
(25) Krumbein, W. E. (1979) Geomicrobiol. J. 1, pp. 139-203.
(26) Krumbein, W. E. (1979) In: Biogeochemical Cycling of Mineral-forming Elements (Trudinger P. A. and Swaine D. J., Ed.), pp. 47-68, Elsevier, Amsterdam.
(27) Krumbein, W. E. (1983) In: Microbial Geochemistry (Krumbein W. E., Ed.), pp. 1-4, Blackwell, Oxford.
(28) Krumbein, W. E. and K. Jens (1981) Oecologia 50, pp. 25-38.
(29) Krumbein, W. E. and P. K. Swart (1983) In: Microbial Geochemistry (Krumbein W. E., Ed.), pp. 5-62, Blackwell, Oxford.
(30) Krumbein, W. E. and D. Werner (1983) In: microbial Geochemistry (Krumbein W. E., Ed.), pp. 125-157, Blackwell, Oxford.
(31) Krumbein, W. E. and J. N. C. van der Pers (1974) Helgol. Wiss. Meeresunters. 26, pp. 1-17.
(32) Lapo, A. V. (1982) Traces of Bygone Biospheres. Mir, Moskau.
(33) Linck, G. (1930) Chemie der Erde 4, pp. 67-69.
(34) Lovelock, J. E. (1979) Gaia A new Look at Life on Earth. Oxford Univ. Press, Oxford.
(35) Lovelock, J. and L. Margulis (1980) Coevolution 1, pp. 20-32.
(36) Margulis, L. and Lovelock, J. (1983) Coevolution 11, pp. 48-52.
(37) Munch, J. C., T. Hillebrand and J. C. Ottow (1978) Can. J. Soil Sc. 58, pp. 475-486.
(38) Nadson, G. (1903) Die Mikroorganismen als Geologische Faktoren I. Trudy Commissi Islact. Min. Vod. G. Petersburg.
(39) Nealson, K. H. (1983) In: Microbial Geochemistry (Krumbein W. E., Ed.), pp. 159-190, Blackwell, Oxford.
(40) Nealson, K. H. (1983) In: Microbial Geochemistry (Krumbein W. E., Ed.), pp. 191-222, Blackwell, Oxford.
(41) Schopf, J. W. (ed.) (1983) Earth's Earliest Biosphere.

Princeton Univ. Press, Princeton, N. J.
(42) Schröder, H. G. (1982) Dissertation, Göttingen.
(43) Stal, L. and W. E. Krumbein (1981) FEMS Microbial. Lett. 11, pp. 295-298.
(44) Tesic, T. and M. S. Tedorovic (1958) Semjlist I bilka 8, pp. 233-240.
(45) Vernadsky, W. (1929) La Biosphere. Alcan, Paris.
(46) Westbroek, P. (1983) In: Biomineralization and biological Metal Accumulation (Westbroek P. and De Jong E. W., Ed.), pp. 1-14, Reidel, Dordrecht.
(47) Winkler, E. M. (1975) Stone: Properties, Durability in Man's Environment. Springer, Berlin.
(48) Winogradski, S. (1887) Bot. Ztrbl. 45, pp. 489-507.

SOLUBILIZATION, TRANSPORT, AND DEPOSITION OF MINERAL CATIONS
BY MICROORGANISMS - EFFICIENT ROCK WEATHERING AGENTS

Friedrich E. W. Eckhardt

Inst. f. Allgemeine Mikrobiologie
Universität Kiel
D-2300 Kiel
West Germany

INTRODUCTION

Mineral environments such as rocks or building stones are usually thought to contain few living micoorganisms. This might be true for stones that have just been quarried or lava shortly after a volcanic eruption, but once the surface of a stone has been exposed to the environment, considerable numbers of various different types of organisms can be observed. In surface samples of freshly-worked building stones, for example, 10^3 colony forming units (cfu) of heterotrophic bacteria and filamentous fungi per gram of stone were observed after 3 days exposure (39).

"It is virtually impossible to find any region on earth that is sterile" (60). Moreover, microorganisms live and are still physiologically active under the extreme environmental conditions found in Antarctica (23), in hot springs at $90°C$ and avove (9), in deserts (22, 38, 62), and in the oceans to depths of around 5000 m, for example in manganese nodules (17).

NUMBERS AND DISTRIBUTION OF MICROORGANISMS IN SOIL, ROCKS, AND BUILDING STONES

Many data are available on the microbiology and chemistry of soils. On the average, 10^7 bacteria, 10^6 actinomycetes, 10^5 fungi, and 5×10^4 algae (cfu/g dry weight) are found by plate counting in the top layers of cultivated soils (4). Horizontal as well as vertical distributions of these microorganisms may vary by factors of 10 and more (27) depending on the character of the soil, the environmental conditions, and on

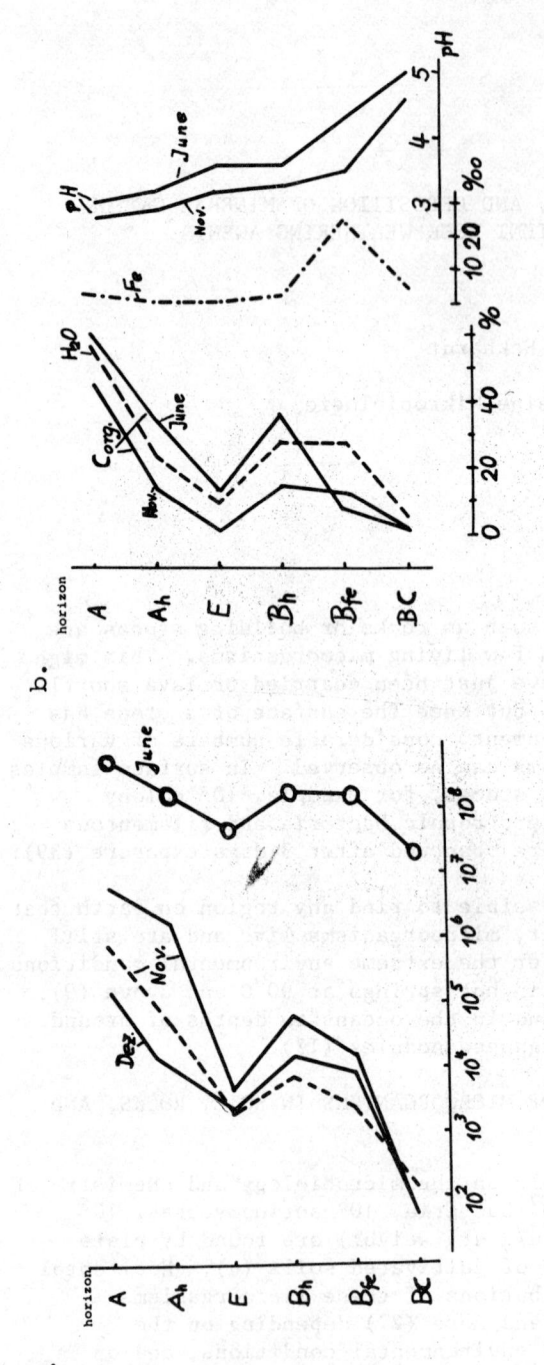

Fig. 1: Analytical data of an iron humic podsol profile

a. Numbers (cfu/g dry wt.) of bacteria (-----), of filamentous fungi (———), and numbers of microbial fluorescent units/g dry wt. (0———0) at the different soil horizons.

b. moisture content (%-fresh wt.), organic matter content (%-dry wt.), total iron content (% -dry wt.), and pH at the different soil horizons

the microbial species contributing to the respective population. Study of the microflora of a podsol showed that the total numbers of heterotrophic bacteria and filamentous fungi were closely correlated with the moisture and organic matter contents of the different horizons (Fig. 1). The total number of microorganisms and species diversity are usually highest in the topsoil layers. However, in the B_s horizon of a brown soil, 10^5 cfu/g iron-precipitating bacteria were found, accounting there for 80% of the total bacterial numbers (75).

Relatively high numbers (10^4-10^6 cfu/g) of bacteria, actinomycetes, and fungi have been reported in the surface layers of various rocks and building stones (6, 13, 14, 16, 30, 37, 39, 73). Along pores, cracks and fissures these microorganisms may invade deeper into the rocks. In the weathering layer of an olivine, for example, the numbers of fungi found 4 cm below the surface were similar to those of the surface layer; the number of actinomycetes increased with depth whereas those of bacteria decreased (Table 1). In porous materials, fractured crystalline bedrock for example, bacteria migrate along with the flow of groundwater; velocities of 28.7 m in 24 hours have been reported (26). Drillings in a biotite gneiss have shown the presence of bacteria at a depth of 160 m (49).

Weathering layer (cm)	Colony forming units (/g) bacteria	actinomycetes	fungi
surface 0-1	1.1×10^5	3.9×10^4	3.5×10^2
subsur- 1-2.5	7.6×10^4	1.0×10^4	1.2×10^3
face 2.5-4	2.9×10^3	2.9×10^5	8.7×10^2

Table 1: Distribution of microorganisms within the weathering layer of an olivine rock (73).

Freshly exposed igneous rocks devoid of organic matter will be "contaminated" first by phototrophic microorganisms (cyanobacteria, algae, lichens) and by chemolithotrophic bacteria followed very soon by heterotrophic microorganisms (bacteria, actinomycetes, fungi). Sedimentary rocks containing diagenetic organic matter may be colonized simultaneously by autotrophic and by heterotrophic microorganisms. The weathering of minerals and rocks by microorganisms has been studied since as early as 1890 (2) and 1912/13 (3).

WEATHERING ACTIVITIES OF MICROORGANISMS

Various bacteria, cyanobacteria, algae, fungi, and lichens have been found to be involved in both the degradation and the formation of minerals and rocks, and in the cycling of mineral-forming elements (18, 40, 67). In addion, different mechanisms have been recognized in the mobilization, transport, and deposition of mineral cations by microorganisms. These include enzymatic reactions, acidolysis by microbially formed organic or inorganic acids, reduction or oxidation, complexation, and microbial adsorption or uptake. More than one of these reactions may be involved in a particular mineral transformation.

One of the most spectacular microbial "weathering" process is the bacterial leaching of metals from ores (7, 8). A great variety of metal sulfides may be solubilized by chemolithotrophic sulfur oxidizing bacteria (Table 2). Leaching of copper and/or uranium oxide has reached industrial importance. This technique was already used empirically in the 18th century in the copper mines of Rio Tinto, Spain (and probably a long time before that in other Mediterranean countries); is was recognized later on as resulting from sulfur oxidizing bacteria (8). In 1971, about 15% of the copper production of the United States was by bacterial leaching (68).

The most important microbial "miners", Thiobacillus thiooxidans and T. ferrooxidans, enzymatically oxidize reduced sulfur compounds (eg metal sulfides) to water-soluble metal sulfates. Metal sulfide leaching is accelerated several-fold by applying T. ferrooxidans in comparison to the chemical (sulfuric acid) leaching technique (Table 3). In addition, T. ferrooxidans transforms ferrous iron to ferric iron which oxidizes, for example, uranium(IV) oxide to the soluble uranium(VI) oxide (Fig. 2). This microbial weathering works best under acid conditions. At pH 2-4.5 T. ferrooxidans can enhance oxidation rates by factors of 10^5-10^6 compared to abiotic mechanisms (44).

A decrease in environmental acidity increases the solubility of most metal cations. In agriculture, the pH of alkaline soils is lowered by fertilization with elemental sulfur, thus providing the crop with sufficient trace elements in a plant-available form. The mechanism is microbial sulfur oxidation causing a decrease in pH (57, 64). The simultaneous presence of gypsum and thiobacilli has been reported in lesions and weathering layers on calcareous building stones and monuments (29, 37, 53, 66). Up to 10^6 cfu/g sulfur oxidizers were found below the weathering crust, but only up to 10^3 cfu/g were found in the unweathered surface stone (41). Whether the gypsum is formed by the bacteria and/or from sulfate input by immission or acid rain is still under discussion.

As another example, concrete building structures have been severely corroded by thiobacilli, especially in sewage systems (21, 57). Exhalation of H_2S from sewage resulting from

mineral	formula	mineral	formula
arsenopyrite	FeAsS	orpiment	As_2S_3
bornite	Cu_5FeS_4	pyrite/marcasite	FeS_2
chalcocite	Cu_2S	sphalerite	ZnS
chalcopyrite	$CuFeS_2$	stibnite	Sb_2S_3
covellite	CuS	tetrahedrite	$Cu_8Sb_2S_7$
enargite	Cu_5AsS_4	gallium sulfide	Ga_2S_3
galena	PbS	cobalt sulfide	CoS
millerite	NiS	molybdenite	MoS_2

Table 2: Metal sulfides biodegradable by direct action of <u>Thiobacillus ferrooxidans</u> (18).

mineral	formula	factor
arsenopyrite	FeAsS	7
sphalerite	ZnS	7
chalcopyrite	$CuFeS_2$	12
covellite	CuS	18
bornite	Cu_5FeS_4	18

Tab. 3: Acceleration factor of the dissolution process by <u>Thiobacillus ferrooxidans</u> in comparison with a conventional leaching technique (61)

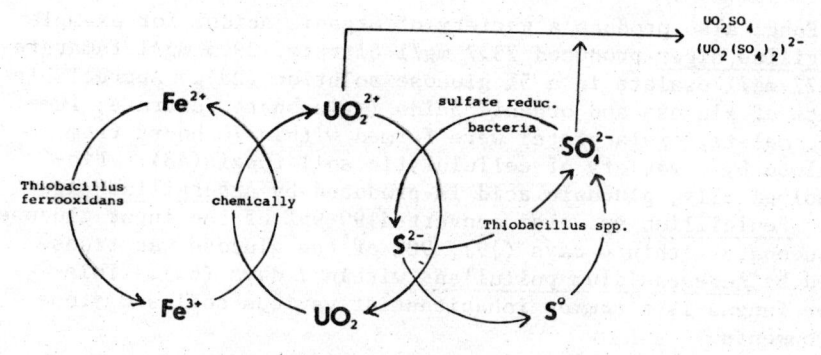

Fig. 2: Biogeochemical uranium cycle

bacterial sulfate reduction, and subsequent sulfuric acid production by Thiobacillus species on the walls of the sewage pipes caused extensive damage to the concrete (47, 54).

In general, colorless sulfur bacteria of various different genera (42) as well as phototrophic sulfur bacteria (52) generate sulfate from reduced sulfur compounds in the appropriate aerobic or anaerobic environments respectively. In contrast, a number of anaerobic bacteria (Desulfovibrio spp., Desulfotomaculum spp.) reduce sulfate to sulfide and may thus induce large-scale deposition of metal sulfides and ores, in marine and lake sediments, for example (18, 33, 78).

Nitrifying bacteria, another group of chemolithotrophic acid-producing microorganisms, oxidize ammonium ions to nitrite and to nitrate under neutral to alkaline conditions. Bacteria of this group have been documented in decaying limestone of buildings and monuments (36), and in weathering serpentinized ultrabasic rocks (45). In field experiments, nitric acid formation is correlated with the solubilization of calcium and magnesium from granite sands (5). Ammonium pollution of water in a cooling tower enables the nitrifying bacteria to produce nitric acid which, in one example, destroyed the concrete parts of the cooling system within only 4 years (34).

Kunze (43) documented corrosion of marble, limestone, granite, and basalt by filamentous fungi and by the mycorrhiza of plants through the action of excreted oxalic and citric acid. Moreover, soil bacteria producing ketogluconic acid degraded a variety of silicate and phosphate minerals (12, 46). Gluconate, glucuronate and other acid glucose derivatives are common constituents of the capsules and extracellular slime layers of most bacteria, cyanobacteria, and algae (20). Gluconate accounts for 10-25% of bacterial slimes (10). Additionally, oxalic, citric, or lactic acid may be excreted by these bacteria, which solubilize carbonate as well as phosphate minerals (70).

Fungi also produce a variety of organic acids, for example Aspergillus niger produced 23.7 mg/l citrate, 29.5 mg/l fumarate, and 322 mg/l oxalate in a 5% glucose solution (28). Appreciable amounts of glucose and organic acids (gluconate, citrate, fumarate, malate, oxalacetate) were formed within 70 hours from cellulose by a variety of cellulolytic soil fungi (48). Biotechnologically, gluconic acid is produced by Aspergillus niger (55). Penicillium sp. also converted 90-95% of the input glucose to gluconate within 4 days (]9); 90% of the glucose was transformed by Aureobasidium pullullans within 2 days (65). This latter fungus is a common inhabitant of various soil and stone environments.

In one study of a natural environment (11, 25), the top layer of a forest soil contained an average of 7 mg/g oxalate, with the oxalate crystals adhering to fungal hyphae demonstrating the source of the organic acid. Ca- and Mg-oxalate crystals have also been found in lichen thalli grown on basalt and serpentinite (32, 76). In culturing the mycobiont of the lichen, Ca-oxalate crystals from excreted oxalic acid formed in the agar below the colonies (76). During my studies I found the weathering crust of a sandstone monument colonized by a variety of bacteria, yeasts, and filamentous fungi. The presence of oxalic and citric acids in the crust was demonstrated by thin-layer chromatography (13).

Experiments to evaluate the weathering capacity of these fungi showed solubilization of up to 60% of the cations in biotite (14,15,16). Batch cultures of A. niger reduced the pH of the medium to 1.8, and solubilized cations from powdered biotite much as 0.1 M oxalic or citric acid does (Fig. 3). Iron gluconate crystals (identified by X-ray diffraction) were formed in the solids during these experiments (14). In addition, fulvic acid-like substances were detected in the culture solutions by sephadex gel chromatography.

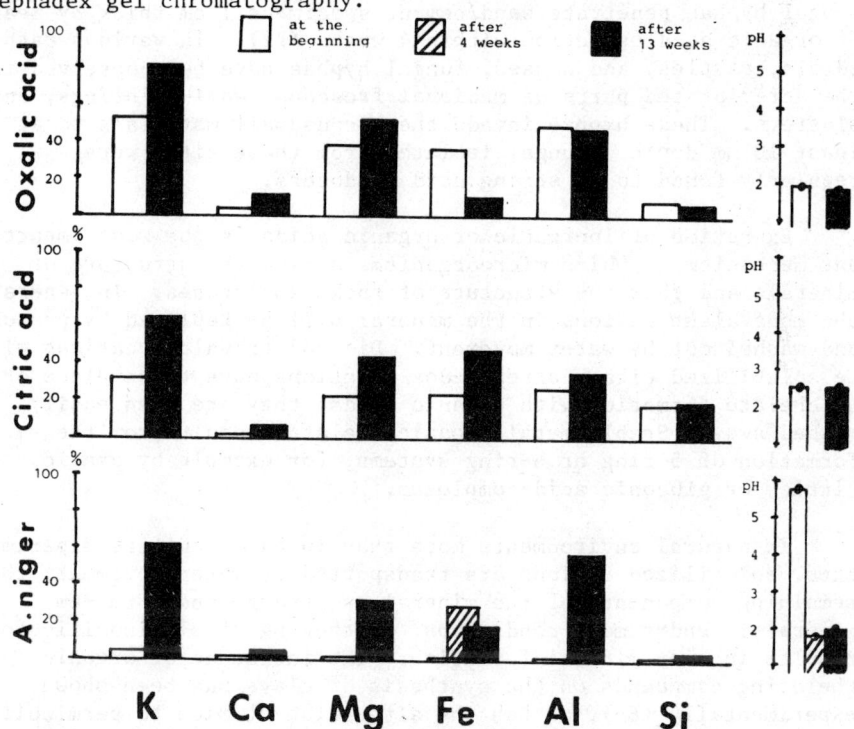

Fig. 3: Solubilization of cations from biotite by 0.1 M oxalic and citric acid, and by the activity of Aspergillus niger in a 0.1% glucose solution.

The results of these experiments show a strong weathering influence of A. niger--and of other filamentous fungi and yeasts --by excretion of organic acids. Similar activities were found in (i) experiments with A. niger, which extracted large amounts of magnesium from chrysotile (76), (ii) the solubilization of Li, Fe, Al, and Si from pegmatites by Trichoderma lignorum and by Penicillium notatum (1), and (iii) leaching Li, Al, and Si from spodumene by filamentous fungi, by T. thiooxidans, and by a slime-producing "silicate" bacterium, Bacillus mucilagenosus (35).

MICROBIAL WEATHERING MECHANISMS

Microbial degradation of minerals, rocks, and building stones affecting sulfides, phosphates, silicates, oxides, carbonates etc. has been observed in vitro as well as in natural sites (67). The limestone coast of the northeastern Adriatic, Yugoslavia, for example, is being removed at a rate of 0.25-1 mm/y by the etching activities of cyanobacteria, algae, and fungi (58). Wood-destroying fungi also etch bricks and concrete; the fungal hyphae penetrate sand/cement specimens 1 cm thick by means of organic acid excretion within 4 weeks (72). In various cathedrals, castles, and houses, fungal hyphae have been observed in the deteriorated parts of medieval frescoes, wall paintings, and plasters. These hyphae invade the porous wall materials to at least 10 mm depth. Fungal isolates from these sites were regularly found to be strong acid producers.

Excretion of inorganic or organic acids is the most important mechanism by which microorganisms attack the structure of minerals and thus the structure of rocks and stones. In general, the monovalent cations in the mineral will be replaced by protons and washed out by water movement. Di- and trivalent cations will be solubilized either after redox reactions have taken place or by chelate formation with organic acids; they are then easily washed away. Stable metal-organic chelates result from the formation of 5-ring or 6-ring systems, for example by oxalic, citric, or gluconic acid complexes.

In natural environments more than in batch culture experiments, solubilized cations are transported by water movement; the remaining components of the mineral re-arrange and form new minerals. Under most conditions, weathering of aluminosilicates results in clay mineral formation. The influence of organic chelating compounds on the synthesis of clays has been shown experimentally (69), as has the alteration of mica to vermiculite by fungi (74). In addition, the microbial counts in an illite deposit (50) support the concept that microorganisms are involved in the transformation of mica and feldspars to clay minerals.

An additional microbial weathering process should also be considered. Heavy metals and metalloids are common constituents of various minerals in rocks and soil (e.g. Table 2), or they may be deposited in locally high concentrations (18). The weathering of rocks and minerals as well as mining processes (and also man-made pollution) increases the concentrations of these generally toxic substances in the environment considerably. In 1897, Gosio (24) already reported the influence of fungi in transforming arsenic in wallpaper pigments to toxic volatile arsines. An increasing number of heavy metals and metalloids which are converted by bacterial and/or fungal oxidation, or methylation, have now been recognized (18, 31, 51, 59, 77).

Various mechanisms have evolved during earth history by which microorganisms can cope with toxic substances in their environment, by which they have "learned" to become resistant. Microbial reduction of mercury ions to metallic mercury makes this element volatile, as does bacterial or fungal methylation to methylmercury. Plasmid-coded methylation is also known for As, Se, Te, Pb, Sn, and Cd (59). In addition, chemical trans-methylation reactions have been detected for a great number of other metals. In addition to Hg, bacteria are able to reduce As, Cr, Se, and Te, whereas other microorganisms can oxidize As and Sb (63). Finally, precipitation of metals by bacteria and fungi may occur: a) indirectly, e.g. in the form of metal sulfides due to bacterial sulfide production, b) by adsorbing Fe, Mn, Cd, Ag, or U onto cell walls by means of polysaccharides etc. (40,67), or c) by uptake into the cells and deposition there, e.g. Fe in the form of magnetite in magnetotactic bacteria. Processes such as precipitation as sulfides have given rise to various ore deposits (18), and may contributed to the accumulation of metals in sediments and soils.

Toxic metal resistance and the capacity of microorganisms to convert heavy metals and metalloids are generally developed in contact with a metal deposit or polluted environment. These microorganisms could therefore be useful as biomonitors in detecting or controlling metal pollution. Attempts to use microorganisms in the bioaccumulation of metals (Ag, Cd, U) from waste waters or dumps have also been reported (59). The activities of these microbes could thus be of interest in managing the heavy-metal problems of sewage sludge, or in cleaning sewage plant effluents.

CONCLUSION

According to Koch's postulates, a microorganism can only be named a "germ of disease" if it (i) is always found in animals suffering from the disease, but not in healthy ones, (ii) can be cultivated in pure culture, (iii) provokes the corresponding disease by inoculation in susceptible animals, and (iv) can be isolated from the induced disease and identified as being the same as the original microorganism. With regard to the processes discussed in this paper, an additional postulate should be added: (v) the microorganism should be present in the weathering stone, soil, etc. in sufficient numbers to cause a substantial effect.

As pointed out in the preceding sections, microorganisms are able to actively attack minerals and rocks. They are present in appropriate numbers at places where weathering occurs. The ability of these organisms to degrade minerals can be demonstrated experimentally. Bacterial leaching of metals in dumps (7, 8) and destruction of concrete by nitrifying bacteria (34) and fungi (72) demonstrate that microorganisms are also able to start the corresponding weathering process anew. An increased use of bacteria and fungi in biotechnology and bioengineering also demonstrates that microorganisms are capable of conducting processes on a large scale. In contrast, mineral weathering processes in nature may proceed more slowly and thus be overlooked by many scientists. However, on a longer time scale, weathering of minerals, rocks, building stones, and monuments by bacteria, cyanobacteria, fungi, algae, and lichens is an important factor-- even in Antarctic sandstones.

REFERENCES

1. Avakyan, Z.A., Karavaiko, G.I., Mel'nikova, E.O., Krutsko, V. S. and Ostrushko, Yu. I., 1981: Microbiology 50, pp. 115-120.

2. Bachmann, E., 1980: Ber. dt. bot. Ges. 8, pp. 141-145.

3. Bassalik, K., 1912/13: Z. Gärungs-Physiol. 2, 1-32; 3, 15-42.

4. Beck, T., 1968: Mikrobiologie des Bodens. BLV, München.

5. Berthelin, J., Gelgy, G., Wedraoggo, F.X., 1981; Coll. Humus et Acote, Reims; Service Science du Sol, Nancy, pp. 112-117.

6. Blöchliger, G., 1931: Mikrobiologische Untersuchungen an verwitternden Schrattenkalkfelsen. Diss. ETH Zürich.

7. Brierley, C.L., 1978: CRC Crit. Rev. Microbiol. 6, pp. 207-262.

8. Brierley, C.L., 1982: Scientific American 247, pp. 42-51.

9. Brock, T.D., 1978: Thermophilic microorganisms and life at high temperature. Springer, New York.

10. Claus, E., Wittmann, H., Rippel-Baldes, A., 1958: Arch. Mikrobiol. 29, pp. 169-187.

11. Cromack, K., Sollins, P., Graustein, W.C., Speidel, K., Todd, A.W., Spycher, G., Li, C.Y., Todd, T.L., 1979: Soil Biol.

Biochem. 11, pp. 463-468.

12. Duff, R.B., Webley, C.M., Scott, R.O., 1963: Soil Sci. 95, pp. 105-114.

13. Eckhardt, F.E.W., 1978: in "Environmental Biogeochemistry and Geomicrobiology" (Krumbein, W.E., Ed.), Ann Arbor Sci. Publ., Ann Arbor, Michigan, pp. 675-686.

14. Eckhardt, F.E.W., 1979: Z. Pflanzernähr. Bodenkd. 142, pp. 434-445.

15. Eckhardt, F.E.W., 1979: Mitt. Dtsch. Bodenkd. Ges. 29, pp. 391-398.

16. Eckhardt, F.E.W., 1980: in "Biodeterioration" (Oxley, T.A. et al., Eds.), Pitman Publ., London, pp. 107-116.

17. Ehrlich, H.L., Ghiorse, W.C., Johnson, G.L., 1972: Dev. Ind. Microbiol. 13, pp. 57-65.

18. Ehrlich, H.L., 1981: "Geomicrobiology", Dekker, New York.

19. Elnaghy, M.A. and Megalla, S.E., 1975: Folia Microbil. 20, pp. 504-508.

20. Finch, P., Hayes, M.B.H., Stacey, M., 1971: in "Soil Biochemistry" Vol. 2 (McLaren, A.D. and Skujins, J. Eds.), Dekker, New York, pp. 257-319.

21. Forrester, J.A., 1959: The Surveyor 118, pp. 881-884.

22. Friedmann, E.I., 1971: Phycologia 10, pp. 411-428.

23. Friedmann, E.I., 1982: Science 215, pp. 1045-1053.

24. Gosio, B., 1897: Ber. 30, pp. 1024-1026.

25. Graustein, W.C., Cromack, K. Jr., Sollins, P., 1977: Science 198, pp. 1252-1254.

26. Hagedorn, C., 1984: in "Groundwater Pollution Microbiology" (Bitton, G. et al., Eds.), Wiley, New York, pp. 181-195.

27. Hattori, T., 1973: "Microbial Life in the Soil", Dekker, New York.

28. Henderson, M.E.K. and Duff, R.G., 1963: J. Soil Sci. 14, pp. 236-246.

29. Jaton, C., 1971: "Contributions à l'étude de l'altération microbiologique des pierres de monuments en France" Thèse, Paris.

30. Jaton, C. and Orial, G., 1978: Int. Symp. Deterioration and Protection of Stone Monuments, UNESCO, Paris, Sess. 4.5.

31. Jernelöv, A. and Martin, A.L., 1975: Ann. Rev. Microbiol. 29, pp. 61-77.

32. Jones, D., Wilson, M.J., Tait, J.M., 1980: Lichenologist 12, pp. 277-289.

33. Jørgensen, B.B., 1983: in "Microbial Geochemistry" (Krumbein, W.E., Ed.), Blackwell, Oxford, pp. 91-124.

34. Kaltwasser, H., 1976: Europ. J. Appl. Microbiol. 3, 185-192.

35. Karavaiko, G.I., Krutsko, V.S., Mel'nikova, E.O., Avakyan, Z.A., Ostroushko, Yu.I., 1980: Microbiology 49, pp. 402-406.

36. Kauffmann, J., 1960: Corrosion-Anticorrosion 8, pp. 87-95.

37. Krumbein, W.E., 1968: Z. Allg. Mikrobiol. 8, pp. 107-117.

38. Krumbein, W.E., 1969: Geol. Rundsch. 58, pp. 333-363.

39. Krumbein, W.E., 1973: Dtsch. Kunst. Denkmalpfl. 31, pp. 54-71.

40. Krumbein, W.E. (Ed.), 1983: "Microbial Geochemistry", Blackwell, Oxford.

41. Krumbein, W.E. and Pochon, J., 1964: Ann. Inst. Pasteur 107, pp. 724-732.

42. Kuenen, J.G., 1975: Plant Soil 43, pp. 49-76.

43. Kunze, G., 1906: Jahrb. wiss. Bot. 42, pp. 357-393.

44. Lacey, D.T., Lawson, F., 1970: Biotechnol. Bioeng. 12, 29-50.

45. Lebedeva, E.V., Lyalikova, N.N., Bugel'skii, Yu.Yu., 1978: Microbiology 47, pp. 898-904.

46. Louw, H.A. and Webley, D.M., 1959: J. Appl. Bacteriol. 22, pp. 216-226.

47. Milde, K., Sand, W., Ebert, A. Bock, E., 1980. Forum mikrobiol. 3, pp. 107-108.

48. Moore, S., Stapelfeldt, E.E., 1976: in "Proc. 3, Int. Biodegr. Symp." (Sharpley, J.M., Kaplan, A.M., Eds.) pp. 711-718.

49. Neher J., Rohrer E., 1958: Eclog. Geol. Helvet. 52, 619-625.

50. Oberlies, F., Zlatanovic, J., 1964: Ber. Dtsch. Keram. Ges. 41, pp. 691-695.

51. Perlman, D., 1965: Adv. Appl. Microbiol. 7, pp. 103-108.

52. Pfennig, N., 1975: Plant Soil 43, pp. 1-16.

53. Pochon, J., Jaton, C., 1967: Chem. Indust., pp. 1587-1589.

54. Pomeroy, R., 1960: Water Sewage Works 107, pp. 400-403.

55. Rehm, H.J., 1980: "Industrielle Mikrobiologie", 2. Aufl., Springer, Berlin.

56. Rigdon, J.H., Beardsley, C.W., 1958: Corrosion 14, pp. 60-62.

57. Rupela, O.P., Tauro, T., 1973: Soil Biol. Biochem. 5, 899-901.

58. Schneider, J., 1976: Contrib. Sedimentol. 6, pp. 1-112.
59. Silver, S., Misra, T.K., 1984: Basic Life Sci. 28, pp. 23-46.
60. Silverman, M.P., 1979: in "Biogeochemical Cycling of Mineral-Forming Elements" (Trudinger, P.A. and Swaine, D.J., Eds.), Elsevier, Amsterdam, pp. 445-463.
61. Solzhenkin, P.M., 1980: Abstr. Int. Conf. "Use of Microorganisms in Hydrometallurgy", Pecs, Hungary, pp. 7-11.
62. Staley, J.F., Palmer, F.R., Adams, J.B., 1982: Science 215, pp. 1093-1095.
63. Summers, A.O., Silver, S., 1978: Ann. Rev. Microbiol. 32, pp. 637-672.
64. Swaby, R.J., Vitolins, M.I., 1968: Trans Int. Congr. Soil Sci., Adelaide, pp. 673-681.
65. Takao, S., Sasaki, Y., 1964: Agr. Biol. Chem. 28, pp. 752-756.
66. Thiebaud, M., Lajudie, J., 1963: Ann. Inst. Pasteur 105, pp. 353-358.
67. Trudinger, P.A., Swaine, D.J., Eds., 1979: "Biogeochemical Cycling of Mineral Forming Elements", Elsevier, Amsterdam.
68. Tuovinen, O., Kelly, D.P., 1972: Z. allg. Mikrobiol. 12, pp. 311-346.
69. Violante, A., Violante, P., 1980: Clays and Clay Minerals 28, pp. 425-434.
70. Wagner, M., Schwartz, W., 1967: Z. allg. Mikrobiol. 7, 33-52.
71. Wainright, M., 1981: Z. Pflanzenernähr. Bodenkd. 144, 41-63.
72. Wazny, J., 1980: in "Biodeterioration" (Oxley, T.A. et al. Eds.), Pitman, London, pp. 59-62.
73. Webley, D.M., Henderson, M.E.K., Taylor, J.F., 1963: J. Soil Sci. 14, pp. 102-112.
74. Weed, S.B., Davey, C.B., Cook, M.G., 1969: Soil Sci. Soc. Amer. Proc. 33, pp. 702-706.
75. Wenzel, A., Schweisfurth, R., 1979: Mitt. Dtsch. Bodenknd. Ges. 29, pp. 493-506.
76. Wilson, M.J., Jones, D., McHardy, W.J., 1981: Lichenologist 13, pp. 167-176.
77. Wood, J.M., 1975: Naturwiss. 62, pp. 357-364.
78. Zajic, J.E., 1969: "Microbial Biogeochemistry", Academic Press, New York-London.

CHEMICAL WEATHERING AND SOLUTION CHEMISTRY IN ACID FOREST SOILS: DIFFERENTIAL INFLUENCE OF SOIL TYPE, BIOTIC PROCESSES, AND H$^+$ DEPOSITION

Christopher S. Cronan[*]

Department of Botany & Plant Pathology
University of Maine at Orono
Orono, Maine 04469

ABSTRACT

In this investigation, weathering rates were calculated for three eastern North American forest soils using five separate estimation techniques. In addition, leaching experiments were performed to examine the influence of selected environmental variables on the weathering process. Weathering rates varied five-fold between soils, ranging from approximately 0.5 Keq ha^{-1} yr^{-1} in a sandy Adams Spodosol, to 1.2 Keq ha^{-1} yr^{-1} in a sandy loam Becket Spodosol, to 2.7 Keq ha^{-1} yr^{-1} in a silty Unadilla Inceptisol. Inter-soil differences in weathering rate were inversely correlated with mean soil particle size and positively correlated with total exchangeable bases in the soil profile. Study results also demonstrated major differences in weathering rates between soil horizons.

[*]Supported by NCSU/EPA Contract APP0026-1980.

INTRODUCTION

There are a number of compelling reasons for studying weathering processes in natural systems. From a geochemical perspective, weathering reactions are critical to the distribution and transport fate of a significant number of elements. Without improved understanding of weathering reaction thermodynamics, kinetics, and responses to environmental variation, it will be impossible to develop effective models for predicting the geochemical behavior of these elements. From an ecological perspective, there is an additional critical need to quantify the variable contribution of weathering to base supply (Ca + Mg + K + Na) in natural ecosystems. For example, base supply rate may determine long-term site quality in areas that are subjected to whole-tree harvesting of the forest resource. Likewise, base supply rate from primary weathering may control the fate of terrestrial and aquatic systems that are exposed to chronic acid deposition from the atmosphere.

The objective of this investigation was to estimate weathering rates for representative acid forest soils in northeastern North America and to examine the influence of selected environmental variables on the weathering process. This field and laboratory study was conducted using three forest soils from the northeastern U.S. Conceptually, the research was designed to address a framework of testable hypotheses. The primary null hypothesis for the investigation was as follows: there are no differences in weathering rates between separate acid forest soils in northeastern North America. Besides this general hypothesis, the investigation also considered the following kinds of more specific hypotheses.

* Weathering rate in the soil is inversely porportional to mean grain size and increases with surface area.

* Weathering is a rate-limited non-equilibrium process which increases per unit area with decreasing soil permeability and increasing water residence time.

* The weathering rate of base cations is proportional to the concentration of bases in the

primary mineral substrate.

* Weathering varies as a function of vegetative cover and its influence on dry deposition and uptake-coupled proton release.

* Weathering rate is controlled by the leachate chemistry of the biologically-derived forest floor. Weathering increases as a function of organic ligand concentrations and their influence on metal complexation and free metal activities.

* Weathering increases as a function of H^+ inputs from atmospheric deposition and internal generation by biota.

METHODS

This investigation was conducted as a soil microcosm leaching study and involved simulated precipitation inputs to undisturbed soil column lysimeters collected from three contrasting forest soils: a Becket series Haplorthod from Old Forge, N.Y., an Unadilla series Dystrochrept from Hanover, N.H., and an Adams series Haplorthod from Ossipee, N.H. These soils were selected to represent a range of contrasting physical-chemical properties with potentially different sensitivities to acidic deposition. In addition, one soil -- the Adirondack Becket -- was chosen to coincide with on-going field investigations by the author and other collaborators on the ILWAS project (The Integrated Lake-Watershed Acidification Study). The comparative environmental, chemical, and physical characteristics of each soil are presented in Tables 1-4.

At each of the three forested sites, ten intact 12 cm diameter soil columns were removed from each of three soil depths: 5 cm (O horizon), 20 cm (O+A horizon), and 50 cm (O+A+B horizon). These ninety soil columns were then equipped with plastic tension plates and were installed in a greenhouse at 12-25°C. Over the next 60 weeks, the microcosms were each subjected to a weekly 3.5 cm simulated rain event. Initially, the soil columns were all exposed to a three month baseline period of pH 5.7 simulated rain events. After completion of the pretreatment

TABLE 1
COMPARATIVE ENVIRONMENTAL CHARACTERISTICS OF STUDY SOILS.

PARAMETER	ADAMS	BECKET	UNADILLA
LOCATION	71°8'W, 43°47'N	75°0'W, 43°40'N	72°17'W, 43°42'N
VEGETATION	OAK-PITCH PINE	N. HARDWOOD	WHITE PINE
PARENT MATERIAL	OUTWASH	TILL	LAKE SEDIMENT
SOIL SUBORDER	TYPIC HAPLORTHOD	TYPIC HAPLORTHOD	TYPIC DYSTROCHREPT
SOIL TEXTURE	SAND	SANDY LOAM	SILT/SILT LOAM
% SAND	HIGH	MEDIUM	LOW
% SILT	LOW	MEDIUM	HIGH
PERMEABILITY	HIGH	MEDIUM	LOW
SOIL CHEMISTRY			
pH	INTERMEDIATE	LOW	HIGH
% CaO	LOW	HIGH	INTERMEDIATE
EXCH. Ca	LOW	LOW	HIGH
EXCH. Al	LOW	HIGH	INTERMEDIATE
B.S.I.	INTERMEDIATE	LOW	HIGH

TABLE 2
COMPARATIVE PHYSICAL PROPERTIES OF THE THREE STUDY SOILS. VALUES REPRESENT AVERAGES FOR THE UPPER 50 CM OF SOIL.

SOIL	% SAND 2-0.05 mm	% SILT 0.05-.002 mm	% CLAY <0.002 mm	GRAPHIC MEAN mm	MEDIAN mm	PERMEABILITY cm sec^{-1}
ADAMS	92.6	6.2	1.2	0.43	0.41	5.60×10^{-3}
BECKET	62.2	31.3	6.5	0.21	0.12	2.47×10^{-4}
UNADILLA	14.7	79.2	6.1	0.03	0.023	1.74×10^{-4}

TABLE 3

COMPARATIVE BULK XRF CHEMISTRY FOR THE THREE STUDY SOILS. VALUES ARE MEANS FOR UPPER 50 CM.

SOIL	Fe_2O_3	MnO	TiO_2	CaO	K_2O	P_2O_5	SiO_2	Al_2O_3	MgO	Na_2O	TOTAL
ADAMS	1.06	0.02	0.17	0.47	2.76	0.09	83.99	9.26	0.11	2.00	99.93
BECKET	5.40	0.02	1.01	1.74	3.25	0.12	74.11	11.74	0.94	1.70	100.03
UNADILLA	6.10	0.04	0.96	0.99	2.42	0.23	70.48	15.87	1.13	1.68	99.90

VALUES ARE WT. %

TABLE 4

COMPARISON OF EXCHANGEABLE ELEMENT POOLS

SOIL	MASS MT/ha	Ca	TOTAL BASE CATIONS
		------Keq/ha------	
ADAMS			
O	310	6.1	9.0
O+A	2,040	8.0	13.2
O+A+B	5,370	8.4	14.6
BECKET			
O	140	3.8	4.9
O+A	1,330	7.8	10.8
O+A+B	4,000	9.8	14.1
UNADILLA			
O	100	5.5	7.3
O+A	1,530	12.9	18.2
O+A+B	4,760	27.2	40.6

period, the columns were distributed in a stratified random fashion into three treatment groups (pH 5.7, pH 4.0, and pH 3.5) corresponding to the following respective annual H^+ loading rates: 36 eq ha^{-1}, 1,840 eq ha^{-1}, and 5,520 eq ha^{-1}. These weekly treatment storm events continued for eleven months.

During each simulated rain event, soil column leachates were collected under tension in plastic bottles and were then analyzed for pH, AE-pH, Ca, Mg, K, Na, NH_4^+, monomeric Al, SO_4^{2-}, NO_3^-, Cl^-, F^-, and DOC. At the termination of the experiment, the soil columns were dissected into depth increments for analysis and the samples were air-dried. These treatment soil samples and the pretreatment field samples were sieved and the <2mm fraction from each sample was analyzed for exchangeable cations using 1 $\underline{N}$ NH_4Cl and AAS. Exchangeable acidity was measured by extraction with 1$\underline{N}$ KCl, followed by titration. Soil horizons from each soil were analyzed for grain size distributions using standard methods of sieving and hydrometer analysis. Each soil horizon (<2mm) was also analyzed for bulk soil chemistry by XRF and for soil mineralogy using x-ray diffraction. Final statistical analyses were performed using the SAS analytical program. Further methodological details are provided elsewhere (1,2).

RESULTS

Weathering Rates

Mass balance data from the soil columns were used to calculate weathering and biological mineralization rates for the three soils and to evaluate how these rates of element replenishment are affected by differences in soil properties and H^+ deposition. For these calculations, the data were plugged into the following equation:

Weathering and/or Mineralization =
Leachate Flux - Input Flux + ΔExchange Pool + Plant Accumulation

Then, the equation was solved using two alternate boundary assumptions: (1) because there were virtually no statistically significant changes in exchangeable cation pools over the experimental period, the ΔExchange Pool term was set to zero; and

(2) it was assumed that the ΔExchange Pool term could be as large as half the 95% confidence interval on the exchangeable cation measurement. Using this approach, it was possible to produce high and low estimates for each soil column and each separate horizon.

Results from mass balance experiments with the Adams, Becket, and Unadilla soils are summarized in Tables 5-7. These estimates illustrate that cation denudation rates differed significantly between the three contrasting forest soils; thus, the null hypothesis could not be substantiated even for three soils from the same general region of eastern North

TABLE 5

COMPARISON OF CATION DENUDATION RATES (Eq ha^{-1} yr^{-1})

DEPTH	UNADILLA	BECKET	ADAMS
O-HORIZON	3150 (420)	900 (40)	350 (90)
A-HORIZON	4690 (570)	870 (260)	380 (40)
B-HORIZON	3190 (800)	650 (260)	780 (70)
TOTAL A+B	7880	1510	1160
TOTAL O+A+B	11030 (1790)	2410 (550)	1510 (200)

STANDARD ERROR IN PARENTHESIS

BASED UPON AN ANNUAL WATER INPUT OF 120 cm.

America. As shown in Table 5, weathering/mineralization rates between soils varied ten-fold in the A-horizon, ranging from 4,690 eq ha^{-1} yr^{-1} in the Unadilla soil to 380 eq ha^{-1} yr^{-1} in the Adams soil. Normalized rates in the same horizon ranged from 3.3 meq Kg^{-1} yr^{-1} in the Unadilla soil to 0.2 meq Kg^{-1} yr^{-1} in the Adams soil (Table 6). In comparison, B-horizon weathering rates ranged from 3,190 eq ha^{-1} yr^{-1} in the Unadilla

TABLE 6

NORMALIZED AVERAGE WEATHERING/MINERALIZATION RATES
EXPRESSED IN Meq kg^{-1} yr^{-1}.

DEPTH	UNADILLA	BECKET	ADAMS
O-HORIZON	31.5	6.4	1.3
	(4.2)	(0.3)	(0.4)
A-HORIZON	3.3	0.7	0.2
	(0.4)	(0.2)	(0.03)
B-HORIZON	1.0	0.2	0.2
	(0.2)	(0.1)	(0.02)

STANDARD ERROR IN PARENTHESIS

TABLE 7

MINIMUM AVERAGE WEATHERING/MINERALIZATION RATES
CORRECTED FOR ERRORS IN THE ESTIMATION OF SOIL
EXCHANGEABLE CATION CONCENTRATIONS. UNITS: Eq $ha^{-1} yr^{-1}$.

DEPTH	UNADILLA	BECKET	ADAMS
O-HORIZON	170	10	50
A-HORIZON	2410	30	4
B-HORIZON	390	50	540
TOTAL A+B	2800	80	540
TOTAL O+A+B	2970	90	590

profile to less than 800 eq ha^{-1} yr^{-1} in the Becket and Adams soils; similarly, normalized weathering rates varied four-fold between the Unadilla B-horizon at 1.0 meq Kg^{-1} yr^{-1} and the Becket and Adams soils at 0.2 meq Kg^{-1} yr^{-1}. Overall, the analysis presented in Table 5 indicated that total mineral soil profile cation denudation rates ranged seven-fold from 7.9 Keq ha^{-1} yr^{-1} in the fine-textured Unadilla Inceptisol to 1.2 Keq ha^{-1} yr^{-1} in the sandy Adams Spodosol.

Using the alternate mass balance boundary assumption described earlier, it was also possible to estimate minimum weathering/mineralization rates for the three soils (Table 7). Based upon these values, it is apparent that much of the base cation flux from all three soils could potentially be accounted for by statistically undetectable changes in the soil exchangeable element pools. Thus, weathering/mineralization rates could be as low as 2,800 eq ha^{-1} yr^{-1} in the Unadilla mineral soil profile, 80 eq ha^{-1} yr^{-1} in the Becket mineral profile, and 540 eq ha^{-1} yr^{-1} in the Adams profile.

The study also provided an important opportunity to compare weathering estimates obtained by several independent methods. In Table 8, the high and low mass balance estimates for each soil are compared with weathering rates based upon sodium and silica fluxes from each soil. The table also includes two additional estimates for the Becket Spodosol derived from field studies by April and Newton (3) and Likens et al. (4). As such, the data illustrate several significant points. First, the different estimation techniques provided weathering rates that varied by as much as 20X for a given soil. In comparing techniques, the silica flux method tended to yield the lowest weathering rates, the sodium flux method produced intermediate estimates, and the mass balance approach provided the highest weathering rates. Yet, in spite of the great variation in results, there were indications of convergence among some procedures. In the Becket soil, results of mass balance, sodium flux, and watershed input-output budget calculations indicated that the soil profile weathering rate lies between 1.0 - 1.5 Keq ha^{-1} yr^{-1}. In comparison, the minimum mass balance estimate and the sodium flux estimate for the Unadilla soil indicated that the weathering rate for this soil profile lies between 2.6 - 2.8

TABLE 8

COMPARISON OF WEATHERING RATES FOR EACH SOIL BASED UPON SEVERAL DIFFERENT ESTIMATION TECHNIQUES. WEATHERING IS TAKEN TO BE THE SUM OF Ca + Mg + K + Na RELEASED FROM PRIMARY MINERALS IN Eq ha^{-1} yr^{-1}.

METHOD	ADAMS	BECKET	UNADILLA
Mass Balance (Maximum)[a]	1,160	1,510	7,876
Mass Balance (Minimum)[a]	540	80	2,800
Sodium Flux Ratio[b]	320	1,030	2,590
Silica Flux Ratio[c]	60	250	330
Mineral Depletion[d]	--	620	--
Watershed Budget			
ILWAS[e]	--	1,480	--
Hubbard Brook[f]	--	2,000	--

[a] From the soil column mass balance experiments (Tables 5 & 7)

[b] Because sodium is considered to be relatively conservative, the solution Na flux and bulk XRF chemistry were used to calculate the release of Ca + Mg + K necessary to account for the observed Na flux.

[c] This technique makes the tenous assumption that solution silica flux is directly related to the rate of release of Si + Ca + Mg + K + Na from primary minerals.

[d] This estimate from April and Newton (1984) is based on a long-term weathering rate determined from mineral depletion in the Becket soil profile of Woods watershed.

[e] This estimate is based on data from the ILWAS watersheds where the Becket Spodosol was obtained. The estimate combines the annual stream export of cations from Woods Lake watershed (April and Newton 1984) with an estimate of the annual biomass increment of cations (Cronan, unpubl. data).

[f] The Hubbard Brook Experimental Forest in New Hampshire (Likens et al. 1977) contains the same Becket Spodosol as the Adirondack soil used in this study.

Keq ha^{-1} yr^{-1}. Finally, although the Adams soil exhibited less clear-cut convergence, the intermediate sodium flux estimate and minimum mass balance estimate indicate that the weathering rate for this soil probably falls in the range of 0.3-0.6 Keq ha^{-1} yr^{-1}.

With soil depth included in the experimental design, it was possible to examine differential leaching fluxes and element turnover rates through the separate soil horizons. As shown in Table 6, this analysis revealed a number of distinct differences in weathering/mineralization rates between soil horizons. First, the normalized forest floor (O-horizon) rates of base cation mineralization were all 5-10X higher than base cation release rates in the mineral horizons. In addition, the Unadilla and Becket A horizons exhibited unit mass weathering rates that were approximately 3X higher than B-horizon rates of weathering. As such, this confirms and quantifies the pattern of cation denudation that one would expect in the zone of intense weathering beneath the acid forest floor. In fact, if one adds the flux of aqueous aluminum from the A-horizon, the inter-horizon differences in weathering rates would increase to approximately 4-5X. In comparison to the Unadilla and Becket soils, unit base cation weathering rates in the Adams soil were low and nearly identical in the A and B horizons. Again, however, the addition of aqueous aluminum to the A horizon cation flux would boost the overall cation denudation rate in the A horizon above the B horizon rate.

Controls on Weathering: Soil Properties

As might be expected, the three contrasting study soils exhibited distinct differences in mineral weathering rates. These differences were reflected in leachate solution chemistry (Table 9) and in calculated total ion flux estimates (Tables 5-7). Given these differences, one of the prime objectives of this investigation was to exploit the three soils to examine the influence of selected physical, chemical, and biological parameters in controlling the patterns of chemical weathering in acid forest soils.

TABLE 9

EXAMPLES OF TREATMENT PERIOD (pH 3.5 TREATMENT) LEACHATE CHEMISTRY FROM THE O, A, AND B HORIZONS OF EACH STUDY SOIL. ALUMINUM DATA ARE IN ug l^{-1}, WHILE OTHERS ARE IN ueq l^{-1}

HORIZON	pH	Ca	Mg	K	Na	Al$_a$	SBC	SO$_4^{2-}$	NO$_3^-$	Cl$^-$	F$^-$
RAIN INPUT	3.52	20	10	25	5	0	60	330	30	10	0
ADAMS
O-HORIZON	4.30	39	13	87	13	290	152	428	61	20	1
A-HORIZON	4.55	61	42	161	24	850	288	444	76	34	1
B-HORIZON	5.50	169	47	154	29	0	399	18	368	31	1
BECKET
O-HORIZON	3.62	85	25	44	8	140	162	435	195	19	1
A-HORIZON	4.01	190	75	71	31	3,870	367	238	1,155	23	4
B-HORIZON	4.15	212	102	64	44	3,870	422	74	1,819	41	10
UNADILLA
O-HORIZON	3.76	474	118	55	7	810	654	414	450	18	1
A-HORIZON	4.13	1,154	276	85	31	3,690	1,546	161	2,385	23	19
B-HORIZON	4.64	1,520	538	20	122	390	2,200	117	2,280	49	9

SBC= SUM OF BASE CATIONS

The initial focus of this analysis was to examine the differential influence of soil type and varible physical-chemical properties on mineral weathering rates. To begin, the data were used to test the hypothesis that weathering is inversely proportional to soil particle size and increases with increasing surface area. As shown in Fig. 1, the data from the three soils fit a negative decay curve and thus supported the hypothesis that mineral weathering rate in the bulk soil profile is strongly related to grain size. In a related test, the data were used to evaluate the hypothesis that weathering is a rate-limited non-equilibrium process which increases per unit area with decreasing soil permeability and increasing soil water residence time. This auto-correlated test provided essentially the inverse relationship of the previous case, indicating that over a range of soil water

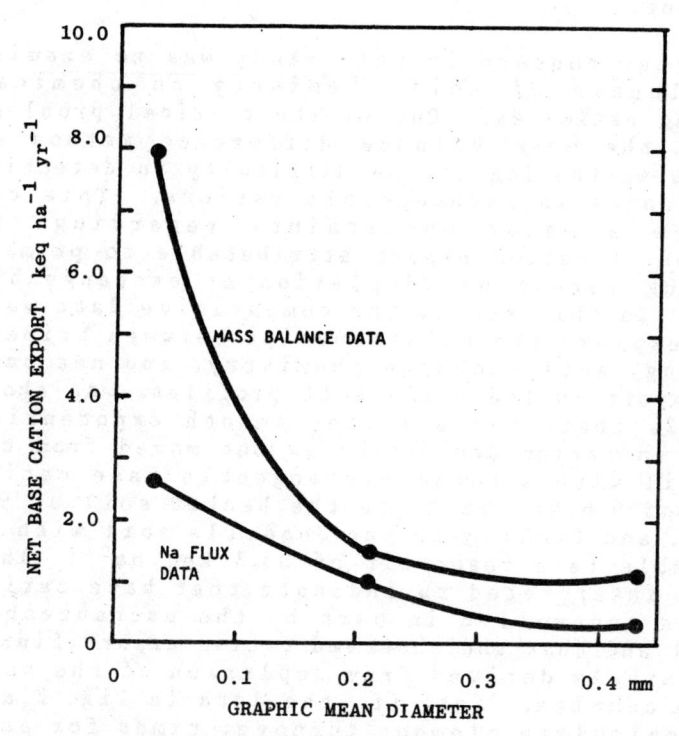

FIG. 1. RELATIONSHIP BETWEEN WEATHERING RATE AND MEAN SOIL PARTICLE SIZE IN EACH STUDY SOIL.

residence times from 2-80 hr., cation export increased with increasing residence time.

Despite close similarities in the mineralogical characteristics of the three soils, an attempt was made to test the hypothesis that weathering rate is controlled by source mineralogy and is proportional to cation concentration in the primary mineral substrate. This hypothesis assumes that simple congruent dissolution of a unit of mineral will yield soluble cation concentrations in proportion to their mineralogical abundance. When the data were examined, the differences in base cation equivalents in the bulk soil proved to be relatively small and were not clearly related to weathering rates. This may indicate: (1) that there are other overriding or intermediate processes (e.g. selective ion exchange) which control solution composition, or (2) that weathering rate is controlled by minor minerals that are poorly represented by an XRF bulk chemistry measurement.

Another concern in this study was to examine the influence of soil chemistry on chemical weathering estimates. One of the critical problems in using the mass balance difference method to calculate weathering is the difficulty in detecting small changes in exchangeable cations. This can introduce a major uncertainty regarding the proportion of cation export attributable to primary weathering versus net depletion of exchangeable cations. In this study, the comparative data were used to explore the relationships between primary weathering, soil exchange chemistry, and net base cation export in the three soil profiles. As shown in Fig. 2, there was a rather smooth exponential increase in cation denudation as one moved from the Adams soil with a total exchangeable base cation reserve of 5.6 keq ha^{-1}, to the Becket soil at 9.2 keq ha^{-1}, and finally to the Unadilla soil with an exchangeable base reservoir of 33.3 keq ha^{-1}. This could be interpreted to indicate that base cation export is controlled in part by the exchangeable base pool and that the observed cation export fluxes may be largely derived from depletion of the soil exchange complex. Yet, if the data in Fig. 2 are used to calculate element turnover times for each soil, it is clear that these observed cation export rates are high enough to deplete the exchangeable bases in each soil within 4-6 yrs (or 8-12 yrs. if the water flux estimate is halved to a more conservative level).

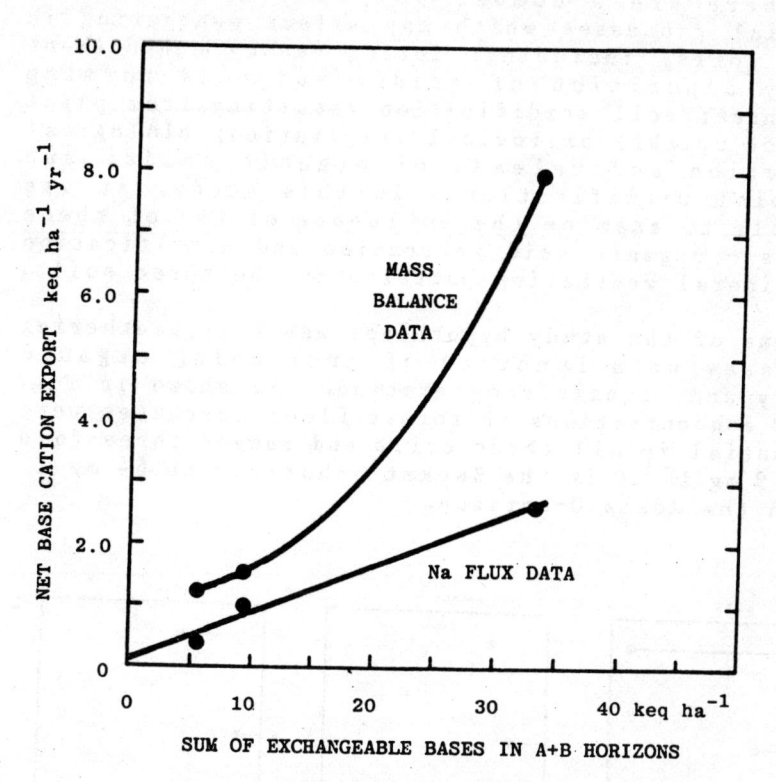

FIG. 2. RELATIONSHIP BETWEEN WEATHERING RATE AND TOTAL EXCHANGEABLE BASES IN EACH SOIL.

Since this depletion obviously could not be sustained and because all other evidence indicates a relatively stable exchange pool, it can be concluded that (1) the exchange pool acts as the immediate response reservoir and as a modifier of solution ionic chemistry; and (2) while high rates of cation export draw upon exchangeable base supplies as a proximal source, sustained cation export is ultimately possible because the base cation weathering rate balances the rate of exchangeable base depletion through leaching. Thus, base cation exports correlate with soil exchangeable base supply because both are dependent on the same master variable, the base cation weathering rate.

Controls on Weathering: Biotic Processes

There are a number of plant-mediated and microbial processes which may affect weathering in forest soils, including: forest canopy enhancement of dry deposition of acidic and acid-forming substances; soil acidification resulting from plant nutrient uptake; biological respiration; biological production and release of organic acids; and microbial nitrification. In this study, it was possible to examine the influence of two of these factors - organic acid generation and nitrification - on mineral weathering patterns in the three soils.

One of the study hypotheses was that weathering increases as a function of increasing organic acidity and organic complexation. As shown in Fig. 3, DOC concentrations in forest floor leachates were substantial in all three soils and ranged three-fold from 19 mg l^{-1} C in the Becket O-horizon to 64 mg l^{-1} C in the Adams O-horizon.

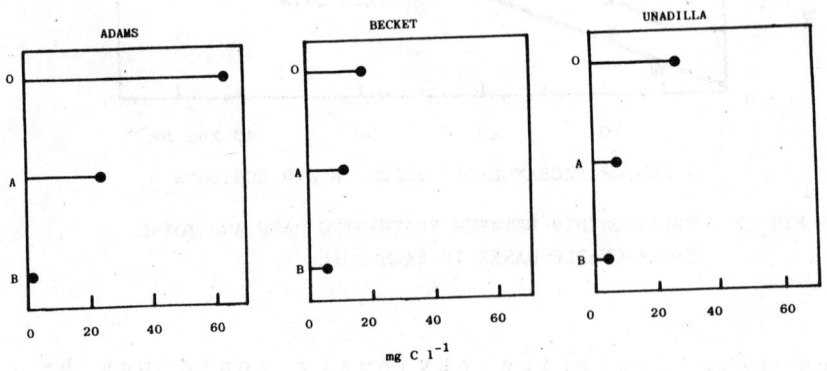

FIG. 3. COMPARATIVE LEACHATE DOC CONCENTRATIONS IN THE O, A, AND B HORIZONS OF THE THREE STUDY SOILS.

Previous studies with the same soils (5) have shown that much of this DOC is comprised of acidic humic substances characterized by an average carboxyl group content of ca. 6 meq gm^{-1} C, with an average pK_a of approximately 3.8 (Table 10). If one extrapolates with these figures to an annual flux estimate, one can calculate that each Kg of soil in the uppermost A-horizon of each soil is exposed to

0.2 - 2.0± meq of organic acid derived protons each year. Furthermore, these organic acids may exert a strong influence on weathering reactions through metal complexation. Comparing just the proton input from the organic acids with the normalized weathering rates in Table 6, one can see that biologically-derived organic acidity and complexation can make a significant, if not dominant, contribution to weathering in the upper soil zone. Thus, within a given soil, it is likely that increases in organic acid release would produce increased weathering. However, in comparing different soils, it appears that fluxes of organic acidity are secondary to other factors in controlling weathering.

Nitrification, like plant uptake, can act as a net acidifying process in soils. As such, it can potentially contribute to accelerated cation export from the soil. In most systems with intact plant communities, there appears to be little direct net contribution from nitrification to the proton budget of the soil. However, where plant uptake is interrupted or where a forest reaches nitrogen saturation (6), nitrification potential may be expressed as a strong pulse of acidity. In this study, nitrification was pronounced in the soil columns and this permitted an examination of the influence of internal acid generation by nitrification upon cation export rates of different soils. This analysis showed that differences in cation export fluxes between the three soils were not clearly related to differences in nitrification and the resulting nitrate flux. In fact, the Becket and Unadilla soils shared the same nitrate flux, but differed 2-5X in cation export. Hence, inter-soil differences in weathering rate seemed to be dominated by other factors than internal acidification by nitrification.

TABLE 10

FUNCTIONAL GROUP ANALYSIS AND ESTIMATION OF AVERAGE pK_a
FOR O/A HORIZON SOIL SOLUTIONS COLLECTED DURING THE SUMMER

pH	3.55
H^+	284 µeq/l
Acidity to pH 7	510 µeq/l
Acidity to pH 9	635 µeq/l
Al_a (Monomeric aluminum)	51 µeq/l
Ratio of Al_o/Al_a	0.79
Fe	9 µeq/l
DOC	52 mg/l
COOH	166 µeq/l
Total COO^- (Anion Deficit)	134 µeq/l
Metal Bound COO^-	49 µeq/l
Free COO^-	85 µeq/l
Total Carboxyl Group Content	5.77 meq g^{-1} C
Estimated Average pK_a	3.84

Controls on Weathering: H^+ Deposition

One of the final objectives of this investigation was to test the effects of strong acid deposition upon weathering rates within contrasting forest soils. This question was of interest because of concern that acidic deposition may deplete soils of available bases faster than they are replenished by weathering. The experiment also offered an important opportunity to examine general patterns of weathering response to increased H^+ inputs. In Fig. 4, weathering rates have been plotted for each soil as a function of the three treatment H^+ deposition rates. To account for baseline differences between soil columns, the soil columns were compared statistically using net mean differences in element fluxes between treatment and pretreatment periods.

These net changes in cation denudation are plotted on the graph and indicate that the Adams and Unadilla soils both exhibited significantly increased base cation exports in response to increased H^+ deposition. In contrast, the Becket soil provided a pattern that was not directly related to H^+ inputs and which may have been dominated by inter-column differences in nitrification, sulfate adsorption, labile aluminum, or the like.

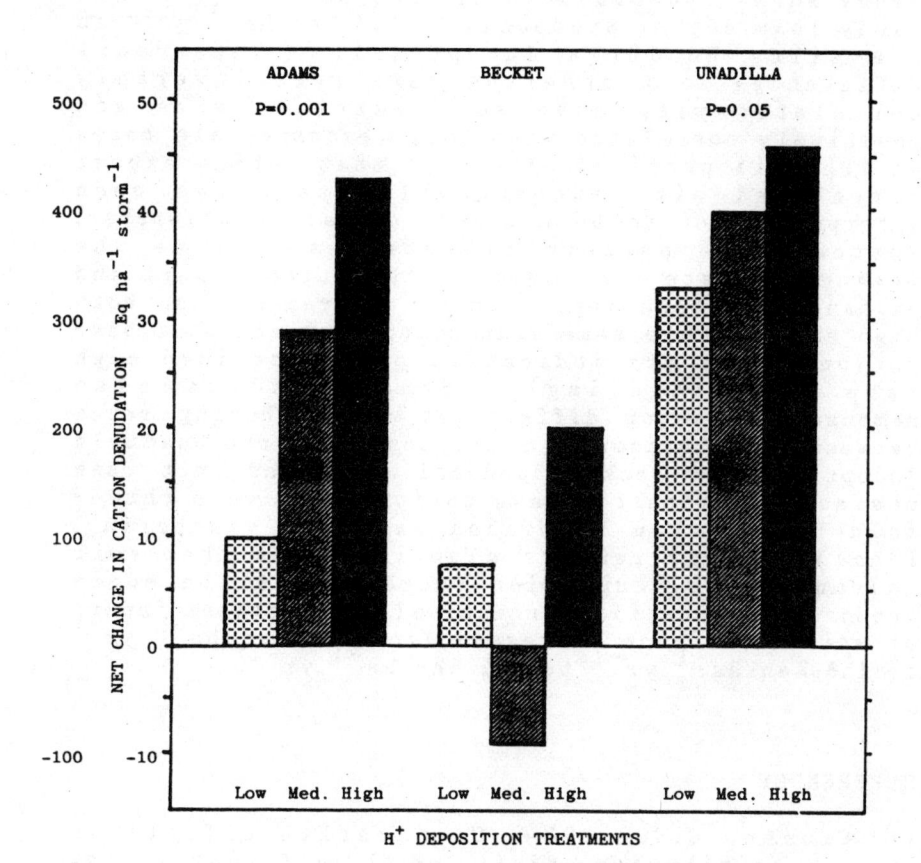

FIG. 4. COMPARISON OF TREATMENT H^+ DEPOSITION EFFECTS ON RATES OF CATION EXPORT FROM EACH SOIL. LOW, MEDIUM, AND HIGH TREATMENTS REFER TO 1, 35, and 106 Eq ha^{-1} storm^{-1}, RESPECTIVELY. TO NORMALIZE EACH TREATMENT GROUP, SOIL COLUMNS WERE COMPARED STATISTICALLY USING NET MEAN DIFFERENCES IN ELEMENT FLUXES BETWEEN TREATMENT AND TREATMENT PERIODS. THE LOWER RANGE ON THE Y-AXIS APPLIES TO THE ADAMS AND BECKET SOILS.

SUMMARY

In this investigation, weathering rates were estimated for three eastern North American forest soils and leaching experiments were performed to examine the influence of selected environmental variables on the weathering process. Based upon multiple estimation techniques, mineral soil weathering rates varied five-fold between soils, ranging from approximately 0.5 Keq ha^{-1} yr^{-1} in the sandy Adams Spodosol, to 1.2 Keq ha^{-1} yr^{-1} in the sandy loam Becket Spodosol, to 2.7 Keq ha^{-1} yr^{-1} in the silty Unadilla Inceptisol. Inter-soil differences in weathering rate were inversely correlated with mean soil particle size and positively correlated with total exchangeable bases in the soil profile. The fact that cation export correlated with exchangeable bases has been interpreted not to be a direct causal relationship. Instead, the pattern illustrates that as the weathering rate increases, both cation export and exchangeable base replenishment increase (i.e. both high exchangeable base status and high rates of base cation export are indicative of a sustained high rate of weathering). Study results also demonstrated major differences in weathering rates between soil horizons. A-horizons from the Unadilla Inceptisol and Becket Spodosol exhibited unit mass weathering rates for base cations that were three-fold higher than B horizon rates of weathering. Finally, study results also indicated that soil column cation denudation increased in the Adams Spodosol and Unadilla Inceptisol as treatment inputs of acid deposition increased from .04 Keq ha^{-1} yr^{-1} to 1.8 Keq ha^{-1} yr^{-1} to 5.5 Keq ha^{-1} yr^{-1}.

REFERENCES

(1) Cronan, C.S. 1984. Comparative effects of precipitation acidity on three forest soils: impacts on nutrient availability. In review.
(2) Cronan, C.S. 1984. Comparative effects of precipitation acidity on three forest soils: carbon cycling responses. Plant and Soil :submitted.
(3) April, R. and Newton, R.M. 1984. Geology and geochemistry of the ILWAS watersheds. Water, Air, Soil Pollut. :submitted.

(4) Likens, G.E., F.H. Bormann, R.S. Pierce, J.S. Eaton, and N.M. Johnson. 1977. Biogeochemistry of a Forested Ecosystem. Springer-Verlag. N.Y. 146 p.

(5) Cronan, C.S.and G.R. Aiken. 1984. Chemistry and movement of dissolved humic substances in ILWAS watersheds, Adirondack Park, N.Y. In review.

(6) Ulrich, B. 1983. Stabilität von waldökosystemen unter dem einfluβ des sauren regen. Allgemeine Forst Zeitschrift 26/27 pp. 670-677.

(7) Johnson, N.M., C.T. Driscoll, J.S. Eaton, G.E. Likens, and W.H. McDowell. 1981. Acid rain, dissolved aluminum, and chemical weathering at the Hubbard Brook Experimental Forest, New Hampshire. Geochim. Cosmochim. Acta 45, pp. 1421-1437.

PROTON CONSUMPTION RATES IN HOLOCENE AND PRESENT-DAY
WEATHERING OF ACID FOREST SOILS

Horst Fölster

Institut für Bodenkunde und Waldernährung
Büsgenweg 2, 34 Göttingen, Germany

ABSTRACT

Results from weathering budgets and input-output budgets of sealed catchments and soil columns are reviewed in order to estimate naturally occuring rates of silicate weathering. Both approaches show limitations, the former because of uncertain methodology and homogeneity of soils, the latter because of disequilibria in the cation store of vegetation, humus layer and exchange pool. Results indicate that Holocene and present-day rates of basic cation release from weathering appear to be similar (0.4 - 1.0 $kEq.ha^{-1}.yr^{-1}$).

INTRODUCTION

The rate of silicate weathering has become an important parameter in estimating the capacity of weathering mantles to neutralize the acid load of air pollution. Present estimates vary from 0.2 to 10 $kEq.ha^{-1}.yr^{-1}$ of proton equivalents consumed during this process. The rather wide range may result from an equally wide range of hydrological and chemical properties of the study site (actual load of dry and wet deposition as well as internal proton production, depth and extension of the weathering mantle, speed and pathways of seepage) but also from differences in the methodological approach. The present paper discusses results from weathering budgets and inflow-outflow budgets of sealed catchments or soil columns.
The former method is expected to provide cumulative means for the whole period of soil formation, the latter estimates actual rates of weathering under present conditions of more or less

accelerated proton input.
Weathering of silicates consumes protons in a two-step process, firstly during the release of basic cations, and secondly during the transformation of hydroxy-compounds of cation acids (mainly Al) into ionic, mobile forms. Under acid conditions, both steps may proceed simultaneously.

PRESENT-DAY WEATHERING RATES FROM INPUT-OUTPUT BUDGETS

Taking the release of basic cations as the most important proton-consuming quantitative indicator of silicate weathering, their input-outflow budgets in sealed catchments have been used to arrive at estimates of weathering rates. Some data have been listed in Table 1. A broader compilation of results from catchment and lysimeter studies by CLAYTON (7) included maximum values of annual release from 0.3 (K) to 4.3 (Ca,Mg) kEqu.ha^{-1}.

Table 1: Release of basic cations from input-outflow budgets of sealed catchments

author			kEq.ha^{-1}.yr^{-1}	comment
LIKENS et al.	(12)	USA	2.00	net outflow, uptake
HAUHS	(9)	FRG	2.34	", no uptake
REID et al.	(20)	UK	1.72	" "
ROSÉN	(21)	Sweden	0.64	", uptake
WRIGHT and JOHANNESSEN	(34)	Norway	0.64	net outflow
BRICKER et al.	(5)	USA	0.24	"

Differences in rates of release of basic cations are to be expected with changing site factors in different catchments. Their influence cannot easily be assessed as information on important parameters is often incompletely recorded or difficult to provide. This concerns data on dry deposition and the impact of vegetation, climate and the history of land use on internal proton production (31,22), the thickness of the weathering mantle and the speed of seepage, which influences residence time and thus the probability of equilibrium being established between seepage and soil (1,9,10,21) as well as the mineral assemblage and the actual state of acidification reached in different layers of the weathering mantle.
Of more immediate concern are the possible systematic errors which may distort the balance. The input-output budget approach normally assumes that all conditions remain constant in the ecosystem except for the loss of silicates. This assumption of equilibrium may in most cases be wrong. Underestimates result

from the neglect of net cation uptake by the vegetation (10), overestimates from the neglect of changes that may occur in the cation store of the humus layer and the exchange store during the observation period.
Such changes may be induced by manipulation, which is a special handicap in studies with lysimeters and extracted soil columns (accelerated mineralisation). However, they are likely to occur in all ecosystems today as a response to changing environmental conditions.
TROEDSSON (29) found in a large scale inventory repetition in Sweden (period 10 years) a significant reduction of exchangeable bases (Ca,Mg,K). STUANES (28) and BJOR and TEIGEN (3) determined a significant drop of base saturation in minicatchments and soil columns treated with artificially loaded rain. At pH 3 (soil pH 3.5), an annual outflow of Ca, Mg, and K of 0.7 kEq.ha^{-1}.yr^{-1} (experimental period 4-5 years) was derived in equal portions from the exchangeable pool and silicate weathering (including mineralisation).
In Germany, cation stores were inventoried in the upper 50 cm of a Inceptisol (loess over sandstone, Solling) at intervals between 1968 and 1983. (all data from: MATZNER, ULRICH (30,13), MATZNER, pers.comm.). Under spruce, cation stores (Ca,Mg,Na) were reduced by 1.0 kEq.ha^{-1}.yr^{-1}, under beech the same cations were lost at even higher rates (1.58 kEq.ha^{-1}.yr^{-1}) during the period 1973-83. Loss of Ca, Mg, and Na totalled about 15 kEq.ha^{-1} over 15 and 10 years respectively, which corresponds to about 50% of the total original store of basic cations.
Release of basic cations from the exchange pool of the order of 1 to 1.5 kEq per ha and year is well in the range of data given for weathering release in Table 1. One should be aware that this rate corresponds to about 5 - 10% of the total basic cation store. With an error in the determination of the individual cations between samples of 5 to 40% of the mean, it is difficult to prove or even quantify these changes except over prolonged observation periods.
Another approach was followed by HAUHS (9) in a catchment of the Harz mountains. Net outflow of basic cations increases from 1.17 kEq.ha^{-1}.yr^{-1} below the rooting zone (80 cm) to 2.3 in the streamlet. Ca and Mg dominate with a molar ratio of 1 to 2. A similar ratio exists in the pool of exchangeable cations while in the rock and the soil minerals (quarz, muscovite, sericite chlorite) the ratio is close to 0.1. Potassium as a dominant mineral component hardly contributes to the basic cation release. The author concludes that exchange reactions account for most of the buffering.
Inconsistencies between basic cation ratios in outflow and the mineral assemblage of the soil mantle have often been recorded and are usually explained by differential weathering (12,21), i.e. the greater resistance to weathering of Na- and

K-containing minerals. However, the weathering budgets of MAZZARINO (Tab. 2) show a loss ratio of (Ca,Mg) to (Na,K) close to 1, though this ratio is bound to vary with the types of minerals in the soil. Mineralogical studies (see below) indicate a substantial loss of Na- and K-felspar and equally substantial losses of K during the weathering of mica to illite and Al-chlorite. Therefore, the conclusions of HAUHS (9) have to be accepted in principle. Ratios of differential weathering cannot be accepted on the base of simple input-output budgets which do not take into account the quantitative impact of exchange reactions.

In spite of the already low base saturation of the soils in both study areas, Solling and Harz (6 - 8% of sum of cations at pH of the soil in NH_4-extract), the present store of basic cations is being rapidly depleted. Similar unbalanced budgets may exist also in the humus layer (9).

Actual disequilibrium is also documented by changes in the input-output budgets of the Solling soil during the period 1969(73) - 81 (Fig. 1, data from MATZNER and ULRICH, 14). While the soil layer (90 cm) retains a constant 1-2 $kEq.ha^{-1}.yr^{-1}$ of protons, net outflow of Al, S, Ca and Mg (as also Na and Mn) increased strongly since 1976. The difference between spruce and beech is explained by the much higher annual acid load of the former ecosystem: 7.0 as against 3.1 $kEq.ha^{-1}.yr^{-1}$ of proton equivalents of which 70% are derived from deposition and 30% from internal proton production (13).

The outflow pattern has to be interpreted as a cumulative effect of vertically differentiated processes: Weathering under less acid conditions released Al-hydroxy-compounds, temporarily also $AlOHSO_4$; growing acidity converted these compounds into mobile Al^{3+}. The process started in the surface soil and progressively extended to the subsoil where this cation acid contributed to acidification. By 1976, the subsoil had been sufficiently acidified to keep Al in mobile form resulting in a simultaneous outflow of Al as well as of basic cations from exchange reactions.

In 1981, outflow of total acidity (Al, H, NH_4, Mn) was three times higher (15.7 $kEq.ha^{-1}.yr^{-1}$) than the respective input (14). Acidification thus appears as a process gradually moving downward in the profile, and not until it breaks through to the surface waters does its impact become apparent from input-output budgets. This also explains why constancy of the budget over several years does not necessarily indicate equilibrium, neither of the whole ecosystem nor between dissolved and exchangeable cations. Such an inference would be misleading. It is equally apparent that the obvious influence of the factors thickness of the soil, length of slope, pathway and speed of seepage not only depends on residence time necessary to establish weathering equilibrium but also on the time required for the acidity front to break through to the surface waters (9,10,19).

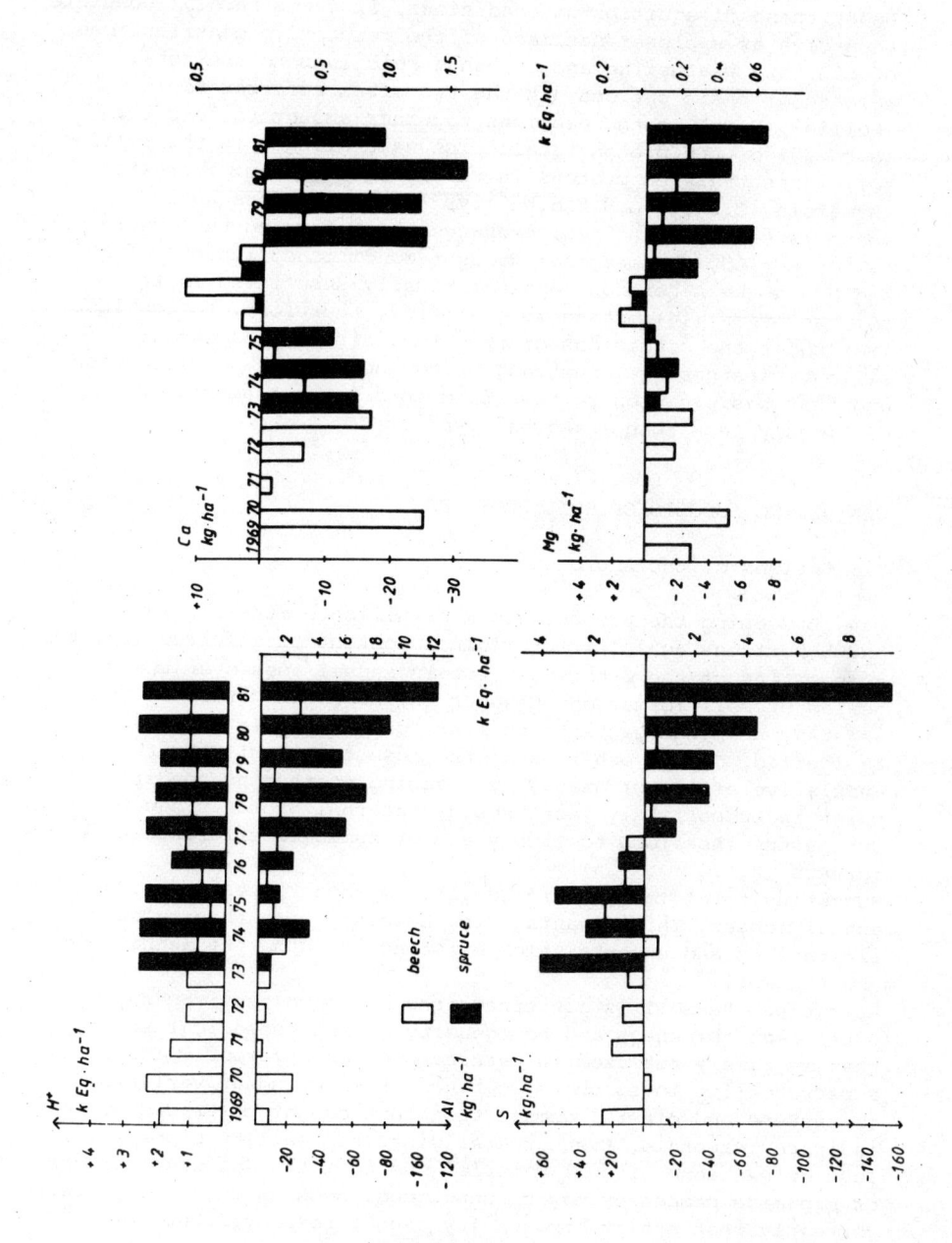

Fig. 1: Annual input-output balance (H, Al, S, Ca, Mg) of 90 cm mineral soil (Solling) below spruce and beech in the years 1969 - 1981.

Under these disequilibrium conditions, it seems hardly possible to arrive at a closer estimate of the respective contributions of silicate weathering and exchange reactions to the total release of basic cations. In the two sites referred to (Solling, Harz), both reactions together amount to less than 15 to 25% of the proton total to be neutralized. In the Solling soil, unneutralized protons in the outflow (90 cm) increased threefold (0.25 to 0.8 $kEq.ha^{-1}.yr^{-1}$). The surface soil accumulated 20 $kEq.ha^{-1}$ of exchangeable H^+ during the observation period, and seasonal drought-conditioned flushes of NO_3^- and H^+ in the A-horizon remain virtually unbuffered in the soil solution (14). It seems justified, therefore, to conclude that under the conditions of the study sites the speed of silicate weathering cannot adjust to the rate of acidification, and that basic cation release from this process amounts to distinctly less than 1 $kEq.ha^{-1}.yr^{-1}$.

WEATHERING BUDGETS OF SOIL PROFILES

Limitations of the Approach

Budgeting the present stock of silicate minerals in the A- and B-horizons against the original content as inferred from the composition of the C-horizon, totals annual loss over the whole period of soil formation. Changes in the cation store of the vegetation or the exchange pool are comparatively small and can be ignored. On the other hand, the loss total reflects the cumulative effect of gradually changing weathering conditions which includes likely increases of internal proton production during the last 1000 to 1500 years of human interference with forests (31).
Budgeting total loss should ideally improve the estimates of annual losses. This advantage is, however, more than matched by limitations and uncertainties existing in both the methodology and the soil.
As far as the soil is concerned, the approach requires a sufficient thickness and homogeneity of the geological stratum that is rarely realized in nature. Even an apparently ideal substrate like loess shows combined or independent variations of texture and mineral composition that cannot be traced by indicator minerals. Even greater disturbances like superficial loss or exchange of soil material due to intra-Holocene natural or man-made processes may go unnoticed. BRONGER et al. (6) state correctly that strict homogeneity cannot be proved but only implied considering all available data.
The base data for weathering budgets may be mineralogical or chemical (2). The latter have the advantage of providing the required data on element loss directly, and that their analytical error is smaller. Determining mineral composition of soil

fractions by microscopy and X-ray diffraction is generally
handicapped by the unsolved problem of transforming grain
counts or reflection intensities into weight percentages. In the
diameter range 20 µm, the estimates become rather semiquantitative, and chemical alterations of minerals by cation
replacement go unnoticed.
On the other hand, mineralogical data provide a more
satisfactory base to control homogeneity and to obtain more
specific data on clay formation and dissolution. Still more
important is the quartz content which represents the best
reference base for calculating the gains and losses of the
original mineral- or element content of A- and B-horizon.
MAZZARINO (15) working with a chemical data base, used geochemical methods to calculate the mineral composition (32). This
method, no doubt, also introduces certain inaccuracies and
operates with some unchecked assumptions though in general it
seems applicable.

Results

Because of these limitations, weathering budgets have to be
treated cautiously and can at best be accepted as approximations.
STAHR (26), working with SW-German soils in stratified material,
determined annual losses of basic cations of the order of
0.4 to 1.3 kEq.ha^{-1}. With homogeneity lacking, the author
inferred the original chemical composition of the strata from
pure material and the degree of admixture.
MAZZARINO (15,16) published weathering budgets of three Holocene
soils and a fossil soil from the last Interglacial, formed in
175 to 300 cm deep loess deposits. Soil formation was assumed to
have lasted 10 000 and 40 000 years respectively. Unweathered
C_{ca}-horizons were not observed. The least weathered B-(B/C)
horizons were selected as reference horizons. The budgets in
Table 2 are based on chemical data using quartz as reference
mineral.
Compared with previous mineralogically based studies of soils
on glacial till, loess, and glaciofluvial sands (4,8,23,24,25,
27) the soils show a normal tendency of clay formation from
silt-size mica, a comparable degree of felspar dissolution while
clay dissolution appears to be more advanced than in soils still
containing C_{ca}-horizons. The observed differences may represent
the natural variation of weathering intensities and the factors
controlling it (original carbonate content, types of silicates,
etc.) but may equally well be influenced by differences in the
methodology.
Assuming that the budgets of MAZZARINO represent fair approximations, minimum proton consumption rates can be calculated
from basic cation release which yields values between 0.4 to
1.1 kEq.ha^{-1}.yr^{-1}. Carbonates have been neglected because their
concentrations are unknown and may have been low. Of the

Table 2: Weathering budgets of 4 loess profiles based on chemical composition and quartz as reference mineral (15,16).

	SPAN	WEST	HOF	DAS
Loss of silicates in soil >0.6 μm				
kg/m^2	170	112	155	296
% of original soil >0.6 um	20	18	29	27
Fine clay (< 0.6 um) formation from mica				
kg/m^2	110	55	58	141
% of original mica	38	32	40	35
Fine clay dissolution				
kg/m^2	51	51	82	176
% of original soil	2.4	3.1	5.7	6.2
Total loss				
kg/m^2	111	108	179	331
% of original soil	5.2	6.6	12.4	11.7
Loss of				
Na$^+$ (kEq.m^{-2})	–	0.10	0.34	0.21
K$^+$ "	0.45	0.16	0.36	0.75
Ca^{2+} "	0.08	0.11	0.13	0.16
Mg^{2+} "	0.59	0.21	0.26	0.60
sum of basic cations	1.12	0.58	1.09	1.72
proton consumption due to basic cation release kEq.ha^{-1}.yr^{-1}	1.11	0.59	1.09	0.43
Al released kmol.m^{-2}	0.59	0.57	1.26	1.56
protons consumed kEq.ha^{-1}.yr^{-1}				
minimum estimate	0.3	0.3	0.6	0.2
maximum estimate	0.3	0.8	2.1	0.7

4 profiles, HOF is the most uncertain because of a discontinuity between the reference horizon and the upper horizons.

Problems arise from Al as the most important cation acid released during weathering. MAZZARINO et al. (16) calculated an annual proton consumption of 0.2 - 0.6 kEq.ha^{-1} assuming a mean net negative charge of 0.5/Al. This estimate may be too low as part of the released Al was leached beyond the reference horizon. It is not known whether this Al was able to move in a hydroxilized form, as Al^{3+} or as an organic complex. Table 2 gives a minimum and a maximum estimate: the former is based on a proton consumption per Al of 0.5, the latter assumes one of 3 for the Al that was leached beyond the lower limit of the soil column.

Considering the loss of basic cations only, the ranges given by MAZZARINO and STAHR compare well (see also 2). The interglacial soil with 0.4 kEq.ha^{-1}.yr^{-1} (Table 2) yields the lowest value. The higher release rates of the Holocene soils may already reflect the impact of higher internal proton production since human interference with the forests (31).

From mineralogical studies of related soils (25,24,6) one can calculate a loss of basic cations released from weathering of felspars of the order of 0.1 - 0.4.kEq.ha^{-1}.yr^{-1}. Loss of K due to alteration of mica to clay minerals runs up to another 0.05 - 0.2 kEq.ha^{-1}.yr^{-1}. This shows that a conservative estimate of 0.4 kEq.ha^{-1}.yr^{-1} of basic cation released may also represent a close approximation for soils which still contain a C_{ca}-horizon and are possibly less strongly weathered than the soils of Table 2.

Original content of fast-reacting carbonate buffer is probably the most important single factor, certainly more important than differences in the original composition of silicate minerals. Even with andesitic glass as the dominant mineral, annual release of basic cations did not exceed 1.7 - 2.0 kEq.ha^{-1} during 3500 to 10 000 years of soil formation on volcanic ash in Mexico (18).

CONCLUSIONS

With due considerations for the uncertainties discussed above, the range of annual rates of release of basic cations derived from weathering budgets (0.4 - 1.1 kEq.ha^{-1}.yr^{-1}) may tentatively be accepted as representing the variation caused by differences in site factors and the progress of acidification during the soil formation period.

Actual rates of basic cation release derived from input-output budgets of catchments and soil columns fall within a range of 0.2 to 2.3 kEq.ha^{-1}.yr^{-1}. In most cases this value includes the effect of exchange reactions due to disequilibria of present soil conditions. The respective contributions of weathering and exchange are unknown but from the information available (see above) one may conclude that actual weathering rates may not exceed the range calculated from weathering budgets.

In order to explain the apparently lacking difference in Holocene and actual silicate weathering, one might argue that differential weathering increased the mean weathering resistance of the mineral assemblage during the course of soil formation, and that this tendency was compensated by growing acidity. However, even if this assumption is correct the impact of acidity cannot have been very marked. During experiments with artificially acidified rain, STUANES (28) observed a distinct step-up of weathering intensity only when the pH of the rain was lowered from 3 to 2 (soil pH 3.5 and 2.5 respectively). Basic cation release increased from 0.35 to 1.0 (exchangeable) and 0.35 to 3.2 (silicate weathering, including mineralisation) kEq.ha^{-1}.yr^{-1}. Thus, the response of silicate weathering to acidification in the pH-range actually observed in nature, appears to be rather slow, and this explains why weathering seems to play a minor role in the neutralisation of acid rain compared to the mobilisation of Al-hydroxy-compounds and proton storage.

Because of the uncertainties connected with H$^+$-consumption of Al, it is not possible to compare Holocene and actual proton consumption rates. Calculated total loss of Al during Holocene weathering may amount to several thousand kmol.ha^{-1}. Present outflow rates of 0.5 to 12 kEq.ha^{-1}.yr^{-1} (Solling 1968 - 1981, Fig. 1) are too low for this loss to be conceived as a result of actual acidification. Al must have moved previously in a less proton-consuming mobile form, or the soil experienced earlier phases of acidification possibly induced by temporary flushes of internal proton production.

REFERENCES

(1) Bache, B.W. 1982, "The implications of rock weathering for acid neutralization." In: Ecological Effects of Acid Deposition. Series PM, Swedish Environmental Protection Board.

(2) Barshad, I. 1964, "The chemistry of soil development." In: Bear, F.E. (Ed.): Chemistry of the Soil. Reinhald, N.Y., pp. 1-70.

(3) Bjor, K. and O. Teigen 1980, "Effects of acid precipitation on soil and forest." 6. Lysimeter experiments in greenhouse. Proc. Int. Conf. Ecol. Impact Acid Precip. Norway 1980. SNSF project, pp. 200-201.

(4) Bosse, I. 1964, "Verwitterungsbilanzen von charakteristischen Bodentypen aus Flugsanden der nordwestdeutschen Geest (Mittelweser-Gebiet)." Diss. Univ. Göttingen.

(5) Bricker, O.P., A.E. Godfrey and E.T. Cleaves 1967, "Mineral-water interaction during the chemical weathering of silicates." In: A.C.S. Adv. Chem. Series No. 73, Trace Inorganics in Water, pp. 128-142.

(6) Bronger, A., E. Kalk und D. Schroeder 1976, "Über Glimmer- und Kalkverwitterung sowie Entstehung und Umwandlung

von Tonmineralen in rezenten und fossilen Lößböden."
Geoderma 16, pp. 21-54.
(7) Clayton, J.L. 1979, "Nutrient supply to soil by rock weathering." In: Impact of Intensive Harvesting on Forest Nutrient Cycling. Proc. Symposium, State University N.Y., Syracuse, pp. 75-96.
(8) Fölster, H., B. Meyer und E. Kalk 6963, "Parabraunerden aus primär carbonathaltigem Wurmlöß in Niedersachsen." Z. Pflanzenernähr. Bodenkd. 100, pp. 1-11.
(9) Hauhs, M. 1984, "Stoffbilanzen von Ökosystemen als Mittel zur Beschreibung von Versauerungstendenzen." Status-Seminar, München. UBA. (in press).
(10) Johnson, N.M., C.T. Driscoll, J.S. Eaton, G.E. Likens and W.H. McDowell 1981, "Acid rain dissolved aluminum and chemical weathering at the Hubbard Brook Experimental Forest, New Hampshire." Geochimica et Cosmochimica Acta 45, pp. 1421-1437.
(11) Kundler, P. 1965, "Waldbodentypen der Deutschen Demokratischen Republik." Neumann-Verlag.
(12) Likens, G.E., F.M. Bormann, R.S. Pierce, J.S. Eaton and N.M. Johnson 1977, "Biogeochemistry of a Forested Ecosystem." Springer, New York.
(13) Matzner, E. and B. Ulrich 1983, "The turnover of protons by mineralization and ion uptake in a beech (Fagus silvatica) and a Norway spruce ecosystem." In: B.Ulrich and J.Pankrath (Eds.): Effects of Accumulation of Air Pollutants in Forest Ecosystem. D. Reidel Publishing Company Dordrecht, Netherlands, pp. 93-104.
(14) Matzner, E. and B. Ulrich 1984, "Raten der Deposition, der internen Produktion und des Umsatzes von Protonen in zwei Waldökosystemen." Z. Pflanzenernähr. Bodenkd. (in press).
(15) Mazzarino, M.J. 1981, "Holozäne Silikatverwitterung in mitteldeutschen Waldböden aus Löß." Diss. Univ. Göttingen
(16) Mazzarino, M.J., H. Heinrichs and H. Fölster 1983, "Holocene versus accelerated actual proton consumption in German forest soil." In: B.Ulrich and J.Pankrath (Eds.): Effects of Accumulation of Air Pollutants in Forest Ecosystems. D. Reidel Publishing Company, Dordrecht, Netherlands, pp. 113-123.
(17) Mazzarino, M.J. und H. Fölster 1984, "Freisetzung und Verteilung von Al- und Si-Oxiden in mitteldeutschen Löss-Böden unter Wald." Catena (Braunschweig) 11, pp. 27-38.
(18) Miehlich, B. 1984, "Chronosequenzen und anthropogene Veränderungen andesitischer Vulkanascheböden in drei Klimastufen eines randtropischen Gebirges (Sierra Nevada de Mexico)." Habilitationsschrift, Hamburg.
(19) Puhe, J. 1982, "Chemische Eigenschaft von Quellwässern im Kaufunger Wald." Dipl. Arbeit, Forstl. Fakultät,

Univ. Göttingen.
(20) Reid, J.M., D.A. McLeod and M.S. Cresser 1981, "Factors affecting the chemistry of precipitation and river water in an upland catchment." J. Hydrology 50, pp. 129-145.
(21) Rosén, K. 1982, "Supply, loss and distribution of nutrients in three coniferous forest watersheds in Central Sweden." Reports in Forest Ecology and Forest Soils, 41. Uppsala.
(22) Rosenquist, I.TH., P. Joergensen and H. Rueslatten 1980, "The importance of natural H^+ production for acidity in soil and water." Proc. Int. Conf. Ecol. Impact Acid Precip. Norway, 1980. SNSF project, pp. 240-241.
(23) Scheffer, F., B. Meyer und H. Gebhardt 1966, "Pedochemische und kryoklastische Verlehmung (Tonbildung) in Böden aus kalkreichen Lockersedimenten (Beispiel Löß)." Z. Pflanzenernähr., Bodenkd. 114, pp. 77-89.
(24) Schlichting, E. und P. Blume 1961, "Art und Ausmaß der Veränderungen des Tonmineralbestandes typischer Böden aus jungpleistozänem Geschiebemergel und ihrer Horizonte." Z. Pflanzenernähr., Bodenkd. 95, pp. 227-239.
(25) Schroeder, D. 1955, "Mineralogische Untersuchungen an Lößprofilen." Heidelberger Beit. z. Mineral. u. Petr. 4, pp. 443-463.
(26) Stahr, K. 1979, "Die Bedeutung periglazialer Deckschichten für Bodenbildung und Standorteigenschaften im Südschwarzwald." Freiburg, Bodenkund. Abhand. Heft 9.
(27) St.Arnaud, R.J. and M.D. Sudom 1981, Canadian J. Soil Sci. 61, pp. 79-89.
(28) Stuanes, A.O. 1980, "Effects of acid precipitation on soil and forest. 5. Release and loss of nutrients from a Norwegian forest due to artificial rain of varying acidity." Proc. Int. Conf. Ecol. Impact Acid Precip. Norway 1980. SNSF project, pp. 198-199.
(29) Troedsson, T. 1980, "Ten years acidification of Swedish forest soils." Proc. Int. Conf. Ecol. Impact Acid Precip. Norway 1980. SNSF project, pp. 184.
(30) Ulrich, B. 1983, "Soil acidity and its relations to acid deposition." In: B.Ulrich and J.Pankrath (Eds.): Effects of Accumulation of Air Pollutants in Forest Ecosystems. D. Reidel Publishing Company, Dordrecht, Netherlands.
(31) Ulrich, B., R. Mayer und P. Khanna 1979, "Deposition von Luftverunreinigungen und ihre Auswirkungen in Waldökosystem im Solling." Schriften aus der Forst. Fak. Göttingen u. der Niedersäch. Forst. Versuchsanst., 58.
(32) Van der Plas, L. and van Schuylenborgh, J. 1970, "Petrochemical calculations applied to soils-with special reference to soil formation." Geoderma 4, pp. 357-385.

(33) Webb, A.H. 1980, "The effect of chemical weathering on surface waters." In: Proc. Int. Conf. Ecol. Impact Acid Precip. Norway 1980. SNSF project, pp. 278-279.
(34) Wright, R.F. and M. Johannessen 1980, "Input-output budgets of major ions at gauged catchments in Norway." Proc. Int. Conf. Ecol. Impact Acid Precip. Norway 1980, SNSF project, pp. 250-251.

EQUILIBRIUM AND DISEQUILIBRIUM BETWEEN PORE WATERS AND MINERALS IN THE WEATHERING ENVIRONMENT

WOLFGANG OHSE, GEORG MATTHESS, ASAF PEKDEGER

Institute of Geology and Paleontology,
Kiel University, Olshausenstr. 40-60,
D-2300 Kiel, West Germany

ABSTRACT

Pore waters from the unsaturated zone of Quaternary glacial outwash sediments were extracted by centrifugation. Mineral stabilities of the major and minor minerals with respect to dissolved inorganic species were calculated on the basis of chemical thermodynamics with the computer program WATEQF (1) and an extended version of PHREEQE (2). The calculations show extents of mineral disequilibria, which are related to corroded grain surfaces or to precipitates detected by scanning electron microscopy (SEM) and X-ray energy spectrometry (XES). Below the weathered zone the pore waters approach saturation with respect to feldspar (K-feldspar, plagioclase), mica, amorphous silica and calcite, whereas quartz oversaturation exists throughout the profile. The undersaturation of the pore waters with respect to feldspar in the top horizons is related to a relative enrichment of kaolinite in these horizons. After a weathering period of about 60,000 years a general decrease of the corrosion state of feldspars with depth is observed, which correlates with an increase in the saturation state of associated pore waters with depth.

INTRODUCTION

Geochemical interactions between pore solutions and rocks can be evaluated by the use of thermodynamic models to calculate the dissolved species in the pore solutions and the mineral saturation indices (SI). This is one possible method to evaluate the existing mineral equilibria or disequilibria in the three-phase system water-rock-gas. There are several limitations to this approach, and caution is required when interpreting the results of model calculation of aqueous speciation and saturation indices. This is necessary because these calculations depend on a theoretical model for the aqueous solution, on correct thermodynamic constants, the accurate analysis of all dissolved species, of the physico-chemical parameters and of the composition of mineral phases, which together control the state of saturation. Thus it is necessary to apply various chemical models to a particular environment and to check the results by other methods, such as scanning electron microscopy (SEM) coupled with X-ray energy spectrometry (XES) which verify precipitation or dissolution phenomena in the sediment.

METHODOLOGY

In Quaternary glacial outwash sediments of the Segeberger Forest, Schleswig-Holstein, relatively undisturbed sediment samples from the unsaturated zone were obtained by a tube core drilling method. The samples were welded into plastic bags to prevent evaporation and kept at the in-situ temperature. The pH and redox potential (Eh) were measured immediately after recovering the tube cores with special punch-in electrodes. After centrifugation (3,500 r.p.m.) in the laboratory, about 10-20 ml of each 30-cm-long sample were extracted, filtered through a 0.2 um membrane filter and acidified with HNO_3 for cation analyses. The extraction method and subsequent chemical analyses were worked out as a compromise between the small volume of the samples and the number of elements which were to be analysed. Na and K were analysed by emission spectrophotometry, Ca, Mg, Fe, Mn and Al by atomic absorption spectrometry, Zn and Pb by argon plasma jet emission spectrophotometry (ICP); Si, SO_4 and NO_3 were determined photometrically, Cl by an ion sensitive electrode and HCO_3 by an automatic microtitration with HCl.

For the SEM and XES investigations, small amounts of dry sediment were mounted on aluminium stubs. The undisturbed field-fresh material was air dried and the coherent sand particles were fixed by a two-component adhesive on the stubs or the minerals from fine and medium sand size-fractions were mounted on the stubs in the same way (3). The heavy minerals of the fine-sand fraction (0.063 - 0.2 mm) were separated by tetrabromoethane. After mounting and coating with carbon the samples were ready for investigation by SEM.

The major minerals in the sand fraction were determined quantitatively by point counting grain mounts. A simple staining method was used to distinguish between K-feldspar and plagioclase under the binocular microscope; quartz and mica are easily recognized by shape. Clay minerals were determined by X-ray diffractometry (3).

CHEMICAL MODELING OF PORE WATERS

The distribution of species was calculated from the chemical analyses of the pore waters by the WATEQF computer model (1) and an extended version of PHREEQE (2) which includes the main heavy minerals. The additional thermodynamic data were selected from the tabulation in WATEQ3 (4). The calculations are based on chemical equilibrium thermodynamics (5, 6). The programs also calculated the saturation state of the solutions with respect to various mineral phases. Generally the saturation state of a solution is defined as the saturation index (SI):

$$SI = \log(IAP/KT)$$

with IAP = ionic activity product and KT = equilibrium constant (dependant on temperature). The SI is zero if the mineral is in equilibrium with the aqueous solution, less than zero if the solution is undersaturated and greater than zero if the solution is oversaturated with respect to the particular mineral. There are no kinetic considerations involved. Hence this type of speciation/saturation calculation indicates whether a particular mineral would tend to dissolve or precipitate.

RESULTS

The tube core drilling on the Radesforder Berg penetrated 32 m of Saale glacial outwash sands (3,7). The ground water table was encountered at 31.2 m depth. Three thin till layers are present at 9 - 10, 24 - 25 and 27 - 29 m below ground level (GL). The mineralogical composition of the sediments is primarily determined by the mineralogy of the crystalline glacial drift originating from Scandinavia. Hence the deposits vary depending on the origin and the amount of Cretaceous, Tertiary and Pleistocene material taken up during transport, and on the effects of degradation within about 60,000 years of weathering. Up to 15 m under GL the sediment is decalcified. Below this depth some scattered Bryozoa occur. The main minerals are quartz (85.9 vol-%), K-feldspar (4.1 vol-%), plagioclase (2.6 vol-%) and mica (1.6 vol-%). The rest (6.8 vol-%) consists of heavy minerals such as goethite, pyroxene, amphibole, cerussite, smithsonite, ilmenite, rutile and zircon, and bryozoa and some undefined organic material and amorphous phases (Fig. 1).

In the upper part of the profile (1-9 m) illite and kaolinite dominate among the clay minerals (Fig. 2). With depth illite increases, kaolinite decreases, while the amount of chlorite varies; expandible clays were not detected.

The chemical analyses of the extracted pore waters show (3) that the pore waters close to GL have dissolved solid concentrations between 100 and 200 mg/l. The waters from 20-30 m under GL have a dictinctly higher solute content (250 - 500 mg/l) due to the increased Ca, HCO3, SiO2, Cl, Na and K concentrations. Correlated to that the pH rises from 5 to 8, while Eh decreases with depth (Fig. 3). Zn and Pb were analysed at three different depth in the upper part of the glacial sands (at 3, 4 and 5 m under GL). For the thermodynamic calculations mean heavy metal background concentrations were used (Zn: 0.35 mg/l and Pb: 0.06 mg/l).

The thermodynamic speciation/saturation calculations indicate calcite undersaturation down to 17 m (Fig. 6) and quartz saturation down to the calcite dissolution depth. The pore waters from deeper horizons are slightly oversaturated with respect to quartz (Fig. 4) and oversaturated with respect to calcite (Fig. 6). Down to 15 m below GL all calcite minerals and bryozoa are dissolved. From a depth of 17 m until 31 m the sediment contains calcite.

Amorphous SiO2 in contrast to quartz, shows a general undersaturation which decreases with depth (Fig. 4). The feldspars are unstable due to their formation under higher pressure and temperature conditions. They are now subject to near-surface conditions, and dissolution and transformation into clay minerals takes place (Fig. 6).

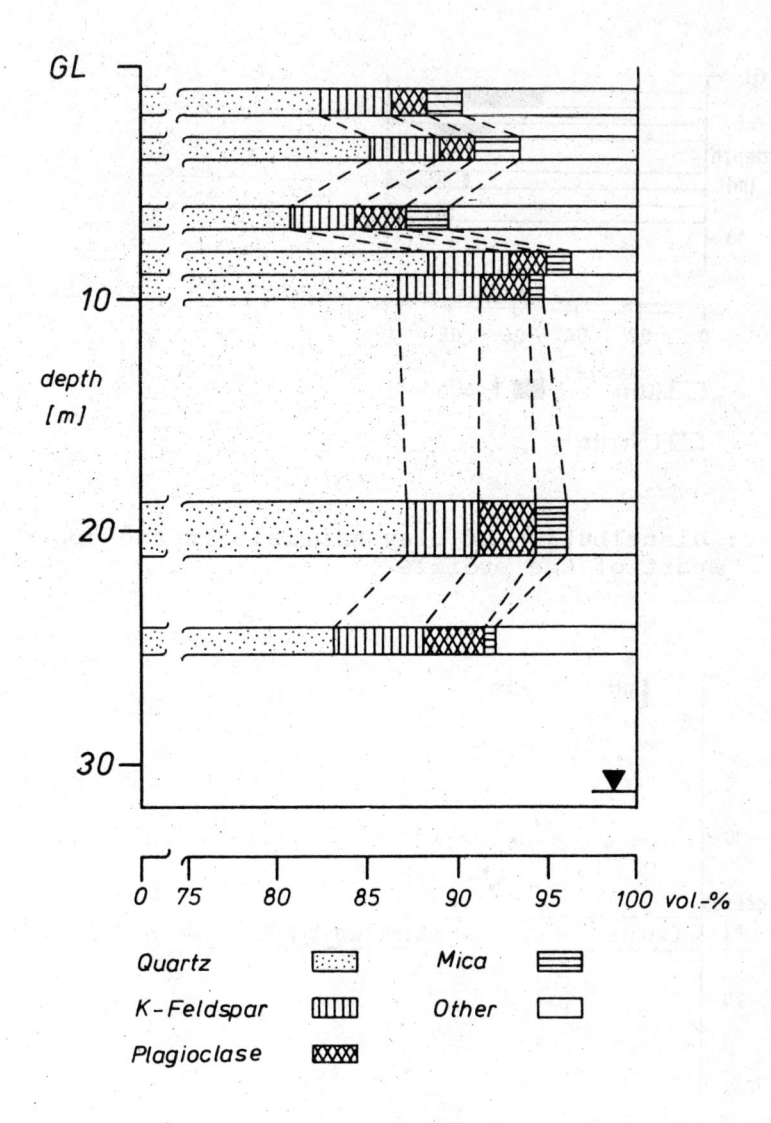

Fig. 1: Distribution of minerals in the sand fraction (drilling: Radesforder Berg).

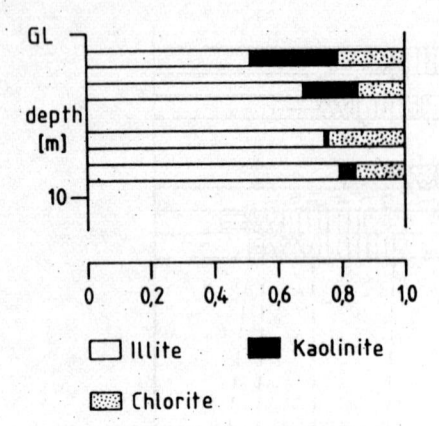

Fig. 2: Distribution of clay minerals in the upper part of the profile.

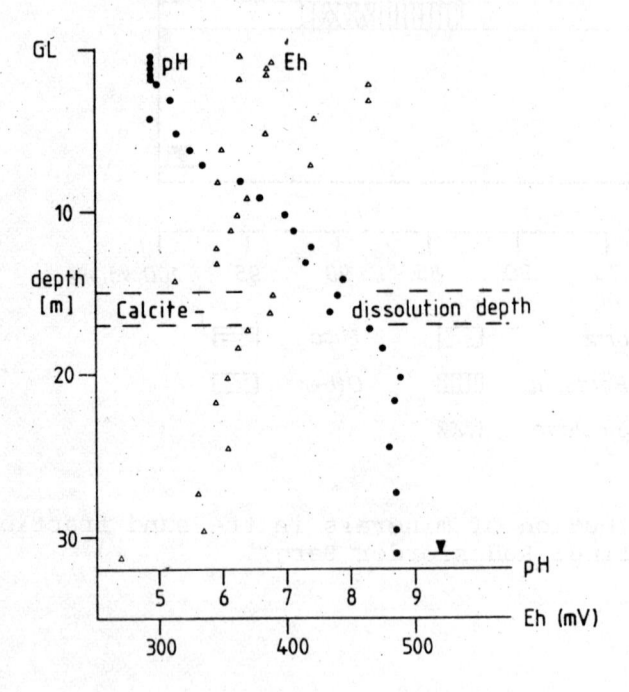

Fig. 3: pH-Eh-profile of the drilling.

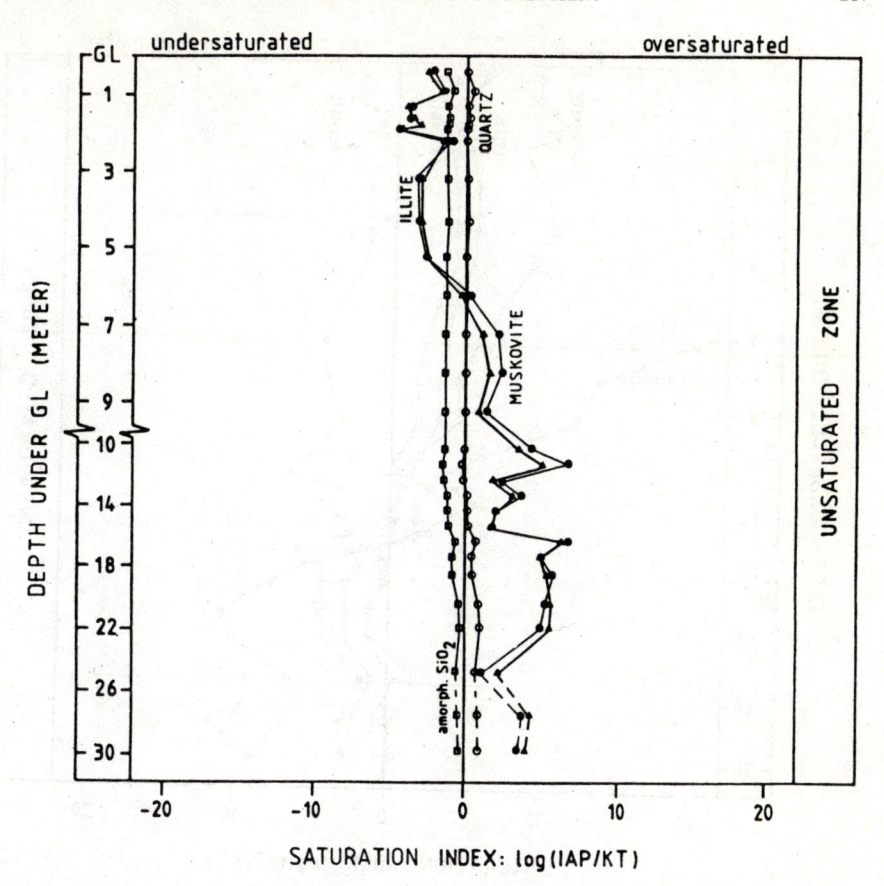

Fig. 4: Saturation of pore waters with respect to quartz, amorphous silica, illite and muscovite.

The pore waters are undersaturated with respect to K-mica (muscovite) and illite until 6 - 7 m under GL (Fig. 4). Physical weathering of mica in connection with potassium release and inclusion of hydrated cations in peripherial parts of the sheets favors formation of hydromuscovite and illite (8). The thermodynamic calculations show that illite is as unstable as K-mica in this environment. Kaolinite is thermodynamically the most stable phase over the whole profile, whereas chlorite is highly undersaturated in the upper part of the profile and gradually becomes highly oversaturated in the deeper part (Fig. 5).

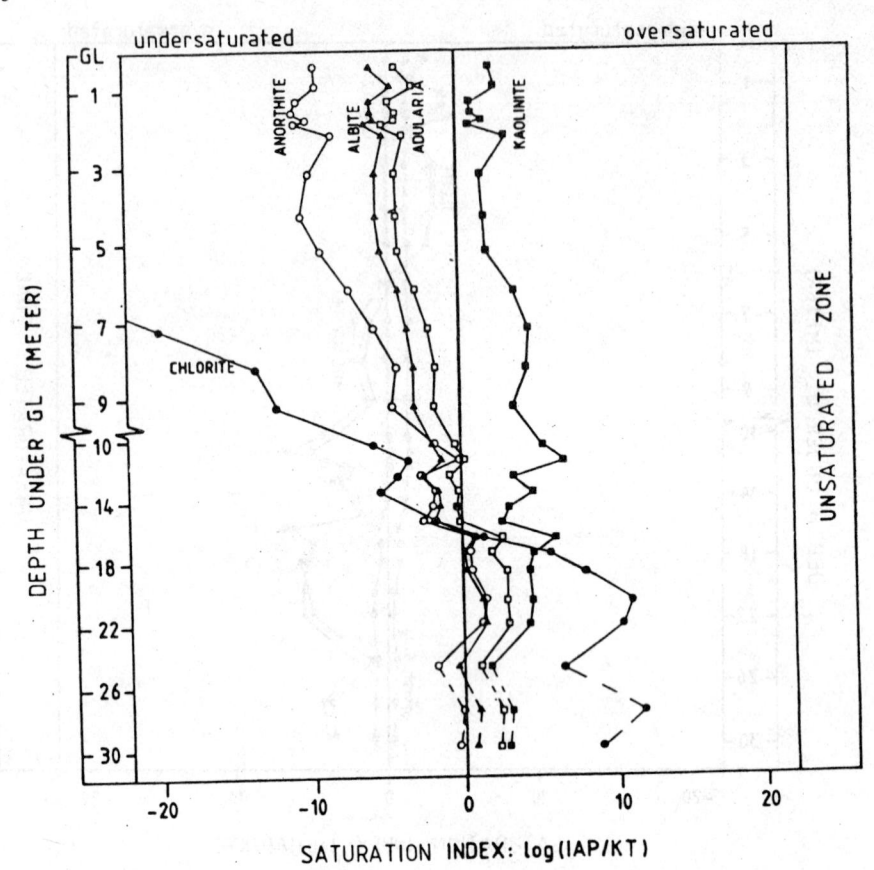

Fig. 5: Saturation of pore waters with respect to feldspars, kaolinite and chlorite.

Preliminary calculations for heavy minerals indicate undersaturation with respect to cerussite and smithsonite. The degree of undersaturation decreases with depth (Tab. 1).

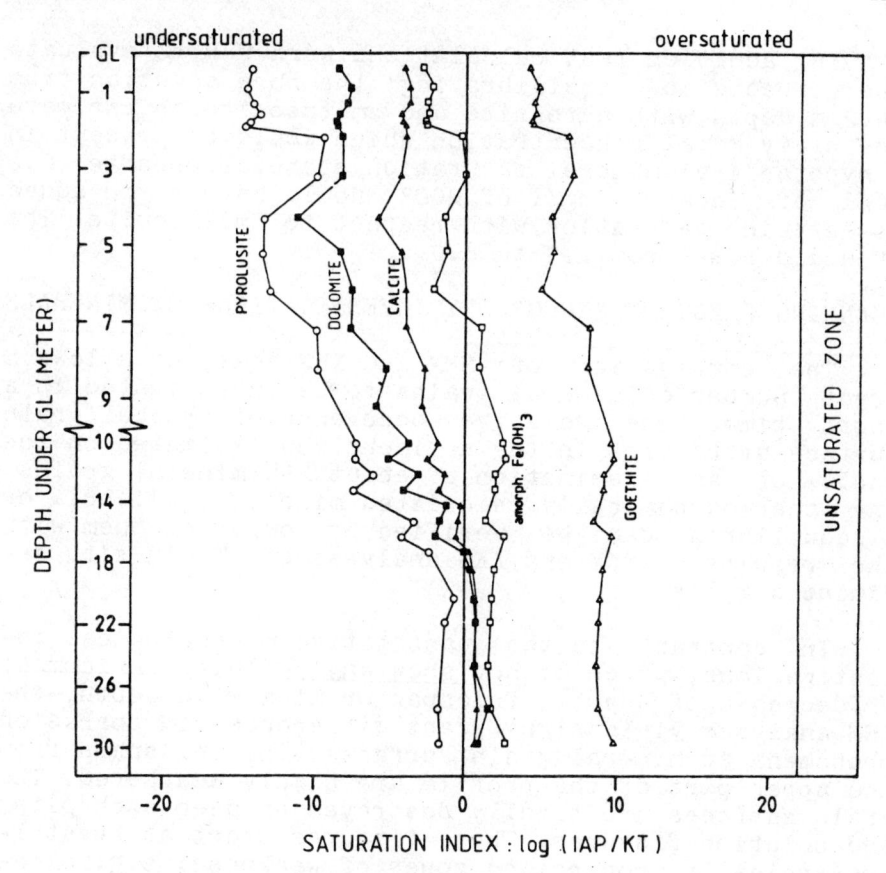

Fig. 6: Saturation of pore waters with respect to calcite, goethite and amorphous Fe(OH)$_3$.

depth	Cerussite	Smithsonite
(m)	log SI	
0.4	- 2.5	- 4.2
5.2	- 2.4	- 4.1
14.2	- 0.7	- 1.1
24.6	- 0.7	- 1.0

Tab. 1: Saturation indices of Corussite and Smithsonite in Radesforder Berg.

In addition test calculations were run to evaluate the effects of equilibrating the pore solution from 14.2 m depth with cerussite and smithsonite to estimate the heavy metal concentration which would be present in a hypothetical mineral saturation state. Thereafter 0.2 mg/l of Pb and 2 mg/l of HCO_3^- would have to be added to attain saturation with respect to smithsonite. The pH would rise from 7.9 to 8.0.

SCANNING ELECTRON MICROSCOPY INVESTIGATIONS OF MINERALS

The combination of SEM and XES analyses allows a great number of mineral grains to be investigated in a short time. The relative occurence of typical grain surface structures in the sediment was estimated on the basis of an examination of about 250 mineral grains. The thermodynamically calculated mineral equilibria or disequilibria can be verified by comparing them with the results of SEM and XES analyses of the in situ sediment samples.

In contrast to the quantitative mineralogical investigations, which do not show any definite enrichment or decrease of quartz, feldspar or mica with depth, the SEM analyses yield significant differences in corrosion phenomena at mineral grain surfaces. The feldspars from the upper part of the profile are highly weathered. The grain surfaces are totally destroyed by deep etch pits and solution fissures. These fissures start at crystallographically controlled zones of weakness, e.g. cleavage, twinning or perthite zones (Fig.: 7.1, 2, 3) and develop solution furrows aligned by crystal symmetry, or they develop honeycomb structures.

HOLDREN & BERNER (9) carried out artificial weathering of feldspars in the laboratory using $HF-H_2SO_4$ solutions. In accordance with their results, a range of dissolution phenomena have been observed on grain surfaces of the sediment, which vary with depth depending on the acidity of the pore waters and the time of reaction. The least weathered state shows only tiny elongated microcracks. Then chemical widening of those microcracks yields rhomboedral or oval shaped solution fissures, often following the 001 or 010 direction. Finally the deeply weathered state is represented by the samples close to GL with honeycomb structures or a net of solution furrows (Fig.: 7.1, 3). Feldspars from the deeper part of the profile are less corroded. This correlates with the decreasing feldspar undersaturation of the pore waters with depth.

The pore waters are undersaturated with respect to amorphous silica and even under near-surface conditions they are saturated with respect to quartz. Therefore quartz dissolution and silica precipitation are not to be expected. The sediment samples show both dissolution and precipitation phenomena. Dissolution of quartz grains in the upper part of the sediment is apparently related to organic acids produced by microorganisms in the soil (Fig.: 7.4, (10)). The organic acids, however, are not considered in the speciation calculations. On the other hand, silica, iron and aluminum precipitates were observed on quartz grains at specific locations favorable for precipitation. Hence microenvironmental processes either retard, enhance or neutralize the thermodynamic tendencies to establishing equilibrium because the saturation states in microenvironments may be different from those in the bulk pore solution. Another reason may be a change in solution composition with time -thus the minerals might be inherited from a time of different water composition (Fig.: 8.1-6).

Mica minerals from the upper part of the profile display distinct exfoliations and transformation phenomena (Fig.: 7.5, 6).

At present, heavy minerals have been investigated only from the 15 m depth. A broad spectrum of partially weathered apatite, goethite, ilmenite, pyroxene, amphibole, rutile, biotite, zircon and barite have been observed (Fig.: 9.1-6, Fig.: 10.5,6). Cerussite and smithsonite appear uncorroded (Fig.: 10.1-4), although the SI indices are -0.7 and -1.1 respectively. In contrast, goethite minerals display distinct corrosion features although the solutions are highly oversaturated with respect to FeOOH (Fig.: 5, Fig.: 9.1-3).

SUMMARY AND CONCLUSIONS

The geochemical models WATEQF and the extended version of PHREEQE used in this study describe fairly well the weathering of the major minerals. They do not, however, predict all the observed changes in the water/rock system. Thermodynamically, the weathering of feldspars, muscovite, illite, chlorite, calcite and quartz should be restricted to the upper layers; deeper than 15 - 17 m no weathering of silicate minerals and calcite should occur. In accordance with this, the SEM investigations of mineral grain surfaces show a gradation of weathering corresponding to increasing alteration towards the top of the core.

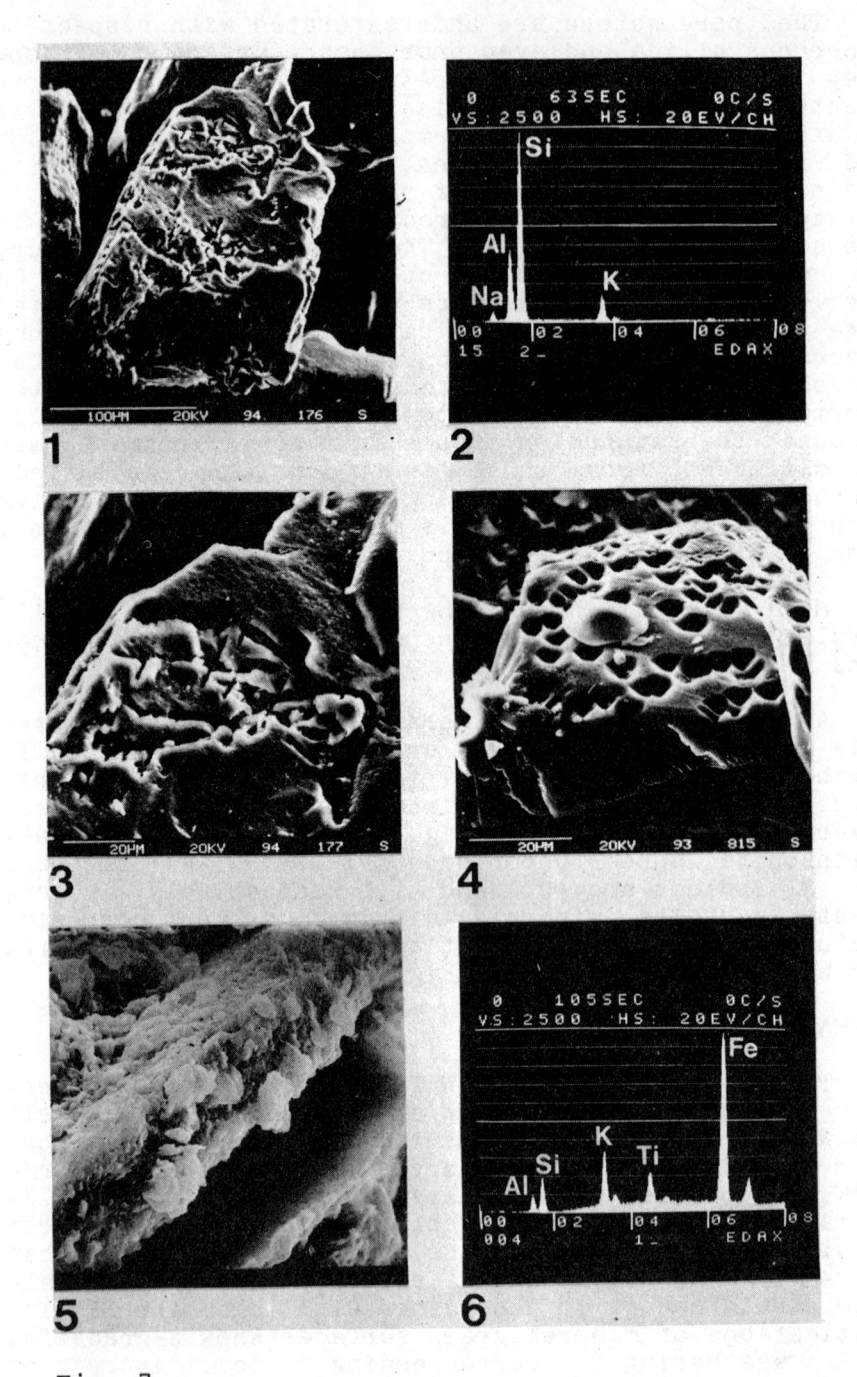

Fig. 7

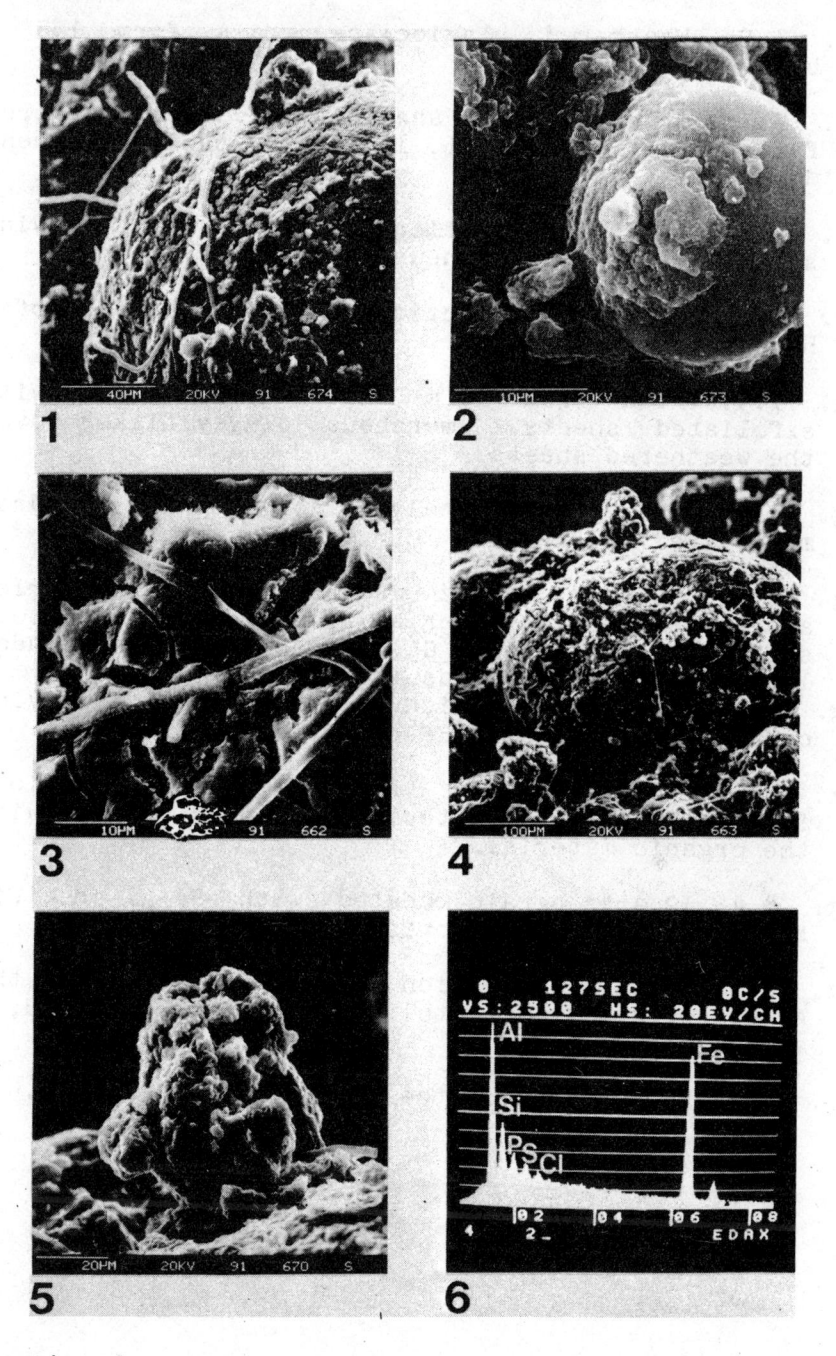

PORE WATERS AND MINERALS IN THE WEATHERING ENVIRONMENT 223

Fig. 8

Fig. 7.1: Weathered plagioclase grain form 1.5 m below ground level.

Fig. 7.2: Integral XES analysis of the weathered feldspar of Fig. 6.1, indicating typical element distribution.

Fig. 7.3: Enlarged section of Fig. 6.1, showing solution fissures aligned by crystal symmetry.

Fig. 7.4: Quartz destruction and development of a honeycomb structure.

Fig. 7.5: Mica from 1 m under ground level with exfoliated sheets. Amorphous clayey flakes cover the weathered sheets.

Fig. 7.6: XES pointe analysis of Fe-rich "hydromuscovite".

Fig. 8.1: Rounded quartz grain, covered with dried amorphous Al-Si-Fe material from Bs-horizon of podsol (35 -40 cm below GL). Organic material (endolithic fungi?) surrounds the grain.

Fig. 8.2: Enlarged section of Fig. 8.1, displaying organic material with Fe-rich precipitate on it.

Fig. 8.3: Drying-cracks (from sample preparation) in the amorphous material, which is covered with the organic material.

Fig. 8.4: Quartz grain coated with Fe-Al precipitate (arrow) from Bs-horizon.

Fig. 8.5: Enlarged section of Fig. 8.4 showing the Al-Fe precipitate with minor amounts of Si, P, S and Cl.

Fig. 8.6: XES analysis of the precipitate in Fig. 8.5.

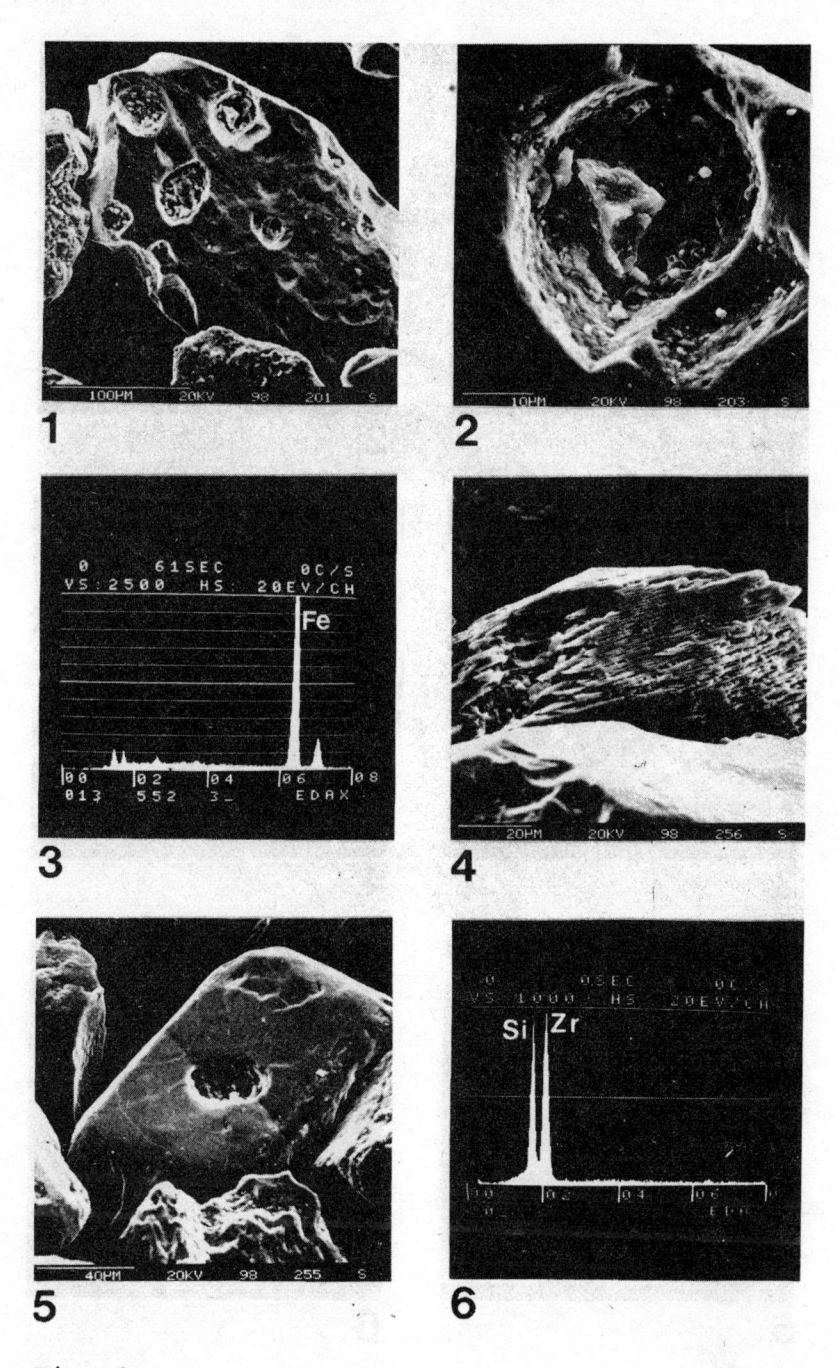

Fig. 9

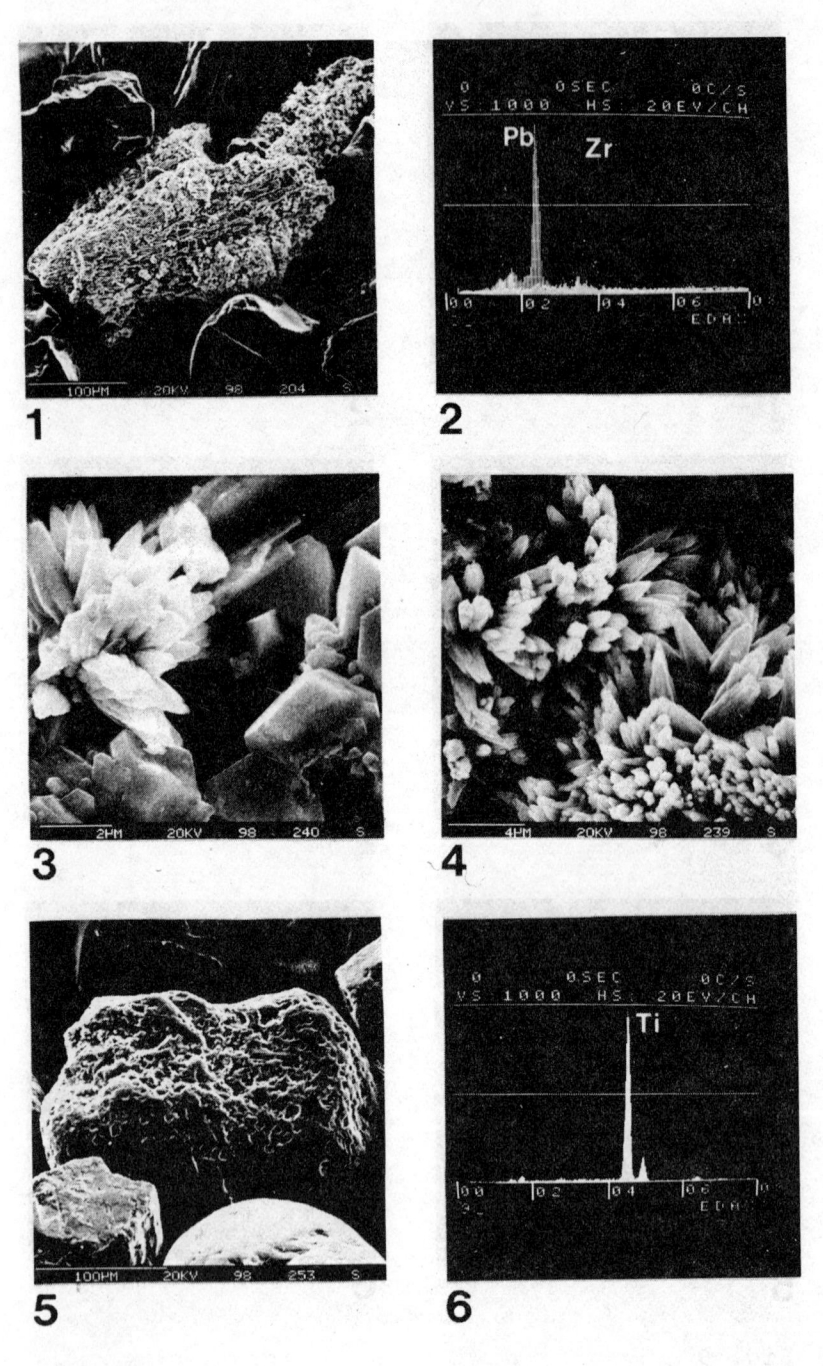

Fig. 10

Fig. 9.1: Weathered goethite.

Fig. 9.2: Enlarged section of Fig. 9.1 showing large corrosion features and tiny V-shaped etch pits.

Fig. 9.3: XES analysis of the goethite.

Fig. 9.4: Weathered amphibole.

Fig. 9.5: Zircon with a central corrosion feature.

Fig. 9.6: XES analysis of the zircon.

Fig. 10.1: Lead-zinc mineralization, probably originating from the crystalline rocks of Scandinavia.

Fig. 10.2: XES analysis of the mineralization in Fig. 10.1.

Fig. 10.3: Enlarged section of Fig. 10.1 showing cerussite (lead carbonate) on the left side and smithsonite on the right side of the figure.

Fig. 10.4: Enlarged section of Fig. 10.1 showing cerussite.

Fig. 10.5: Weathered rutile with solution pits.

Fig. 10.6: XES analysis of the rutile.

LITERATURE

(1) PLUMMER, L.N., JONES, B.F. & TRUESDELL, A.H. 1976, WATEQF - a fortran IV version of WATEQ, a computer program for calculating chemical equilibrium of natural waters.- U.S. Geol. Surv. Water Res. Invest., 76-13, 61 p.; Washington, D.C.

(2) PARKHURST, D.L., THORSTENSON, D.C. & PLUMMER, L.N. 1980, PHREEQE - a computer program for geochemical calculations.- U.S. Geol. Surv. Water Resour. Inv. 80-96, 210 p.; Washington, D.C.

(3) OHSE, W. 1983, Lösungs- und Fällungserscheinungen im System oberflächennahes unterirdisches Wasser/gesteinsbildende Minerale - eine Untersuchung auf der Grundlage der chemischen Gleichgewichts-Thermodynamik.- Diss. Kiel Univ., 242 p.; Kiel.

(4) BALL, J.W., JENNE, E.A. & CANTRELL, M.W. 1981, WATEQ3 - a geochemical model with uranium added.- U.S. Geol. Surv. Open file Report, 81-1183, 18 p.; Washington D.C.

(5) GARRELS, R.M. & CHRIST, C.L. 1965, Solutions, Minerals and Equilibria.- 450 p.; San Francisco (Freeman, Cooper & Co).

(6) STUMM, W. & MORGAN, J.J. 1981,, Aquatic Chemistry.- 780 p.; New York (Wiley & Sons).

(7) SCHULZ, H.D. 1977, Über den Grundwasserhaushalt im norddeutschen Flachland. Teil II: Grundwasserbeschaffenheit der Geest Schleswig-Holsteins.-Bes. Mitt. Dt. Gewässerkdl. Jb. 40: 1-141; Kiel.

(8) SCHEFFER, F. & SCHACHTSCHABEL, P. 1982 ,Lehrbuch der Bodenkunde.- 11. ed., 442 p.; Stuttgart (Enke).

(9) HOLDREN, G.R. & BERNER, R.A. 1979. Mechanism of feldspar weathering.- I. Experimental studies.- Geochim. Cosmochim. Acta 43: 1161-1171; Oxford.

(10) OHSE, W., MATTHESS, G. & PEKDEGER, A. 1983, Gleichgewichts- und Ungleichgewichtsbeziehungen zwischen Porenwässern und Sedimentgesteinen im Verwitterungsbereich.-Z. dt. geol. Ges. 134: 345-361; Hannover.

(11) SPOSITO, G. & MATTIGOD, S.V. 198, A computer program for the calculation of chemical equilibria in soil solutions and other natural water systems.- Department of Soil and Environmental Sciences, 92 p.; University of California, Riverside CA.

HYDROGEOCHEMICAL CONSTRAINTS ON MASS BALANCES IN FORESTED WATERSHEDS OF THE SOUTHERN APPALACHIANS

Michael Anthony Velbel

Dept. of Geology and Geophysics, Yale University (now at: Dept. of Geological Sciences, Michigan State University, East Lansing, MI 48824-1115, U.S.A.)

ABSTRACT

Two variables, parent rock type and flushing rate (the amount of water flushed through the weathering profile per hectare per year), control the long-term average dissolved load of streams in forested watersheds of southwestern North Carolina, U.S.A. The same variables explain qualitative stability relations, as shown by stability field diagrams which are, in turn, consistent with the hydrology and kaolinite-gibbsite clay mineralogy of the profiles. Tardy's Re, a simple semiquantitative mass-balance tool, ranges from 1.36 to 1.65, again qualitatively consistent with the known clay mineralogy of the systems. The consistency of hydrology, aqueous geochemistry and clay mineralogy places useful constraints on more sophisticated geochemical mass-balance models.

INTRODUCTION

Most geochemical research on weathering has been oriented toward explanations of the chemical composition of soil- and groundwaters that drain areas of active rock-water interactions. Many early studies attempted to interpret the composition of groundwater (and resulting streamwater) in terms of chemical reactions between parent minerals and weathering products in near-surface weathering environments. The studies suggest or necessarily assume that soil- and groundwaters are in equilibrium with observed or inferred weathering products or with hypothetical metastable phases. However, a recent study undertaken in the Absaroka Mountains by Miller and Drever (23) has demonstrated that stream waters there are not in equilibrium with weathering products found in the nearby soil. Miller and Drever suggest, after Paces (27,28), that these waters reflect deep incipient alteration of large volumes of rock and that the composition of deep weathering waters are controlled not by equilibrium with weathering products, but by the relative rates of dissolution, precipitation, and water movement through the deep weathering zone. This idea represents an innovative departure from earlier work on chemical weathering, both in its emphasis on non-equilibrium (kinetic) factors, and in its emphasis on the deep sub-soil portion of the weathering profile.

Determining rates and mechanisms of mineral weathering in nature has recently become a major goal of low temperature geochemistry. It is generally agreed that geochemical mass-balance studies are the most reliable approach to determining mineral weathering rates in nature (4). Mass-balance modeling, however, requires careful constraints on mineral reaction stoichiometries, and input and output fluxes.

The purpose of this report is to discuss the effect of lithology and hydrology on the nature and composition of dissolved output fluxes (effluxes) from forested watersheds of the southern Blue Ridge Mountains of eastern North America. The major emphasis of this report is on using hydrological, mineralogical and aqueous geochemical information to constrain the stream output (efflux) terms in more general geochemical mass-balance models of mineral weathering rates in deeply-weathered crystalline rocks (saprolites). Mineral weathering in this area has been or will be discussed elsewhere as will the overall mass-balance model based on these accumulated studies (35-42).

THE STUDY AREA

Topography and Climate

The Coweeta Hydrologic Laboratory of the U.S. Forest Service is located in the Nantahala Mountains 15 km (10 miles) southwest of Franklin, North Carolina. The physiographic Coweeta Basin totals some 1625 hectares, ranging in altitude from over 1585 meters (5200 feet) at its western limit to about 670 meters (2200 feet) in the valley of Coweeta Creek in the east. Slopes within individual watersheds average about 45% (24 degrees). Average annual rainfall is among the highest in eastern North America, from 170 centimeters (80 inches) at lower elevations to 250 centimeters (100 inches) on the upper slopes; this corresponds roughly to decreasing rainfall with distance east from the western boundary of the Laboratory. Less than 5% of the annual precitation falls as snow. More important than precipitation, however, is what Duchaufour (8) calls "climatically controlled drainage," the amount of water actually flushed through the weathering profile, which is always less than rainfall due to evapotranspiration. Annual climatically controlled drainage at Coweeta ranges from 106 to 210 cm (10.6 to 21.0 million liters per hectare per year). The mean annual temperature is 12.8 degrees Celsius (55 degrees F); average maxima and minima are $33°C$ ($92°F$) and $-17°C$ ($1°F$), respectively (6,16,32).

Bedrock Geology

Two major lithostratigraphic units occur in the study area. The Tallulah Falls Formation (9) consists of metagraywackes, pelitic schists, and metavolcanic rocks, which were derived mainly from sedimentary protoliths of low mineralogical maturity (e.g., graywackes). The Coweeta Group (10,12) consists of biotite gneisses, metaarkoses, metasandstones, quartzites, and pelitic and biotite schists, which were derived predominantly from sedimentary protoliths of intermediate to high mineralogical maturity (e.g., arkoses, quartzarenites). The characteristic minerals of these two units include quartz, plagioclase feldspar, biotite and muscovite micas, and almandine garnet, along with minor amounts of staurolite, kyanite and other "heavy" minerals (11,12). From a practical point of view, the qualitative difference between the two main lithostratigraphic units must suffice; the heterogeneity of rock types, combined with the extreme structural complexity of the area, make it almost impossible to estimate quantitatively the absolute or relative abundance of the different minerals and rock types.

Hydrology

"There is another world under this, and it is like ours in everything - animals, plants, and people - save that the seasons are different. The streams that come down from the mountains are the trails by which we reach this underworld, and the springs at their heads are the doorways by which we enter it, but to do this one must fast and go to water and have one of the underground people for a guide. We know that the seasons in the underworld are different from ours, because the water in the springs is always warmer in winter and cooler in summer than the outer air."

Cherokee Origin Myth (25)

Base flow is apparently sustained by prolonged drainage of moist but unsaturated soil and saprolite (15,17). During the "hydrologic survey" of Coweeta in the early 1930's, 28 wells were dug to bedrock (by hand) to depths of 1.5 to 11 meters (5 to 35 feet); average regolith depth at Coweeta is around 6.1 meters (20 feet) (6,32). When pumped dry, 21 failed to recover until heavy rains occurred. From this behaviour, it was concluded that water in these wells reflected "cistern" storage, rather than water levels in an areally extensive aquifer. The remaining seven wells were located near stream channels or in mountain flood plains, and may be connected to local bodies of "groundwater." Hewlett (15) concluded that a saturated groundwater-table-like aquifer does not exist to supply base flow to Coweeta streams.

In a simple and elegant experiment designed to test a hypothetical source of base flow, Hewlett and Hibbert (15,17) constructed giant concrete troughs, which were filled with "subsoil" (C horizon soil), packed to approximately its natural bulk density; moisture tensiometers and neutron-scattering probes were emplaced, as was a "spigot" at the lower end. The experimental hillslopes were then saturated with water to simulate intense precipitation, covered to prevent evaporation, and permitted to drain under the influence of gravity. The results suggested that drainage coninued long after the pores became unsaturated, and that the unsaturated soil zone could contribute sufficient water to sustain observed rates of base flow even after 60 days without recharge. Because recharge is usually much more frequent at Coweeta than in the experiment, Hewlett concluded that prolonged drainage of unsaturated pores is the primary source of base flow to Coweeta streams. There is no evidence that the mountain streams of Coweeta are fed by water from the permanently-saturated water-table.

Storm flow at Coweeta is also dominated by drainage in the unsaturated zone. Hewlett and Hibbert (18) attribute the "flashy" response of Coweeta streams to storms as a result of "subsurface translatory flow, or the rapid displacement of stored water by new rains." (p. 275). "Above the zone of saturation, we may regard such movement as due to thickening of the water films surrounding soil particles and a resulting pulse in water flux as the saturated zone is approached. The process under rainfall is varying everywhere in a most complex way but such movement can be verified in an elementary manner by allowing a soil column to drain to field capacity in the laboratory and slowly adding a unit of water at the top. Some water will flow from the bottom almost immediately, but it will be apparent that it is not the same water added at the top." (p.279). Hewlett and Hibbert (18) cite experimental work using tritium-labelled water to support this notion. Horton and Hawkins (19) found that 87% of the water originally held in pores was "pushed" out of the soil by a plug of tritium-tagged water before any tritium appeared in the effluent. High-runoff episodes therefore involve "the rapid displacement of stored water by new rain" rather than interflow or overland flow. Observations on hydrogen, oxygen and radon isotopes in other natural systems led Sklash and Farnvolden (30) to the same conclusion regarding the major contribution of subsurface water in high-runoff episodes. Winner (44) determined that similar subsurface flow characteristics probably apply to much of the North Carolina Blue Ridge.

A large body of observation, experiment, and theory, based on work at Coweeta and elsewhere, suggests that both base flow and storm flow in streams draining deeply-weathered, saprolitic landscapes are sustained by water from depth in the subsurface (saprolite), and that neither overland flow not interflow contributes directly (or significantly) to the streams.

Clay Mineralogy

Clay minerals in soils and saprolites at the Coweeta area are kaolinite, gibbsite, hydrobiotite, and goethite (35,37,38,40, in prep.). In general, weathering profiles developed on the mineralogically immature Tallulah Falls Formation have a higher kaolinite/gibbsite ratio than those on Coweeta Group rocks. In addition to this lithologic influence on clay mineralogy, there is also a climatic influence -watersheds in the wetter western part of the basin have a lower kaolinite/gibbsite ratio than do dryer watersheds further east on similar rock types. Gibbsite is ubiquitous in soil and shallow saprolite: gibbsite also occurs widely at depth in the saprolite, especially in the more thoroughly flushed western watersheds, but was not detected at depth in poorly flushed watersheds (40, in prep.). Hydrology also exerts an influence: kaolinite increases in relative abundance with depth in all profiles (see Figure 3). Incipiently weathered bedrock in outcrop and at depth contains expandable clays (35). Neither these nor the hydrobiotite weathering products of mica show any systematic variability with rock type or flushing rate (35,37,38,40, in prep.)

AQUEOUS GEOCHEMISTRY

Thus far, this discussion has dealt primarily with evidence from hydrology and clay minerals. There is, however, an other aspect of mineral-water interactions to be considered; the chemistry of Coweeta waters. The dissolved content of the waters often determines the nature of the solid weathering products precipitated from it, and waters carry with them the dissolved products of weathering (all the material released by weathering which is not incorporated in the solid weathering products).

Since mineral nutrient cycling studies were initiated at Coweeta in the late 1960's, a wealth of information has been gathered on elemental concentrations in, and fluxes via, precipitation and stream water. The most comprehensive reports of the results to date is Swank and Douglass (33), wherein are reported elemental concentrations and fluxes for all Coweeta watersheds which had ever been studied up to the date of preparation of the paper. The stream chemistry data suggested to Swank and Douglass (32,33) that higher concentrations of sodium and calcium in south-facing watersheds north of Shope Fork might be due to lithologic variations in bedrock, based on preliminary geologic mapping (10).

The data of Swank and Douglass (33) are weighted means for study intervals of various duration (depending on watershed and element) between 1969 and 1976.

Streamwater Chemistry as an Indicator of Groundwater Chemistry

If we are to use streamwater chemistry data to model weathering reactions in the saprolite, we must first be sure that the streamwater reflects the chemistry of water which has in fact percolated through saprolite. Table 1 compares the chemistry of soil solutions (collected by lysimeter at 25 cm depth; unpublished Coweeta Hydrologic Laboratory data, Swank, written communication), "groundwater" (subsurface water collected from a well at the rock-saprolite interface; unpublished data, Swank, written communication), and streamwater (33) from watershed 6. Within the range of analytical errors and possible minor natural variations, the major-element chemistry of streamwater is indistinguishable from that of subsurface water which has percolated

TABLE 1: COMPARISON OF SOIL-, WELL-, AND STREAMWATER FOR WATERSHED 6

	Concentration (ppm)				
	K	Na	Ca	Mg	pH
Soil	.921	.271	2.418	1.268	6.12
Well	.604	1.081	1.123	.651	5.10
Stream	.591	1.094	1.063	.643	6.64

through the saprolite (Swank, written communication). Furthermore, the streamwater and subsurface water chemistry are profoundly different from those of soil solutions. For the balance of this discussion of majorelements, it is considered established that streams are merely samples of subsurface water which have undergone no significant chemical change (except re-equilibration with atmospheric gases which would affect pH) on leaving the saprolite to feed the streams. The chemical character of streams is apparently determined by the processes which alter the composition of water as it percolates through the saprolite, and the streams effectively sample the solutions which result from weathering processes in the saprolite.

Hydrologic and Lithologic Influences on Stream Chemistry

Johnson and Swank (21) found that, within an individual watershed, concentration is independent of discharge; in other words, high-runoff episodes do not reflect dilution of base flow by interflow or overland flow (see above). High-runoff episodes are caused by addition of rain to the top of the solum, pushing "old" water out of lower levels of the profile (18). This "old" water has been in prolonged contact with soil and rock minerals, and may have reached a "steady-state" composition, as was experimentally shown (3). Consequently, base flow and storm flow have essentially the same composition; storms merely "force out" more of the same water which normally sustains base flow.

Each watershed, however, is different; several different rock units are weathering at Coweeta, and the "mean annual flushing rate" varies within the Coweeta basin. Consequently, each individual watershed possesses an "average" or "steady-state" composition (3) that differs from the "steady-state" composition of other watersheds on different rock types or with different flushing rates. Furthermore, interwatershed variability in "average" streamwater composition is systematic with respect to rock type and discharge.

Shown in Figure 1a are long-term average concentrations of Na, Ca, K, Mg, silica, and pH of streamwater for four control (undisturbed, unmanaged) watersheds underlain dominantly or exclusively by Tallulah Falls rocks, plotted as a function of long-term average discharge (i.e., flushing rate). Several elements (sodium, postassium, and silica) decrease in concentration with increased flushing rate, whereas calcium

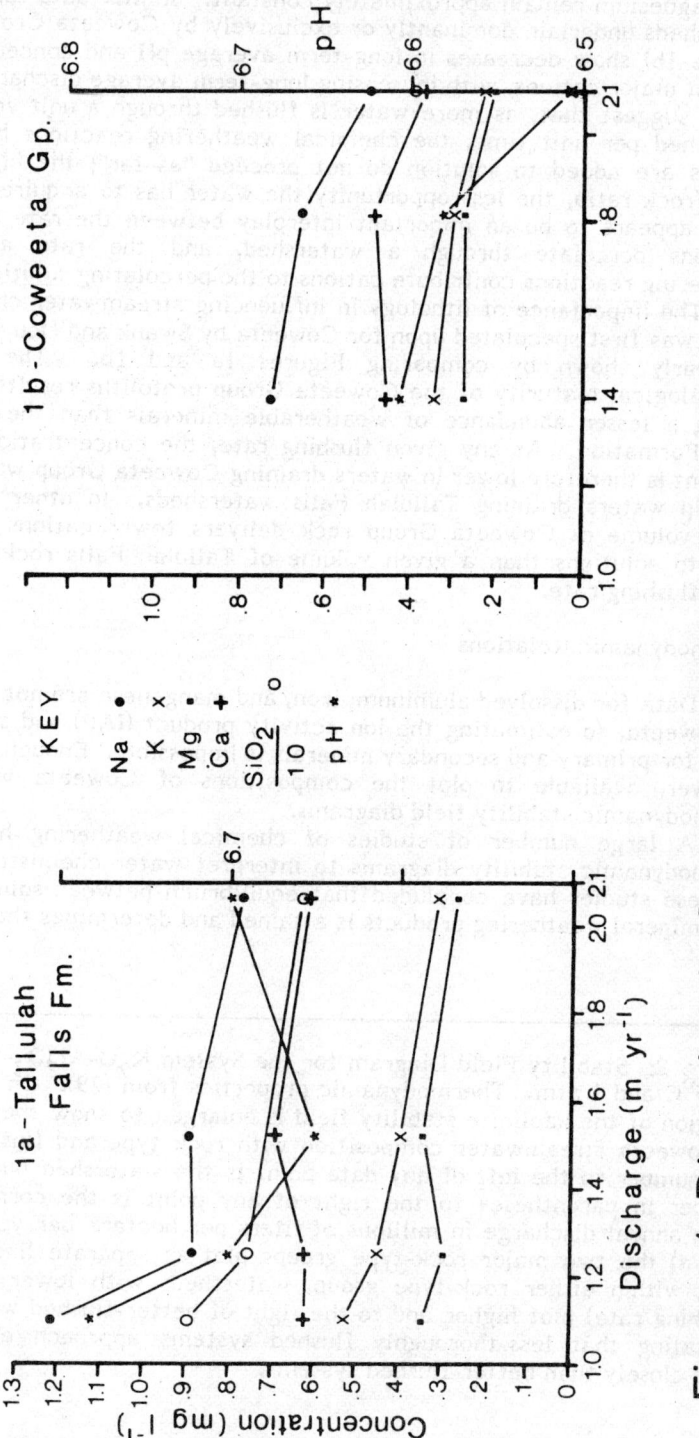

Figure 1: Dissolved Concentration vs. Discharge

and magnesium remain approximately constant. Similar data for control watersheds underlain dominantly or exclusively by Coweeta Group rocks (Figure 1b) show decreases in long-term average pH and concentrations of most major cations, with increasing long-term average discharge. The trends suggest that, as more water is flushed through a unit volume of watershed per unit time, the chemical weathering reactions by which cations are added to solution do not proceed "as far"; the higher the water/rock ratio, the less opportunity the water has to acquire solutes. There appears to be an important interplay between the rate at which solutions percolate through a watershed, and the rate at which weathering reactions contribute cations to the percolating solution.

The importance of lithology in influencing streamwater chemistry, which was first speculated upon for Coweeta by Swank and Douglass (33), is clearly shown by comparing Figures 1a and 1b. The greater mineralogical maturity of the Coweeta Group protoliths results in their having a lesser abundance of weatherable minerals than the Tallulah Falls Formation. At any given flushing rate, the concentration of any element is therefore lower in waters draining Coweeta Group watersheds than in waters draining Tallulah Falls watersheds. In other words, a given volume of Coweeta Group rock delivers fewer cations per unit time to solutions than a given volume of Tallulah Falls rocks, at the same flushing rate.

Thermodynamic Relations

Data for dissolved aluminum, iron, and manganese are not available at Coweeta, so estimating the ion activity product (IAP) and saturation state for primary and secondary minerals is impossible. Enough data are, however, available to plot the compositions of Coweeta waters on thermodynamic stability field diagrams.

A large number of studies of chemical weathering have used thermodynamic stability diagrams to interpret water chemistry. Many of these studies have concluded that equilibrium between solutions and clay-mineral weathering products is attained and determines the nature

---

Figure 2: Stability Field Diagram for the System $K_2O\text{-}Al_2O_3\text{-}SiO_2\text{-}H_2O$ at $25°C$ and 1 atm. Thermodynamic properties from (29). On the right, a region of the kaolinite stability field is enlarged to show the variation of Coweeta streamwater composition with rock type and flushing rate. The number to the left of any data point is the watershed number; the number in parentheses to the right of any point is the corresponding mean annual discharge in millions of liters per hectare per year. Note that a) the two major rock-type groups plot as separate lines, and b) that, within either rock-type group, watersheds with lower discharge (flushing rate) plot higher and to the right of better-flushed watersheds, indicating that less-thoroughly flushed systems approach equilibrium more closely than better-flushed systems.

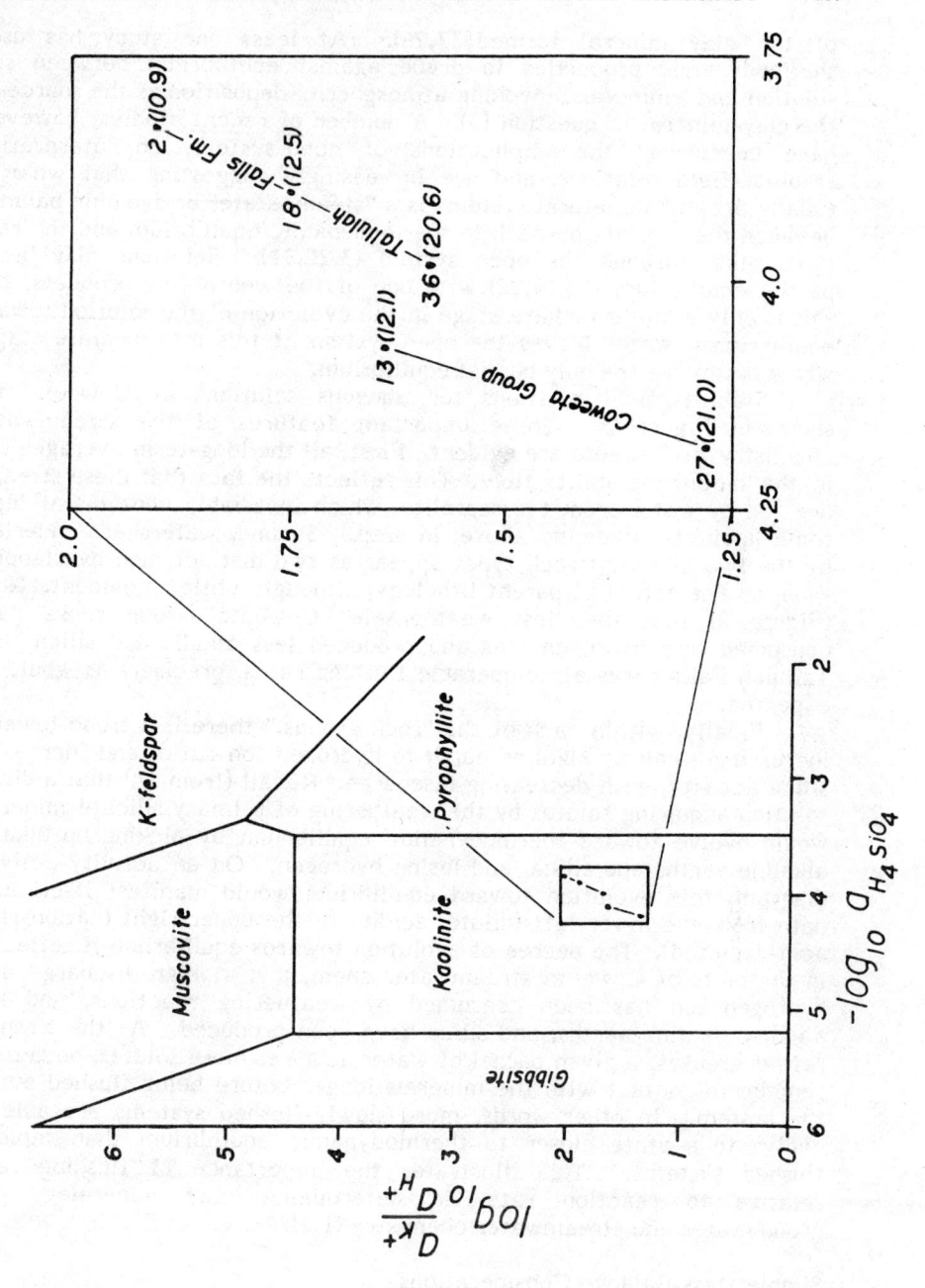

of the clay mineral formed (7,26). At least one study has used thermodynamic properties to argue against equilibrium between soil solution and minerals, invoking atmospheric deposition as the source of the clay-mineral in question (5). A number of recent studies, however, have considered the implications of open-systems on interpreting stability-field relations, and are increasingly suggesting that what is usually present in natural systems is a "steady-state" or dynamic balance between the rate of approach to thermodynamic equilibrium and the rate of flushing through the open system (3,20,31). Solutions may be in partial equilibrium (13,14,22) with one of the weathering products, but this is only an intermediate stage in the evolution of the solution toward equilibrium. Water leaves the open system at this intermediate stage, which is why we see only partial equilibrium.

Stability-field relations for aqueous solutions at Coweeta are shown in Figure 2. Three important features of the streamwater chemistry at Coweeta are evident. First, all the long-term averages fall in the kaolinite stability field. This reflects the fact that these streams are fed by water from the saprolite, which invariably contains at least some kaolinite at depth (above; in prep.) Second, watersheds underlain by the two different rock types appear as two distinct, non-overlapping groups. The effect of parent lithology, although subtle, is evident. Note (Figure 2) that the "less weatherable" Coweeta Group rocks have consumed less hydrogen ions and produced less alkalis and silica than Tallulah Falls rocks at comparable flushing rates, precisely as would be expected.

Finally, within each of the "rock groups," there is a trend towards increasing alkali or alkaline earth to hydrogen ion ratio, and increasing silica activity, with decreasing discharge. Recall (from 14) that a dilute solution acquiring solutes by the weathering of primary silicate minerals would evolve toward thermodynamic equilibrium by picking up alkalis, alkaline earths and silica, and losing hydrogen. On an activity-activity diagram, this evolution toward equilibrium would manifest itself as a path from the lower left (dilute, acidic) to the upper right (cation-rich, acid-depleted). The degree of evolution towards equilibrium is reflected in the plots of Coweeta streamwater chemistry; at high discharge, less hydrogen ion has been consumed by weathering reactions, and less alkalis, alkaline earths, and silica have been produced. As the flushing rate decreases, a given parcel of water acquires more solutes, because it remains in contact with the minerals longer before being flushed out of the system. In other words, more slowly-flushed systems are able to evolve to a state closer to thermodynamic equilibrium than rapidly-flushed systems. This illustrates the importance of flushing rate, relative to reaction rate, in determining clay mineralogy and groundwater and streamwater chemistry (1,2).

Simple Mass-Balance Considerations

Tardy (34) presented a simple method for estimating the ratio of silica to alumina retained in the solid weathering products, from the dissolved concentration of major alkalis, alkaline earths, and silica being

TABLE 2: Re FOR COWEETA CONTROL WATERSHEDS

Watershed	Re
2	1.65
18	1.36
27	1.63
36	1.41

removed from the weathering systems in streams. Several assumptions are involved, the most important of which are: a) that the weatherable minerals all have feldspar or biotite stoichiometries (effectively limiting the method's applicability to granitic and similar crystalline rocks, for which Tardy considers quartz and muscovite unweatherable), and b) that all minerals dissolve stoichiometrically. Given these assumptions,

$$Re = \left(\frac{SiO_2}{Al_2O_3}\right)_{residue} = 2\left(\frac{3K^+ + 3Na^+ + 2Ca^{2+} - SiO_2}{K^+ + Na^+ + 2Ca^{2+}}\right)_{streams}$$

If Re = 0, the weathering products are aluminous (e.g., gibbsite). Weathering systems exhibiting this character are said to be allitic. Where Re = 2, there is a 1:1 atom ratio of silica to aluminum, which corresponds to the 1:1 clay, kaolinite. These weathering products are said to be monosiallitic. If Re = 3 or more, enough silica has been retained in the weathering profile to form 2:1 clays (e.g., smectite), and the profile is said to be bisiallitic. Intermediate cases are also tractable. For instance, equal molar proportions of gibbsite and kaolinite would correspond to a Re of 1.33. Tardy's Re has proven to be a useful qualitative indicator of the characteristic weathering products in drainage basins of various size on crystalline rocks (e.g., 34,43).

Table 2 shows the Re values for Coweeta control watersheds for which sufficient data are available. Values between zero and two indicate the both gibbsite and kaolinite form in Coweeta weathering profiles; more specifically, approximately subequal molar proportions of gibbsite and kaolinite are suggested with kaolinite dominant.

The "Re" approach is not without its conceptual difficulties, but, for the moment, it is sufficient that the "Re" approach is at least qualitatively in accord with the observed clay mineralogy and with the relationship of clay mineralogy and hydrology (i.e., the fact that streams are fed by saprolitic water, which explains its "kaolinitic" character).

SUMMARY

The relationships between hillslope hydrology, clay mineralogy of plagioclase weathering products and aqueous geochemistry are summarized schematically in Figure 3. Feldspar dissolves stoichiometrically (37,38,40,41) releasing all of its calcium, sodium, silica, and aluminum to solution, where they join the alkalis, alkaline earths, silica, and aluminum already in solution. Aluminum is precipitated as gibbsite from this solution in the upper parts of the profile (Figure 3), via the reaction:

$$Al(OH)_2^+{}_{(aq)} + H_2O_{(l)} \rightarrow Al(OH)_{3\,(gibbsite)} + H^+ \quad \text{(Reaction 1)}$$

because rapid flushing keeps silica concentrations low, preventing the attainment of kaolinite saturation. (In some schistose saprolites, the aluminum is consumed as fixed-hydroxy-interlayers and neither gibbsite nor kaolinite form.) Where no aluminum sinks exist (i.e., in the absence of biotite), and where there is prolonged contact between water and weatherable primary minerals, waters deep in the saprolite can acquire enough aluminum and silica to reach kaolinite saturation, and kaolinite precipitates (Figure 3) via the reaction:

$$2Al(OH)_2^+{}_{(aq)} + 2SiO_{2\,(aq)} + H_2O_{(l)} \rightarrow Al_2Si_2O_5(OH)_4 + 2H^+$$

$$\text{(Reaction 2)}$$

Upon reaching the rock-saprolite interface, most of the water is shunted laterally down the slope, ultimately emerging to feed streams. A small fraction of the water enters the system of grain-boundary and grain-traversing fractures in the rock below the rock-saprolite interface (Figure 3), where it continues to attack primary minerals until it has a) transformed biotite to expandable clay, and b) reached saturation with smectite, which is then precipitated as pseudomorphous void-fillings in etched feldspar, and as fracture-fillings. In this manner, the rapidly flushed waters effect the mineral transformations observed petrographically (35-42), produce the observed distribution of clay minerals as summarized Figure 3, and give rise to the character of aqueous solutions.

It is most important to note that, although weathering "microsystems" which form expandable clays exist, the hydrology of the weathering profile limits the amount of water which passes through the "smectitic," subsaprolite portion to an insignificant fraction of the water which passes through the saprolite. The geochemical character of subsurface waters, of the streams they feed, and the stream dissolved effluxes from these profiles, are therefore determined almost exclusively by weathering reactions taking place above the rock-saprolite interface. (This is also suggested by the stability field relations and Tardy's Re.) From a quantitive point of view, rock-water interactions taking place below the rock-saprolite interface do not contribute to elemental budgets, and the complex microfracture

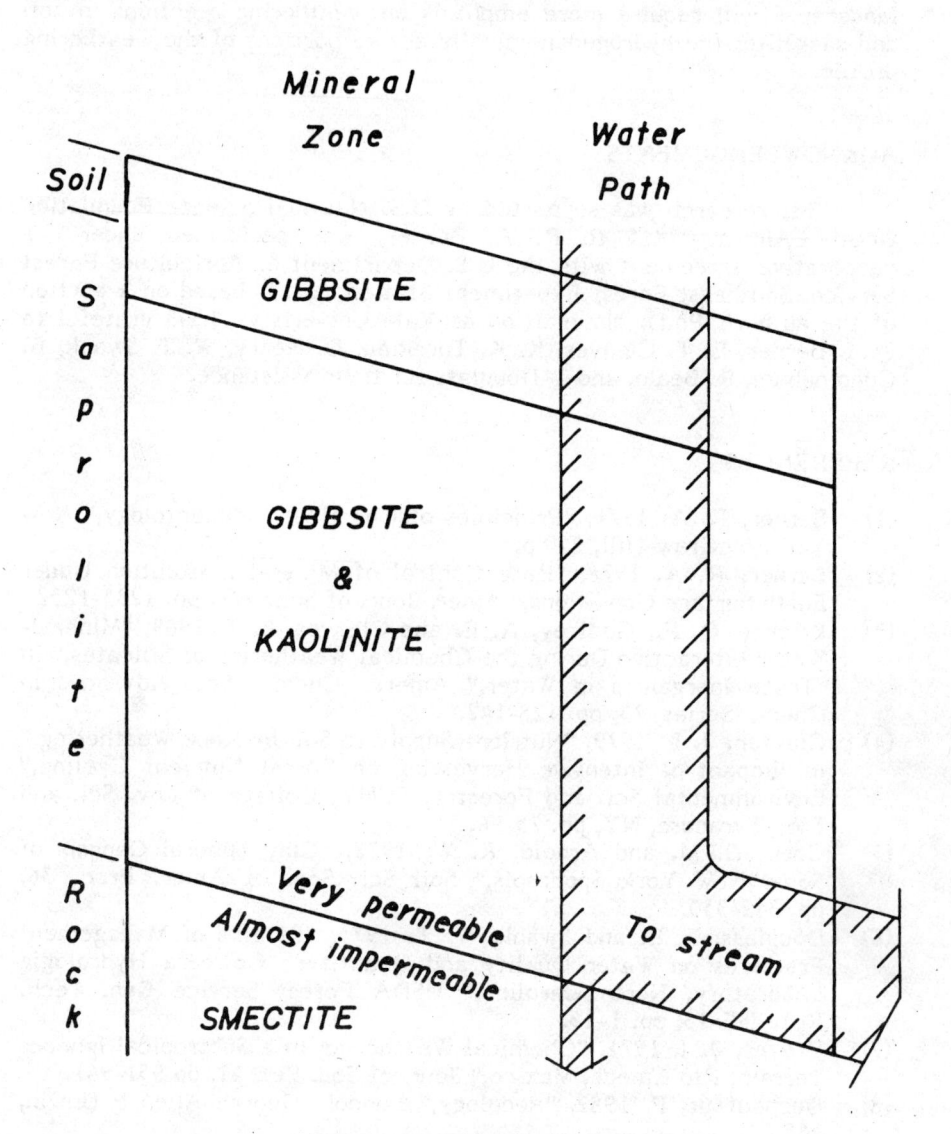

Figure 3 : Schematic Diagram of Watershed 27 Hillslope Hydrology and Weathering Profile

hydrology and stoichiometry of reactions therein can be safely ignored in constructing geochemical mass-balances for the watersheds. Further refinement of geochemical mass-balances in such deeply weathered landscapes will require more emphasis on weathering reactions in soil and saprolite, the hydrogeochemically active portions of the weathering profile.

ACKNOWLEDGEMENTS

This research was supported by U.S. National Science Foundation Grant EAR 80-07815 to R. A. Berner, was performed under a cooperative agreement with the U.S. Department of Agriculture Forest Service, Southeast Forest Experiment Station, and is based on a portion of the author's Ph.D. dissertation at Yale Univeristy. I am grateful to Prof. Berner, E. T. Cleaves, K. K. Turekian, D. Neary, W. T. Swank, B. Cunningham, R. Beale, and J. Douglass for their assistance.

REFERENCES

(1) Berner, R. A. 1971, "Principles of Chemical Sedimentology," New York, McGraw-Hill, 240 p.
(2) Berner, R. A. 1978, "Rate Control of Mineral Dissolution Under Earth Surface Conditions," Amer. Jour. of Sci., 278, pp. 1235-1252.
(3) Bricker, O. P., Godfrey, A. E. and Cleaves, E. T. 1968, "Mineral-Water Interaction During the Chemical Weathering of Silicates," in "Trace Inorganics in Water," Amer. Chem. Soc. Advances in Chem. Series 73, pp. 128-142.
(4) Clayton, J. L. 1979, "Nutrient Supply to Soil by Rock Weathering," in "Impact of Intensive Harvesting on Forest Nutrient Cycling," Environmental Sci. and Forestry, SUNY, College of Env. Sci. and For., Syracuse, NY, pp. 75-96.
(5) Coen, G. M. and Arnold, R. W. 1972, "Clay Mineral Genesis of Some New York Spodosols," Soil Sci. Soc. of Amer. Proc. 36, pp. 342-350.
(6) Douglass, J. E. and Swank, W. T. 1975, "Effects of Management Practices on Water Quality and Quantity: Coweeta Hydrologic Laboratory, North Carolina," USDA Forest Service Gen. Tech. Rpt. NE-13, pp. 1-13.
(7) Drever, J. I. 1971, "Chemical Weathering in a Subtropical Igneous Terrain, Rio Ameca, Mexico," Jour. of Sed. Pet. 41, pp 951-961.
(8) Duchaufour, P. 1982, "Pedology," London, George Allen & Unwin, 448 pp.
(9) Hatcher, R. D., Jr. 1971, "The Geology of Rabun and Habersham Counties, Georgia," Geol. Surv. of Georgia, Bull. 83, 48 pp.
(10) Hatcher, R. D., Jr. 1974, "An Introduction to the Blue Ridge Tectonic History of Northeast Georgia," Georgia Geol. Surv. Guidebook 13-A, 60 pp.

(11) Hatcher, R. D., Jr. 1976, "Introduction to the Geology of the Eastern Blue Ridge of the Carolinas and Nearby Georgia," Carolina Geol. Soc. Field Trip Guidebook, 53 pp.

(12) Hatcher, R. D., Jr., 1979, "The Coweeta Group and Coweeta Syncline: Major Features of the North Carolina-Georgia Blue Ridge," SE Geol. 21, pp. 17-29

(13) Helgeson, H. C. 1968, "Evaluation of Irreversible Reactions in Geochemical Processes Involving Minerals and Aqueous Solutions - I. Thermodynamic Relations," Geochim. et Cosmoch. Acta 32, pp. 853-877.

(14) Helgeson, H. C., Garrels, R. M. and Mackenzie, F. T. 1969, "Evaluation of Irreversible Reactions in Geochemical Processes Involving Minerals and Aqueous Solutions - II. Applications," Geochim. et Cosmoch. Acta 33, pp. 455-481.

(15) Hewlett, J. D. 1961, "Soil Moisture as a Source of Base Flow From Steep Mountain Watersheds," USDA Forest Service, SE Forest Expt. Station, Stat. Paper No. 132, 11 pp.

(16) Hewlett, J. D. and Douglass, J. E. 1968, "Blending Forest Uses," USDA Forest Service Res. Paper SE-37, 15 pp.

(17) Hewlett, J. D. and Hibbert, A. R. 1963, "Moisture and Energy Conditions Within a Sloping Soil Mass During Drainage," Jour. of Geophys. Res. 68, pp. 1081-1087.

(18) Hewlett, J. D. and Hibbert A. R. 1966, "Factors Affecting the Responses of Small Watersheds to Precipitation in Humid Areas," in "International Symposium on Forest Hydrology," Proc. of Nat. Sci. Foundation Adv. Sci. Sem., pp. 275-290.

(19) Horton, J. H. and Hawkins, R. H. 1964, "The Importance of Capillary Pores in Rainwater Percolation to the Ground Water Table," E. I. du Pont de Nemour and Co., Savannah River Plant, DPSPU 64-30-23, 13 pp.

(20) Johnson, N. M. 1971, "Mineral Equilibria in Ecosystem Geochemistry," Ecology 52, pp. 529-531.

(21) Johnson, P. L. and Swank, W. T. 1973, "Studies of Cation Budgets in the Southern Appalachians on Four Experimental Watersheds with Contrasting Vegetation," Ecology 54, pp. 70-80.

(22) Michard, G., Sarazin, G. and Iundt, F. 1979, "Controle des Concentrations D'aluminum et de Silice Dissous Lors de L'interaction des Eaux et des Roches Magmatiques," Bull. Min. 102, pp. 367-373.

(23) Miller, W. R. and Drever, J. I. 1977, "Chemical Weathering and Related Controls on Surface WAter Chemistry in the Absaroka Mountains, Wyoming," Geochim. et Cosmoch. Acta 41, pp. 1693-1702.

(24) Monk, C. D., Crossley, D. A., Jr., Todd, R. L., Swank, W. T., Waide, J. B. and Webster, J. R. 1977, "An Overview of Nutrient Cycling Research at Coweeta Hydrologic Laboratory," in "Watershed Research in Eastern North America," Smithsonian Inst., pp. 35-50.

(25) Mooney, J. 1900, "Myths of the Cherokee," 19th Ann. Rept. Bur. Amer. Ethnology, 1897-1898, Part I.

(26) Norton, D. 1974, "Chemical Mass Transfer in the Rio Tanama System, West-central Puerto Rico," Geochim. et Cosmoch. Acta 38, pp. 267-277.
(27) Paces, T. 1973, "Steady-state Kinetics and Equilibrium Between Ground Water and Granitic Rock," Geochim. et Cosmoch. Acta 37, pp. 2641-2663.
(28) Paces, T. 1976, "Kinetics of Natural Water Systems," in "Interpretation of Environmental Isotope and Hydrochemical Data in Groundwater Hydrology," Data in Groundwater Hydro., Intern. Atomic Energy Agency, Vienna, pp. 85-108.
(29) Robie, R. A., Hemingway, B. S. and Fisher, J. R. 1978, "Thermodynamic Properties of Minerals and Related Substances at 298.15 K and 1 Bar (10^5 Pascals) Pressure and at Higher Temperatures," U.S. Geol. Surv. Bull. 1452, 456 pp.
(30) Sklash, M. G. and Farvolden R. N. 1982, "The Use of Environmental Isotopes in the Study of High-Runoff Episodes in Streams," in "Isotope Studies of Hydrologic Processes," Northern Illinois Univ. Press, DeKalb, pp. 65-73.
(31) Smith, T. R. and Dunne, T. 1977, "Watershed Geochemistry: The Control of Aqueous Solutions by Soil Materials in a Small Watershed," Earth Surf. Proc. 2, pp. 421-425.
(32) Swank, W. T. and Douglass, J. E. 1975, "Nutrient Flux in Undisturbed and Manipulated Forest Ecosystems in the Southern Appalachian Mountains," Pub. No. 117 de l'Ass. Intern. des Sci. Hydro. Sym. de Tokyo, pp. 445-456.
(33) Swank, W. T. and Douglass, J. E. 1977, "Nutrient Budgets for Undisturbed and Manipulated Hardwood Forest Ecosystems in the Mountains of North Carolina," in "Watershed Research in Eastern North America," Smithsonian Inst., pp. 343-364.
(34) Tardy, Y. 1971, "Characterization of the Principal Weathering Types by the Geochemistry of Waters from Some European and African Crystalline Massifs," Chem. Geol. 7, pp. 253-271.
(35) Velbel, M. A. 1982, "Weathering and Saprolitization in the Southern Blue Ridge," XIth Intern. Cong. of Sed. Abst., pp. 171-172.
(36) Velbel, M. A. 1983, "Rate Controls During the Natural Weathering of Almandine Garnet," Geol. Soc. of Amer. Abst. with Prog. 15, p. 712.
(37) Velbel, M. A. 1983, "A Dissolution-reprecipitation Mechanism for the Pseudomorphous Replacement of Plagioclase Feldspar by Clay Minerals During Weathering," in "Petrologie des Alterations et des Sols, Volume I," Memoires Sciences Geologiques, 71, pp. 139-147.
(38) Velbel, M. A. 1984, "Silicate Mineral Transformations During Rock Weathering in the Southern Blue Ridge," Geol. Soc. of Amer. Abst. with Prog. 16, p. 204.
(39) Velbel, M. A. 1984, "Geochemical Mass Balances and Weathering Rates in Forested Cathments of the Southern Blue Ridge," EOS (Amer. Geophys. Union Trans.) 65, p. 211.

(40) Velbel, M. A. 1984, "Mineral Transformations During Rock Weathering, and Geochemical Mass-balances in Forested Watersheds, of the Southern Appalachians," Unpub. Ph.D. Dissert., Yale Univ., 175 p.
(41) Velbel, M. A. 1984, "Weathering Processes of Rock-forming Minerals; Chapter 4," in "Environmental Geochemistry," Miner. Assoc. of Canada Short Course Notes 10, pp. 67-111.
(42) Velbel, M. A. (in press), "Natural Weathering Mechanisms of Almandine Garnet," Geology.
(43) Velbel, M. A. and Dowd, J. F. 1983, "Distribution of Weathering Products in Weathered Bedrock, Western Fairfield County, Connecticut," Geol. Soc. of Amer. Abst. with Prog. 15, p. 200.
(44) Winner, M. D., Jr. 1977, "Ground-Water Resources Along the Blue Ridge Parkway, North Carolina," U.S. Geol. Surv. Water Res. Invest. 77-65, 166 pp.

PAST AND PRESENT SERPENTINISATION OF ULTRAMAFIC ROCKS; AN EXAMPLE FROM THE SEMAIL OPHIOLITE NAPPE OF NORTHERN OMAN

Colin Neal
Institute of Hydrology, Maclean Building, Crowmarsh
Gifford, Wallingford, Oxon OX10 8BB, UK.
Gordon Stanger
The Open University, Dept of Earth Sci., Milton Keynes,
MK7 6AA, UK.

ABSTRACT

Spring waters emerging from partly serpentinised ultramafic rocks in the Semail Ophiolite Nappe of Northern Oman are of a calcium hydroxide type. They have pH values of 10 to 12, are unusually saline and possess anomalous stable hydrogen and oxygen isotopic compositions. The springs, which occur mainly at the contact between "mantle" and "crustal" rock divisions, have low discharge rates and originate by chemical evolution from the recharge of magnesium bicarbonate surface waters. Rare, and in some cases, previously unidentified features associated with these springs include; deuterium depleted hydrogen and nitrogen gases, suolunite, magnesian huntite, brucite, portlandite and amorphous calcium hydroxide. Spring water compositions, petrographic features, and variations in the chemical compositions of the ultramafic rock secondary assemblage, provide strong evidence for the occurrence of both low and high temperature serpentisation processes. These low temperature reactions take place under relatively arid hydrological conditions where groundwater flow is restricted, atmospheric CO_2 and O_2 are completely removed from solution and the normal surface weathering processes are thus inhibited. The unusual isotopic composition of the waters is explained in terms of isotopic exchange between surface waters and hydrous minerals in the groundwater environment related to the low water/rock ratio. The high salt concentrations in the spring waters are derived from the leaching of sea salts introduced during the emplacement of the nappe. Whilst the low temperature serpentinisation processes control the chemical characteristics of the spring waters, it is concluded that the earlier pervasive phase of partial serpentinisation occurred at higher temperatures (up to 350°C) when the ophiolite was obducted from an oceanic to an epicontinental environment.

INTRODUCTION

The foothill areas of the Hajar mountains of Northern Oman constitute an unusual and highly illuminating environment of scientific interest, mainly due to the largest and best exposed of all ophiolites which provide a complete cross section through up to 15km of oceanic crustal and upper mantle material. Consequently the area has been subject to intense geological and hydrochemical studies for the past decade; for example an issue of the Journal of Geophysical research (1981, vol 86, No 84) was dedicated to the topic while several research projects are being or have been undertaken at the Department of Earth Sciences, the Open University (1,2,13,37). Wide-ranging studies on the hydrochemistry of the Oman region have also been undertaken as part of water resource programmes (3). Ground and spring waters in the region are hyperalkaline precipitating unusual hydroxide and carbonate phases (4) and are associated with hydrogen gas evolution (5). They provide a water resource for agricultural use and generate the main 'soils' of the foothill areas (3). Here, results of 6 years detailed field and laboratory studies on the hydrogeochemistry of the region are presented to provide information on the nature of hydration reactions in a unique groundwater setting and provide details on the processes of low temperatures serpentinisation.

Climate and Relief

The climate of Northern Oman is hot and semi-arid. Mean annual air temperature in the foothills averages 25 to 28°C depending upon altitude and proximity to the coast. Monthly mean maxima may reach as high as 46°C whilst the low albedo of "desert varnished" mantle sequence rocks leads to even higher surface temperature during the summer. Average rainfall in the foothills is typically 150 to 200 mm but is highly erratic in intensity and in both spatial and temporal distribution. Consequently, most months are rainless and droughts of up to two years are common. Rainstorms are generally short but intense. Thus invariably steep hillslopes, coupled with the almost total lack of soil and vegetation (ie. giving rise to minimal surface water retention) results in ephemeral flash flood runoff to the adjacent alluvial plains. Potential evapotranspiration exceeds the mean annual rainfall by a factor varying from 12 to over 20.

The entire mountain arc has been subjected to vigorous mid-Tertiary uplift resulting in limestone massifs of up to 3000 m, surrounded by a more extensive foothill region. Most of these

foothills are composed of allocthonous oceanic sediments and
ophiolites which range in altitude from the sea level to no more
than 1500 m despite their higher structural position relative to
the limestone massifs. The mantle sequence, however, is more
susceptible to both chemical and physical degradation than the
associated basic and sedimentary lithologies, and thus seldom
exceeds 1000 m in altitude.

The Geological Setting of the Semail Ophiolite Nappe

30,000 km^2 of Northern Oman and the United Arab Emirates
consists of the Semail Ophiolite Nappe (Fig 1). This ophiolite
has undergone minimal deformation relative to most other "alpine"
ophiolites scattered throughout the Tethian orogenic belt. The
ophiolite geomorphology is distinctive and characteristic of most
of the foothills areas, although the lower Semail nappe (ultramafic "mantle" rocks) exhibits sharp-peaked sierra with dark -
weathering friable surfaces, whereas the more rounded-relief
hills are characteristic of the upper Semail nappe (mainly basic
"crustal" rocks). The terrain is dissected by innumerable structurally controlled secondary wadis draining into major fault -
guided or "superimposed" primary wadi systems. Drainage density
in the mantle areas is substantially greater than in the basic
areas. Steep hill slopes, typical of the arid regime, are
accentuated in some areas by faulting, resulting in incised
precipitous wadi courses. Flat ground within the ophiolite areas
is virtually absent with the exception of occasional remnants of
cemented terraces.

A marked contrast in geological and hydrogeological properties exists between the crustal and mantle sequences. The
former suite is highly variable both texturally and petrologically,
and is dominated by gabbros and basalts (2,6). The structural
relationship between the various lithologies are complex but the
relatively undeformed crystalline units of the crustal sequence
may nevertheless be considered as uniformly impervious. The
great majority of the mantle sequence is partly serpentinised
harzburgite with minor dunite and lherzolite, and has a primary
silicate composition of olivine (Fo>77%) + orthopyroxene
(enstatite) + clinopyroxene (diopside), on average 80, 19 and 1%
respectively. Aluminous magnesiochromite, typically averaging
about 1%, is the only siginificant accessory mineral, but is inert
with respect to the various processes discussed. The silicate
minerals show no consistent chemical variation with depth (1) and
since the scale of petrographic variation is trivial compared with
the extent of the nappe and the length of groundwater pathways,
the ultramafic rock composition may be regarded as homogeneous.
The mantle sequence exhibits intense fracturing and structural
incompetence which has permitted widespread groundwater flow.
Consequently, unlike the crustal sequence, it is appropriate to
consider the mantle sequence as an aquifuge in which fracture
flow of low transmissivity is ubiquitous, at least in the near

Figure 1. The distribution of ophiolites and hyperalkaline springs in N. Oman

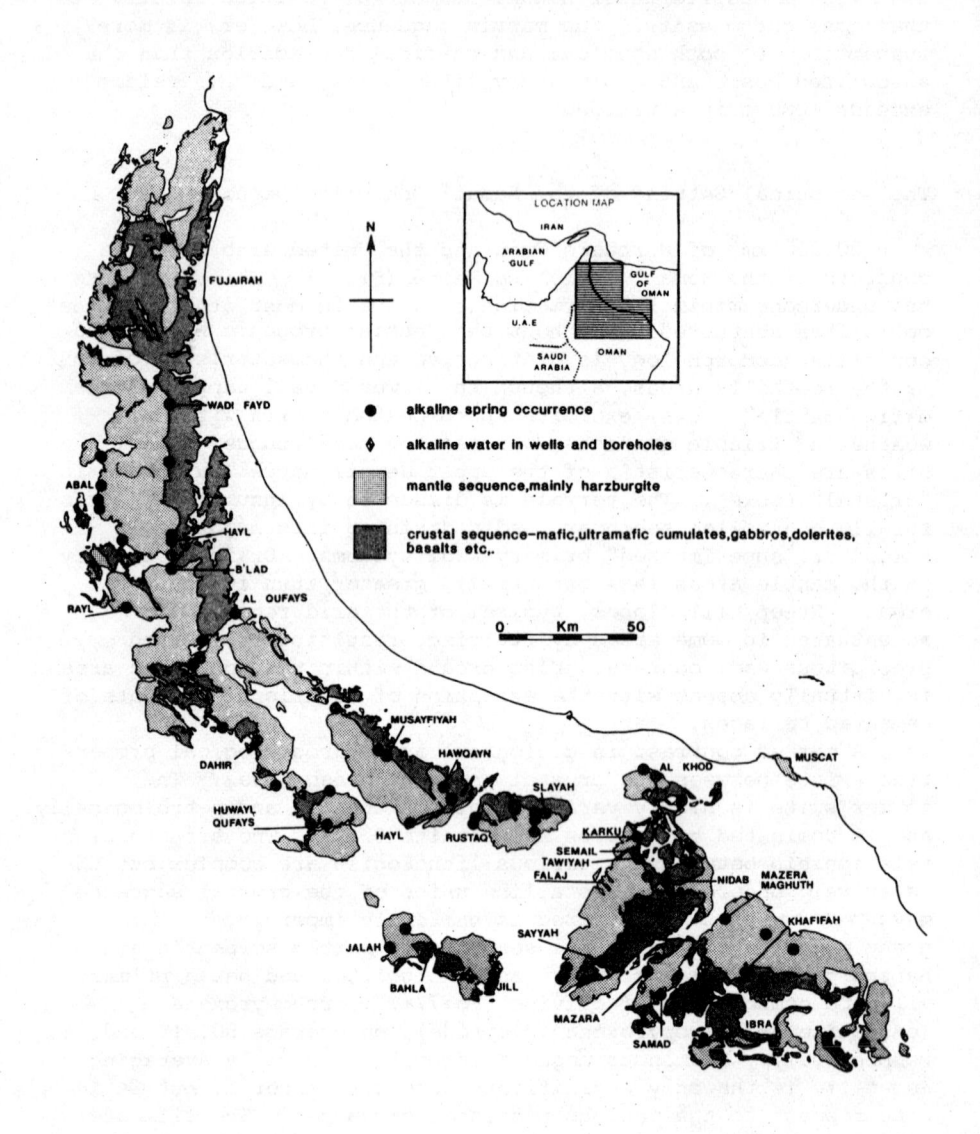

surface environment (ie. up to several hundred metres depth).

ANALYSIS

In the field, water samples were filtered using 0.45 μm membrane filters. HCO_3^-, CO_3^{2-}, OH^- (by acidimetric titration) and pH determinations were made (15) at the time of collection using an Orion pH meter and electrode. Subsequently, samples were stored and analysed by standard colorimetric and spectrographic methods (7) whilst iodine species were determined spectrophotometrically for total iodine (8) and iodate (9). Stable H and O isotope analyses were determined, on samples stored in glass stoppered glass bottles, by isotope ratio mass spectrometry using the methods of Coleman et al. (10) and Epstein and Mayeda (11). Ion activities and saturation indices were determined according to the method of Plummer et al. (57) using the thermodynamic data collated by Truesdell and Jones (58), and Helgeson (61).

THE HYDROGEOLOGY OF THE OPHIOLITE SURFACE AND GROUNDWATER

Crustal Sequence

Boreholes and hand dug wells have seldom been constructed in the crustal rocks owing to the difficulty of construction and the low permeability. In general these have been unsuccessful, although a few low yielding wells have been dug in weathered zone source areas, ie. forming a near surface aquifer of highly fractured breccia or locally derived gravel. Although not discussed in detail in this paper such groundwaters have a large hydrochemical range and most can be considered to be of a Na^+ Cl^- - Ca^{2+}/Mg^{2+} - $2HCO_3^-$ type (Table 1). Their compositions are determined by such factors as residence time, near surface evaporation and equilibrium with aeolian carbonate, rather than by the highly variable intrinsic rock chemistry. In general the Na^+, Cl^- and SO_4^{2-} concentrations are correlated indicating high evaporative concentration of rain water rather than rock dissolution (53).

Mantle Sequence

The hydrochemistry of the surface and near surface waters is relatively constant, being of a Mg^{2+} - $2HCO_3^-$ type characteristic of many ultramafic environments (12). Mg^{2+} - $2HCO_3^-$ type waters are generally present as shallow groundwater in ultramafic wadi gravels and are common as ephemeral and occasionally perennial surface flow, especially at rock bars, shallow bedrock and over "calcrete", where the water table banks up behind hydraulic barriers (this leads to frequent alterations between surface and groundwater flow).
Springs and groundwaters in the mantle sequence are of low flows, varying from minute seepages up to about 10 $l.s.^{-1}$ and are predominantly hyperalkaline. They are of a Na^+ - Cl^-/Ca^{2+} - $2OH^-$

type, as detailed later in this paper, and are highly reactive upon emergence in the surface near surface environment. Upon mixing of hyperalkaline and surface waters in both environments, $CaCO_3$ precipitates (4) by the reaction

$$Ca^{2+} + 2OH^- + 2HCO_3^- \rightarrow CaCO_3 + 2H_2O + CO_3^{2-}$$

As the hyperalkaline water emerges by upward seepage from the mantle rocks, the water saturated gravels become cemented by $CaCO_3$ precipitation, thereby forcing the main flow of alluvial groundwater to the surface. Subaerial emergences of alkaline water from fractures in the mantle rocks also facilitates direct reaction with atmospheric CO_2;

$$Ca^{2+} + 2OH^- + CO_2 \rightarrow CaCO_3 + H_2O$$

Hence alkaline spring occurrences is closely associated both with travertine and calcrete/conglomerate terraces, and wadi beds. The former arise from the "self sealing" tendency of individual spring pathways, resulting in spring migration, often to successively higher outflow points. In consequence, flat sedimentary terraces frequently occur where springs emerge near wadi channels, thereby creating the conditions necessary for village settlement.

Table 1. Groundwater compositions from the crustal sequence (concentrations in $mg\ell^{-1}$) (53)

No.	pH	E.C.	Ca^{2+}	Mg^{2+}	Na^+	K^+	Sr^{2+}	CO_3^{2-}	HCO_3^-	SO_4^{2-}	Cl^-	NO_3^-
127	7.8	1665	83	59	347	5.4	1.6	0	290	363	310	15.1
126	8.4	545	36	25	110	4.1	0.3	12	218	167	50	17.7
190	8.0	545	43	35	31	2.6	0.4	0	268	79	47	4.4
194	7.5	1410	134	57	184	2.6	1.6	0	368	374	193	8.4
195	8.4	460	19	42	16	2.4	0.2	5	171	95	63	0.4
210*	8.9	5000	626	.2	669	3.7	2.0	10	16	527	1620	2.7
239	7.4	3040	128	157	425	6.8	2.1	0	408	526	710	19.0
248	8.3	629	31	40	73	2.3	0.5	0	283	77	40	16.6
264	8.9	2320	63	16	105	3.5	0.8	3	183	622	68	9.9
272	8.1	4410	284	173	720	1.3	4.2	0	254	1594	810	23
273	8.7	2860	109	51	580	0.9	1.4	0	356	815	430	20
274	8.9	1710	73	36	285	4.1	1.2	0	271	279	360	19
330	8.5	963	32	65	78	1.3	0.6	0	346	68	106	3.8
338	8.2	1905	52	28	353	1.6	0.9	0	335	247	370	12.5
339	8.8	826	8	5	178	1.3	0.1	5	307	65	54	12.2
406	8.0	3570	243	115	500	1.9	2.7	0	216	814	360	9.5

(I - locations and site details given in ref 53)
* Water from underlying ultramafic rocks

PAST AND PRESENT SERPENTINISATION OF ULTRAMAFIC ROCKS. 255

Although the low relief intercepts of mantle sequences groundwater pathways coincide with surface wadi flows, there are many examples of alkaline spring emergence both up to 50 m above the associated wadi waters, and isolated from wadi water influences (ie. they are independent of local wadi water recharge). Indeed, the occurrence of hyperalkaline groundwaters in all wells and boreholes within the massive ultramafic rock indicates their almost universal presence in the ultramafic saturated zone. However, the emergence of such groundwater is closely associated with major structural features of the mantle sequence ie. the basal thrust zone of the ophiolite and the crustal - mantle sequence boundaries (Fig 2). The basal sheared zone is the most prominently fractured feature of the ophiolite nappe, being typically composed of up to 800 m of complex, locally imbricated

Figure 2.   Schematic cross section through the ophiolite sequence.

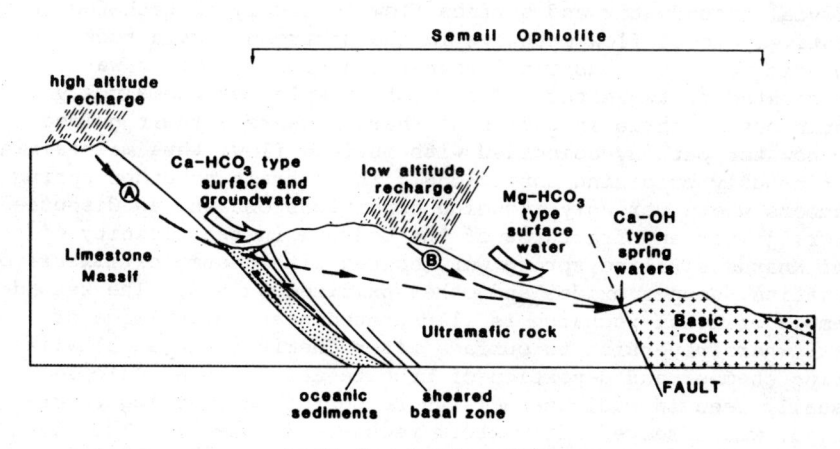

or brecciated serpentine (6,13). Consequently the potential for groundwater seepage through such an intensely sheared unit is great on both macro and meco scales. Near the top of the ultramafic sequence there is a transition from harzburgite to more olivine and clinopyroxene rich rocks (ie. dunite and lherzolite) which in turn are succeeded by olivine depletion and plagioclase enrichment (gabbro). Thus a marked contrast in structural competence arise between the gabbro (and overlying crustal rocks) and the underlying ultramafic suite which, during tectonism, has resulted in increased shearing and local dislocation. These features have been sufficient to produce a major groundwater pathway within the nappe. Apart from these two settings, most of the remaining alkaline springs are closely associated with further complications of the lower nappe macro-

structure such as local convergence of the basal thrust plane
and the top of the mantle sequence, as well as major faulting.
The relationship between the spring occurrences and geological
structure is particularly marked where mantle rocks terminate
at marginal faults against alluvium of lower relief forming
linear travertine terraces up to several kilometers length.

Although the scale of groundwater movement in the mantle
sequence is not amenable to direct observation, a purely local
spring water origin can be discounted. For example, the
differential head between some springs and adjacent wadi waters,
the lack of any significant catchment area in the spring vicinity,
and the slow rate of recession of many springs observed over
3 years, can only be explained by a long groundwater pathway and
consequently a large storage capacity. This is supported by
seven of nine tritium measurements of alkaline waters, which are
sufficiently low to confirm their origin as pre-nuclear bomb
test recharge (ie. base flow residence times in excess of 25
years); the two exceptions were both from sites where large
alluvial groundwater and surface flow is likely to recharge short,
relatively local flow pathways in the upstream mantle rocks.
Sporadic to continuous local recharge by surface water was
illustrated at two sites. The first example was anomalously
continuous recharge in Wadi Bani Kharus, where a fault guided
groundwater pathway coincided with surface flow, thus maintaining
continuously oxidising conditions. In contrast to other spring
chambers where strongly reducing conditions obtain, as discussed
later, joints and fractures of the bedrock in the vicinity of the
Bani Kharus alkaline spring were coated with a complex mixture of
haematite, copper oxides and other oxidised phases. The second
example of local recharge is illustrated near the village of
Karku where reversion to surface water chemistry in an alkaline
spring chamber and deposition of brucite and granular calcite
(usually seen in wadi channels where surface and spring waters
mix, 4) was observed after storm recharge (Table 2). The
importance of this hydrochemical variation is discussed later.

Regarding the depth of ultramafic groundwater circulation,
the only indication is that of spring temperature which varies
from 21° to $41^{\circ}C$ (compared with normal groundwater temperature
in Oman, at equivalent altitude, of $30 - 33^{\circ}C$). The lower spring
temperatures correspond to low discharge springs and seepages
(0 to 1 $l.s.^{-1}$) where evaporative cooling is intense. Otherwise
comparison between alkaline spring waters and surface stream
waters, where they occur in close proximity, shows the former to
be warmer by an average $4.3^{\circ}C$. In some cases this differential
water temperature is also due to evaporative cooling, but many
springs have temperatures exceeding 36° which, together with
differential water temperatures of up to $15^{\circ}C$ clearly indicates
geothermal warming. Assuming a typical geothermal gradient for
the mantle area of $20^{\circ}C$ km^{-1} and isotropic thermal conditions,
these results suggest that hyperalkaline groundwater circulation
varies from near surface to ~ 0.7 km depth.

Table 2. Karku alkaline spring : variations with time (concentrations in mgℓ^{-1})

No	T°C	Date	pH	Ca^{2+}	Mg^{2+}	Na$^+$	K$^+$	Sr^{2+}	OH$^-$	CO$_3^{2-}$	HCO$_3^-$	SO$_4^{2-}$	NO$_3^-$	Cl$^-$	SiO$_2$
411	*	7.01.75	11.4	42	.01	151	7.0	-	36	12	0	141	3.0	120	3.0
	32	18.03.78	11.5	75	.29	264	10.5	-	-	0	0	5	3.0	242	3.0
68	34	4.09.79	10.4	33	.05	333	17.8	.14	0	80	19		0.8	298	122
133	30	17.11.79	9.9	5	12	416	10.8	.04	0	5	118	142	1.5	346	0.4
203	32	17.02.80	11.8	45	.09	310	16.6	.29	102	0	0	30	6.2	370	27.4
233	31	10.04.80	11.6	67	.08	260	12.0	.25	59	0	0	2	8.8	380	1.0
271	36	6.06.80	11.4	48	.27	273	12.1	.32	56	0	0	16	0.6	352	1.2
376	36	7.09.80	11.5	58	.42	254	12.9	.30	81	0	0	15	3.7	317	0.7
394	31	18.10.80	11.0	4	.04	307	12.5	.13	29	0	0	16	0.4	330	0.7
454	34	29.11.80	11.6	54	.07	407	12.7	.28	94	0	0	47	5.5	341	0.8

* Data from reference 3.

THE CHEMICAL EVOLUTION OF SURFACE AND GROUNDWATERS FROM THE MANTLE SEQUENCE

The mantle sequence waters are of remarkably consistent chemistry. The surface waters are all of a Na$^+$ - Cl$^-$/Mg^{2+} - 2HCO$_3^-$ type, with the exception of waters containing a significant groundwater component (Table 3,4). Correspondingly the spring and groundwaters are of a Na$^+$ - Cl$^-$/Ca^{2+} - 2OH$^-$ type of relatively uniform composition, independent of catchment area, areal distribution and position in the nappe sequence (Tables 5,6,7). This is despite the widespread distribution of the mantle sequence waters, the input of Ca^{2+} - 2HCO$_3^-$ type waters from adjacent catchments, the varied near surface water residence times and complex arid zone recharge and evaporative processes.

The generation of the Na$^+$ - Cl$^-$/Mg^{2+} - 2HCO$_3^-$ waters is related to both arid zone evaporative processes and surface/near surface weathering mechanisms in a system open to the atmosphere. The waters might initially be considered as Na$^+$ - Cl$^-$ rich since the rain water of the foothills contains on average 6ppm Cl$^-$ (volume weighted average of 75 determinations). Such Cl enrichment is equivalent to an average evaporation loss of 92%. This high figure is attributable to the low rainfall over the mantle sequence, the high surface temperatures and high potential evaporation rates of up to 19 mm day^{-1}. This is further reflected in the H and O stable isotope data for the surface waters, where evaporation produces higher δD and δ^{18}O values (Fig 3) and a difference in mean isotopic composition between rainfall and surface waters of δD + 10°/oo., δ^{18}O + 3°/oo. The high total I/Cl$^-$ ratio of the surface waters, as compared with sea water, reflects atmospheric cycling of iodine (14) rather than iodine release from the rock.

The high concentrations of Mg^{2+} and HCO$_3^-$ result from the weathering of Mg bearing phases in the surface and near surface zones. Precise classification of the dominant weathering processes involved is difficult owing to the diversity of possible mechanisms (eg. olivine, pyroxene, magnesite, dolomite, brucite decomposition) and the absence of substantial secondary magnesian phases other than serpentine. However, for the surface waters, the high molar ratios of Mg^{2+}/H_4SiO_4 (9.2) and Ca^{2+}/H_4SiO_4 (2.3) differ markedly from the primary mineral assemblage (Mg^{2+}/SiO_2 = 1.63, Ca^{2+}/SiO_2 = 0.007) and suggests that primary mineral dissolution alone is not a significant hydrochemical control.

Olivine breakdown, for example, gives;

$$Mg_{1.84} Fe_{0.16} SiO_4 + 3.68 CO_2 + 4H_2O \rightarrow$$
$$1.84\ Mg^{2+} + 3.68\ HCO_3^- + H_4SiO_4 + 0.16\ Fe(OH)_2$$

where the $[Mg^{2+}]/[H_4SiO_4]$ ratio is 1.84.
For orthopyroxene breakdown;

$$Mg_{0.892} Ca_{0.018} Fe_{0.09} SiO_3 + 1.82 CO_2 + 3 H_2O \rightarrow 0.892\ Mg^{2+}$$
$$0.018\ Ca^{2+} + 1.82\ HCO_3^- + H_4SiO_4 + 0.09\ Fe(OH)_2$$

where the $[Mg^{2+}]/H_4SiO_4]$ amd $[Ca]/[H_4SiO_4]$ ratios are 0.892 and 0.018 respectively.

This apparent discrepancy is partly obviated by considering serpentine formation (oversaturated) in the surface water (Table 8) as the principal reaction product in the mantle sequence. The alternative reactions are;
For olivine breakdown

$$Mg_{1.84} Fe_{0.16} SiO_4 + 1.7 H_2O + 0.92 CO_2 \rightarrow$$
$$0.46\ Mg_3Si_2O_5(OH)_4 + 0.46\ Mg^{2+} + 0.92\ HCO_3^- + 0.16\ Fe(OH)_2$$
$$+ 0.08\ H_4SiO_4$$

where the $[Mg]/[H_4SiO_4]$ ratio is 5.75.
For orthopyroxene breakdown;

$$Mg_{0.892} Ca_{0.018} Fe_{0.09} SiO_3 + 0.036 CO_2 + 1.51 H_2O \rightarrow 0.297\ Mg_3Si_2O_5(OH)_4 + 0.406\ H_4SiO_4 + 0.018\ Ca^{2+} + 0.036\ HCO_3^- +$$
$$0.09\ Fe(OH)_2$$

where the [Mg]/[H$_4$SiO$_4$] and [Ca]/[H$_4$SiO$_4$] ratios are 0 and 0.04 respectively.

Table 3. Examples of mantle sequence wadi water compositions (53) (concentrations in Mg^{-1})

(a) Wadis with upper catchments in Limestone.

No	pH	EC	Ca^{2+}	Mg^{2+}	Na$^+$	K$^+$	Sr^{2+}	CO$_3^{2-}$	HCO$_3^-$	SO$_4^{2-}$	NO$_3^-$	Cl$^-$	SiO$_2$
22	8.7	445	32	40	50	4	.4	12.0	188	145	7	42	
91	8.3	522	24	36	47	5	.2	10	176	46	6	80	
137	8.4	485	29	46	38	4	.2	4	220	59	9	45	19
291	9.0	392	16	36	21	2	.3	0	183	23	5	32	28
314	9.2	409	27	35	15	2	.3	0	171	28	13	27	20

(b) Wadis with catchments wholly in ultramafic areas

No	pH	EC	Ca^{2+}	Mg^{2+}	Na$^+$	K$^+$	Sr^{2+}	CO$_3^{2-}$	HCO$_3^-$	SO$_4^{2-}$	NO$_3^-$	Cl$^-$	SiO$_2$
37	10.4	306	7	37	11	1	.2	16	98	30	10	11	16
38	8.1	336	34	40	17	2	.2	0	150	25	13	30	18
41	8.5	450	4	59	15	5	.4	0	88	71	7	45	4
75	9.6	605	4	17	110	6	.0	26	21	16	1	181	12
92	8.6	747	24	71	83	5	.2	22	231	84	1	120	18
106	8.3	319	18	34	15	2	.2	8	139	11	3	25	9
130	8.5	385	9	40	93	2	.2	10	134	39	5	50	17
246	8.3	415	9	51	16	1	.2	7	193	32	6	45	
278	9.8	471	12	50	17	2	.3	14	167	30	10	33	22

(I - locations and site details given in reference 53)

Table 4. Statistical data on the chemical compositions of surface mantle sequence waters (53)

Parameter* (conc. mgl^{-1})	Mean	Std dev	Skew	Kurt	No of samples	Significant correlations $p \leq 0.001$
Temp (°C)	29.7	3.88	-1.09	0.41	55	Ca
Elec Cond	579.3	246.8	1.23	3.02	70	Ca,Mg,Na,K,Sr,HCO$_3$,SO$_4$,Cl
pH	8.46	1.17	-5.47	40.2	70	
Ca^{2+}	18.8	14.4	1.94	75.6	70	T,F,Cl,SO$_4$,HCO$_3$,Sr
Mg^{2+}	45.5	19.4	0.51	0.99	70	E.C.,Sr,HCO$_3$,SO$_4$
Na$^+$	44.5	38.1	1.53	2.18	70	E.C.,Ca,K,SO$_4$,Cl,F,TI
K$^+$	2.86	1.65	1.18	1.05	70	E.C.,Na,CO$_3$,Cl
Sr^{2+}	0.34	0.25	1.52	2.68	70	E.C.,Ca,Mg,HCO$_3$,SO$_4$,NO$_3$,F
SiO$_2$	12.2	12.0	0.70	-0.46	70	HCO$_3$
CO$_3^{2-}$	7.49	14.0	4.10	23.15	70	E.C.,Ca,Mg,Na,Sr,HCO$_3$
HCO$_3^-$	186.6	88.7	0.47	0.41	70	E.C.,Ca,Mg,Sr,SiO$_2$,SO$_4$,F
SO$_4^{2-}$	49.8	29.0	1.62	4.57	70	E.C.,Ca,Mg,Na,Sr,HCO$_3$
NO$_3^-$	6.8	4.2	1.07	1.27	70	Sr,F
Cl$^-$	76.6	57.3	1.62	3.41	70	E.C.,Ca,Na,K,TI
F$^-$	0.06	0.02	4.91	34.0	70	Ca,Na,Sr,HCO$_3$,NO$_3$
IO$_3^-$	0.016	0.009	0.59	0.51	32	
Total I	0.021	0.010	0.72	0.19	63	Na,Cl
NO$_2^-$	0.03					

Table 5. Hyperalkaline spring compositions·
(concentrations in mgℓ^{-1})

I	pH	Ca^{2+}	Mg^{2+}	Na^+	K^+	OH^-	SO_4^{2-}	NO_3^-	Cl^-	SiO_2
62	11.6	61	0.026	230	12.3	77	8	0.4	317	0.6
63	11.7	54	0.021	250	13.0	76	0	0.2	340	0.4
64	11.6	38	0.352	251	15.6	59	8	0.4	349	1.0
65	11.8	23	0.008	331	15.6	85	45	2.2	411	7.4
66	11.6	23	0.010	260	14.4	78	30	0.0	394	13.0
74	11.0	50	2.200	115	6.5	31	23	0.7	160	1.5
116	11.4	79	0.100	174	7.6	57	21	0.9	255	1.3
205	11.8	52	0.050	317	11.8	105	11	2.9	415	8.2
206	11.9	55	0.023	273	13.3	115	4	5.6	385	7.3
207	11.9	91	0.033	287	13.4	104	32	4.5	380	5.2
208	11.9	89	0.054	245	13.8	104	13	5.1	365	8.2
286	11.4	59	2.600	254	10.7	76	7	2.8	224	1.2
287	11.5	62	0.051	184	8.7	42	6	0.5	256	1.4
288	11.4	55	0.037	191	8.8	68	6	0.7	280	10.5
369	11.2	71	0.290	133	7.4	45	7	2.3	212	0.4
370	11.2	76	0.015	136	6.8	53	12	1.9	221	0.6
422	11.8	66	0.026	66	6.0	39	17	6.0	178	1.0

(I - locations and site details are given in reference 53)

Table 6. Statistical data on the chemical compositions
of hyperalkaline groundwaters (53)
(concentrations in mgℓ^{-1})

Parameter* (conc. mgl^{-1})	Mean	Std dev	Skew	Kurt	No of samples	Significant correlations $p \leq 0.0001$
Temp (°C)	32.4	4.1	-0.79	0.74	57	pH (-vecorr)
Elec.Cond	1580.6	569.1	0.59	0.93	64	pH,Na,K,OH,SO$_4$,Cl,TI
pH	11.4	0.3	-1.14	2.50	64	E.C.,OH, T(-vecorr)
Ca^{2+}	60.8	25.0	0.24	0.37	64	Sr
Mg^{2+}	0.28	0.59	2.95	7.64	64	
Na$^+$	226.6	108.6	1.32	2.46	64	E.C.,SiO$_2$,OH,SO$_4$,Cl,TI,NO$_2$,K
K$^+$	10.6	5.0	1.66	3.77	63	E.C.,Na,SiO$_2$,Cl
Sr^{2+}	0.31	0.20	1.52	2.12	64	Ca, SO$_4$
SiO$_2$	4.92	7.86	2.40	5.71	64	Na,K
OH$^-$	61.6	23.8	-0.10	-0.10	64	E.C.,pH,Na
SO$_4^{2-}$	23.3	26.9	2.81	9.11	63	E.C.,Na,Sr,Cl
NO$_3^-$	2.25	1.98	1.08	0.54	64	
Cl$^-$	306.4	138.5	1.70	3.55	64	E.C.,Na,K,SO$_4$,TI,NO$_2$
F$^-$	0.051	0.005	5.15	26.09	64	
IO$_3^-$	0.003	0.004	2.98	8.87	62	
TI	0.052	0.025	0.79	0.53	64	E.C.,Na,Cl
NO$_2^-$	0.25	0.36	2.70	9.32	62	Cl

If these reactions are important in controlling the chemistry of the surface waters either an additional source of Ca^{2+} or an additional sink for SiO_2 is required to account for the Ca/SiO_2 anomaly. Potential sources of Ca^{2+} are supplied by both aeolian dust from the limestone mountain area and veins in the mantle sequence. Alternatively, a SiO_2 sink would be facilitated by an additional supply of aqueous Mg^{2+} and OH^- from forsterite decomposition, thus leading to further serpentine precipitation. Precipitation of silica itself, though common in wetter ultramafic environments in the form of silica-geothite weathering residuals (54,55), is of minor occurrence in Oman and is restricted, both in time and space, to localised serpentine dissolution under acidic groundwater conditions (55).

If however Mg^{2+} - $2HCO_3^-$ waters are independent of serpentinisation processes, they must be generated by the dissolution of near surface brucite formed during earlier serpentinsation reactions (see below), ie:

$$Mg(OH)_2 + 2CO_2 \rightarrow Mg^{2+} + 2HCO_3^-$$

This reaction is consistent with the low brucite contents of the partly serpentinised rock, the high solubility of the mineral in the surface waters, and would also account for both the generation of Mg^{2+} - $2HCO_3^-$ groundwaters from completely serpentinised ultramafic rock (3) and the formation of highly magnesian carbonate veins (discussed below).

Table 7. Average chemical concentrations ($mg\ell^{-1}$) in mantle sequence ground and surface waters.

	Surface Waters			Spring Waters		
	1	2	3	1	2	3
pH	8.8	9.1	8.4	11.6	11.2	11.4
Na^+	87.1	109.9	44.5	257.6	227.2	226.6
K^+	4.3	4.4	2.86	10.8	8.6	10.6
Mg^{2+}	31.7	37.3	45.5	0.15	0.47	0.28
Ca^{2+}	24.8	14.5	18.8	61.0	57.6	60.8
Cl^-	140.4	178.3	76.6	361.6	317.3	306.4
SO_4^{2-}	30.2	65.8	49.8	21.3	16.8	23.3
CO_3^{2-}	8.8	26.2	7.5	-	0.0	0.0
HCO_3^-	148.8	144.6	186.6	-	0.0	0.0
OH^-	0.0	1.9	0.0	54.7	67.8	61.6
SiO_2	16.4	23.6	12.2	3.5	1.0	4.9
No of samples	13	19	70	13	50	64

Sample collection 1. 1972-1976 (3), 2. 1978, 3. 1980-1981 (53)

Figure 3. Stable hydrogen and oxygen variations in rainfall, surface and spring waters, of the mantle sequence*

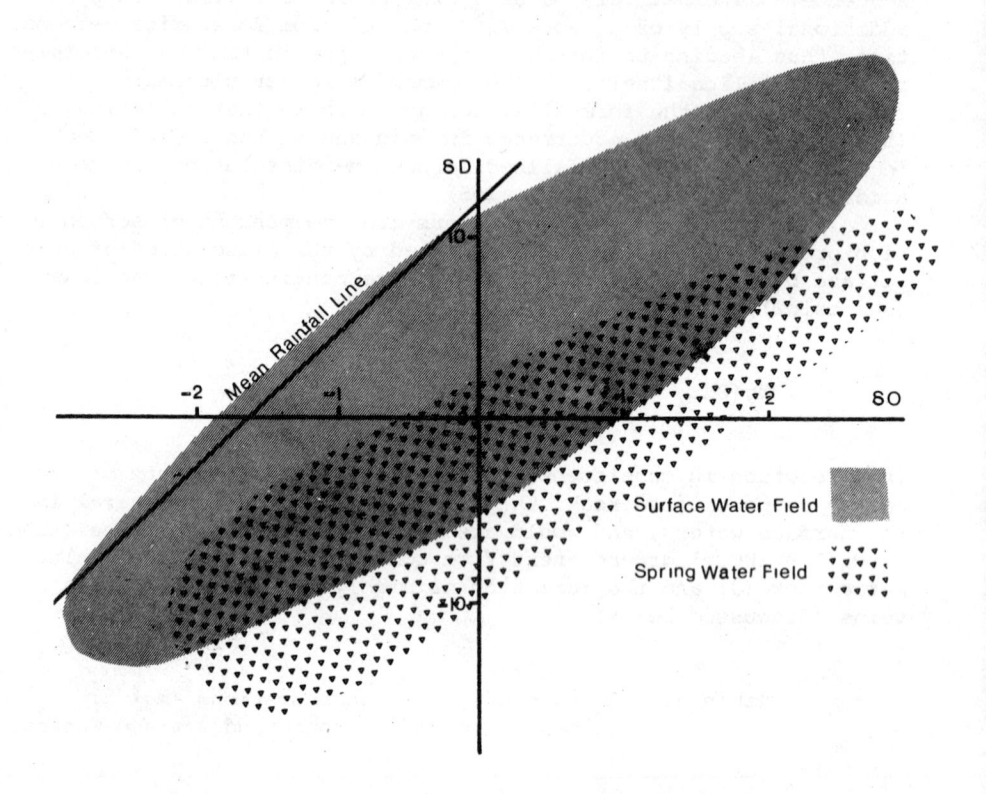

(*data, a compilation of results from 3,53,59 and unpublished data collected at the Institute of Hydrology; surface and spring fields both contain approximately 70 data points)

The hydrology of the mantle sequence shows that the hyperalkaline groundwaters evolve from the surface $Na^+ - Cl^-/Mg^{2+} - 2HCO_3^-$ waters. In so doing, the surface waters become highly oversaturated with respect to serpentine, and slightly undersaturated to saturated with respect to the primary minerals, olivine and orthopyroxene (Fig 4, Table 9). This conforms to a low temperature serpentinisation model whereby primary rock components are hydrated to generate the secondary minerals serpentine ± brucite (15-20). As silicate dissolution is initiated, the release of Mg^{2+}, Ca^{2+} and OH^- to solution results in rising pH, and consequently the conversion of HCO_3^- to CO_3^{2-}. With further OH^- release the CO_3^{2-} itself precipitates in the form of aragonite, calcite or dolomite, thus leaving only the

Table 8. Saturation state of mantle sequence surface waters; (log IAP/KT)

	Primary Phases			Alteration Phases			Carbonate Phases		
	Fo	En	Di	Chr	Br	Ar	Ca	Do	Mg
Well in M.S.	-5.48	-1.33	-0.60	2.81	-2.55	-0.60	-0.34	-0.77	0.81
Musayfiyah	-3.90	-0.27	2.05	5.56	-1.99	0.29	0.55	2.11	1.27
Nidab	-7.92	-2.09	0.20	-0.28	-4.18	0.49	0.75	0.71	-0.33
Hawqayn	-5.61	-1.11	1.01	3.2	-2.76	0.49	0.75	2.07	1.03
Karku	-4.19	-0.22	2.26	5.01	-2.45	0.66	0.91	2.48	1.27
Ibra	-5.22	-0.77	1.68	3.83	-2.77	0.70	0.96	2.37	1.12
Semail	-6.48	-1.21	0.37	1.78	-3.66	0.67	0.93	2.38	1.15
Slayah	-4.17	-0.39	2.05	5.28	-2.10	0.66	0.92	2.70	1.49
Falaj/Buri	-5.44	-1.00	0.47	3.08	-2.88	-0.10	0.16	1.27	0.82

Minerals: Fo = forsterite, En = clinoenstate, Di = Diopside, Chr = chrsotile, Br = Brucite, Ar = aragonite, Ca = calcite, Do = Dolomite, Mg = Magnesite.

Table 9. Saturation states of hyperalkaline groundwaters (log IAP/KT) with respect to primary and secondary mantle sequence phases.

Spring location	Primary Silicates			Alteration Products	
	Forsterite	Clinoestatite	Diopside	Chrysotile	Brucite
Semail	-0.01	2.22	7.05	10.02	0.42
Musayfiyah	-0.23	0.67	6.44	10.18	0.75
Jill	-2.84	1.62	3.21	5.16	0.38
Jalah	-1.47	-0.01	6.53	8.45	0.27
Hawqayn	-1.58	-0.27	5.64	8.64	0.63
Sayyah	-2.70	-0.60	3.59	6.22	-0.55
Karku A	0.00	-0.46	6.73	10.18	1.24
Karku B	-1.95	-1.24	4.05	6.68	0.66
Slayah	-1.12	-0.58	4.65	8.54	1.44

hydrous products (serpentine ± brucite) and only Ca^{2+} and OH^- as the major ions in solution. The Ca^{2+} is normally derived from the orthopyroxene (enstatite) which contains on average 1.0% CaO (2). The only other calcium bearing mineral in the rock is clinopyroxene (diopside $CaMgSi_2O_6$) which is apparently thermodynamically stable in both ground and surface waters (Table 9). Several complex features of the low temperature serpentinisation reactions are important.

1) Although the clinopyroxene is stable in the surface and spring waters the early phase of serpentinisation involves loss of $CaCO_3$ from the surface water to give solutions with very low Ca activity. Hence these partly reacted waters

will become undersaturated with respect to clinopyroxene. Consequently during the early serpentinisation stage diopside decomposition can be invoked. This is supported by field observation since; a) clinopyroxene decomposes in actively serpentinising groundwaters, b) lherzolite layers are $CaCO_3$ coated at rock fractures in contact with near surface waters, and c) the fresh appearance of diopside relative to both olivine and estatite is consistent with the transient period during which diopside is susceptible to hydration (as compared to the relatively long periods of stable high pH conditions during which only olivine and estatite decompose).
2) Despite classical serpentinisation being described in terms of independent olivine and orthopyroxene decomposition the generation of $Ca^{2+} - 2\,OH^-$ waters requires the simultaneous breakdown of both phases since monomineralic hydration of these minerals would give rise to Mg^{2+} or H_4SiO_4 rich solutions; ie. at variance with the results shown in Table 7. For example with olivine hydration

$$Mg_{1.84}Fe_{0.16}SiO_4 + 1.5\,H_2O \rightarrow 0.50\,Mg_3Si_2O_5(OH)_4 + 0.34\,Mg(OH)_2 + 0.16\,Fe(OH)_2$$

the resulting solution being Ca and H_4SiO_4 free and saturated with respect to brucite.
Correspondingly with orthopyroxene hydration;

$$Mg_{0.892}Ca_{0.018}Fe_{0.09}SiO_3 + 1.51\,H_2O \rightarrow 0.289\,Mg_3Si_2O_5(OH)_4 + 0.018\,Ca^{2+} + 0.036\,OH^- + 0.09\,Fe(OH)_2 + 0.404\,H_4SiO_4$$

where the $[Ca]/[H_4SiO_4]$ ratio is 0.04. This compares with the average ratio in the spring waters of 19. Consequently in order to generate $Ca^{2+} - 2OH^-$ waters, brucite and serpentine by the low temperature model, the order in which the reactions proceed is:
a) Initial olivine and orthopyroxene dissolution giving rise to increasing pH, and hence to $CaCO_3$ precipitation and the dissolution of clinopyroxene.
b) Upon reaching high pH, further olivine and orthopyroxene co-dissolution gives serpentine and $Ca^{2+} - 2OH^-$ (in solution). Since the H_4SiO_4 and Mg^{2+} concentrations in solution are both very small the forsterite and enstatite solubilities are approximately equal and consistent;

ie. held in balance by the rate controlling process of hydrous phase precipitation. Although precipitation of both serpentine and brucite is theoretically possible, there is a gross saturation contrast with respect to the two minerals (Table 9) such that the precipitation kinetics probably limits the brucite to trivial amounts. Thus forsterite and enstatite dissolution is essentially in 1.8 : 1.0 stoichiometric proportion to give just serpentine.

$$1.80 Mg_{1.84} Fe_{0.16} SiO_4 + Mg_{0.892} Ca_{0.018} Fe_{0.09} SiO_3 + 3.17 H_2O \rightarrow 1.39 Mg_3Si_2O_5(OH)_4 + 0.375 Fe(OH)_2 + 0.018 Ca^{2+} + 0.036 OH^-$$

c) With the exhaustion of the available orthopyroxene, olivine hydration continues to generate serpentine together with brucite but with the previously formed Ca^{2+} and OH^- remaining stable in solution.

d) With the exhaustion of the available olivine the rock is completely serpentinised, the groundwater reverts to $Mg^{2+} - 2HCO_3^-$ type and the brucite redissolves or reacts to form magnesite.

3) The hyperalkaline groundwaters are unusually saline, being approximately 4 times higher in Na^+ and Cl^- than the surface waters (Table 7). While evaporation of the groundwaters at spring outflow points might at first seem plausible an 80% average evaporation relative to the surface water input is doubtful since a) the waters were sampled from "overtopping" spring channels, b) waters were commonly covered with a surface carbonate film restricting evaporation and c) lack of any soil cover minimises near surface water retention during mantle sequence recharge. Further, isotopic evidence supports this argument since the spring waters are isotopically lighter with respect to H and O than the corresponding surface waters (Fig 3), the reverse being expected if evaporation had occurred (3,21). Alternatively low temperature serpentinisation ("olivine hydration") might be proposed to explain loss of water from the system (given the high rock to water ratio). However this is inconsistent with the stable isotopic evidence (cf feature 6) since the resultant waters should be extremely D rich and ^{18}O depleted. Consequently most of the chloride (on average 75%) of the spring water must be leached from the aquifer. The Cl^-, of marine origin, could have been incorporated into the mantle sequence in several ways; 1) adsorbed onto positively charged crystal surfaces during early serpentinisation (25), 2) precipitated as minor secondary phases in the form of transition metal chlorides/hydroxy chlorides (22,26), and 3) isolated as free salts during fluid circulation. Such

salt enrichment of serpentinised peridotite relative to fresh peridotite (typically 80 to over 400 ppm Cl respectively) has been demonstrated in other ultramafic terrains (22-24). The spring and groundwater salt anomalies seen in Oman must therefore be a legacy of residual sea salts preserved by slow meteoric circulation with a high rock-water ratio.

4) During the serpentinisation process the input water (aquifer recharge by oxygenated surface waters) becomes progressively more reducing as ferrous iron is released from the olivine and pyroxene phases. During this transition the oxidised species in the surface waters (eg SO_4^{2-}, IO_3^-, NO_3^-) become reduced to varying degrees to from HS^-, I^-, NO_2^-. The most remarkable example of the highly reducing nature of the groundwater is the discharging of H_2 gas from the spring sites and the build up of H_2 gas pockets in the ultramafic rock (5). In contrast to the IO_3^- rich surface waters, the groundwaters are dominated by the I^- with a four-fold increase in total iodine due to iodine release from the fractured mantle sequence. For most springs and groundwaters the redox potential is low (-165 to -630mV) and SO_4^{2-} - NO_3^- concentrations are lower than in surface waters. The decrease in NO_3^- may result from N_2 gas loss and the generation of NH_4^+ (not determined here) and NO_2^-. Lower concentrations of SO_4^{2-} in the ground compared to the surface waters (Table 7) indicates a sulphur "sink" within the ultramafic rock but so far this has not been identified. The nature of this sink is probably sulphide in the from of finely disseminated or otherwise concealed pyrite since (a) HS^- and S^{2-} as the alternative species are insufficiently concentrated (<10mgl^{-1}) to account for the SO_4^{2-} depletion; H_2S degassing would not occur under strongly reducing groundwater conditions, (b) elemental S in not thermodynamically stable under the groundwater conditions encountered (27) and (c) iron (II) is generated during the serpentinisation process (eg. in the form of oxides and hydroxides). For both N and S species non equilibrium processes seem to be operative since NO_3^- and SO_4^{2-} concentrations would be expected to be very low under these groundwater conditions (60).

Although there is little available data on trace reduced phases in the Oman mantle sequence, detailed analysis in other alpine ultramafic rocks has demonstrated the widespread and possibly universal association between partly to completely serpentinised ultramafic rocks (never the fresh peridotites), and reduced species such as native metals, intermetallic compounds and sulphides low in sulphur (29-35). Such products and especially those of the less reactive metals (Ni, Co, Cu, Pb, Zn, Ag) would, if present as sulphides, be produced by reactions of the form:

$$NiFeS_2 + 2H_2 + 2OH^- \rightarrow NiFe + 2H_2O + 2HS^-$$

Many similar reactions could be proposed for the other common trace minerals such as native nickel, awaruite (Ni_3S_2), wairauite (CoFe), heazlewoodite (Ni_3S_2), millerite (NiS) etc. Mechanisms for the potential precipitation of sulphide from groundwater therefore arise in both recharge and discharging groundwater environments; 1) the reduction of SO_4^{2-} to HS^- as groundwater enters an environment of active serpentinisation 2) dissolved H_2 in groundwater flow from a serpentinising to non-serpentinising environment would precipitate any complexed sulphur species. The latter was proposed by

Figure 4. Rock-water phase diagram for the system MgO - SiO_2 - H_2O (saturation boundaries at 25°C/1bar total pressure and a_{H_2O} = 1.0 (20).

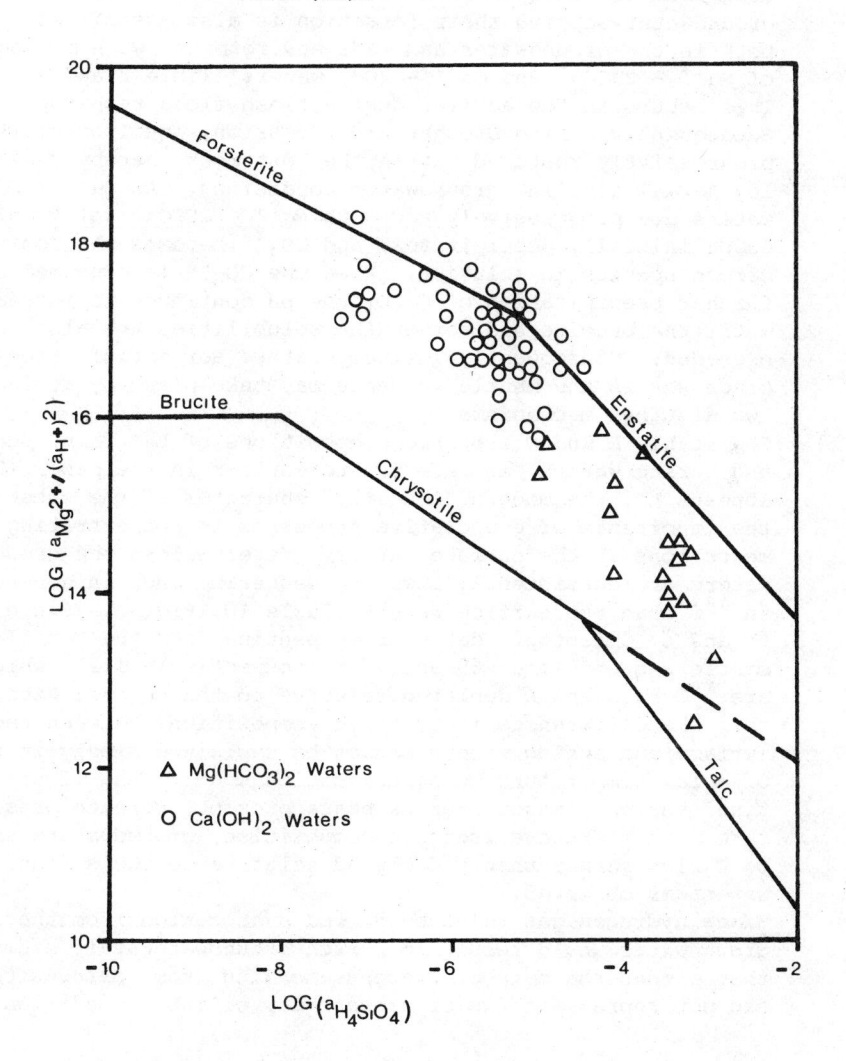

Coveney (36) to account for gold in Californian serpentine;

$$2AuS^- + H_2 \rightarrow 2Au^o + 2HS^-$$

in which at least 80% of the ore bodies were situated next to or no greater than 30 metres from the serpentine wall rocks. Thus, whilst recognising that most extensively "high level" sulphides are of undoubted hydrothermal origin (37), these low temperature mechanisms also seem to play a part in the redistribution of minor sulphides, mainly pyrite, at the basic/ultramafic fault bounded interface (38) and in wholly serpentinised mantle sequence rocks (39).

5) Although $CaCO_3$, $Mg(OH)_2$ and serpentine formation can be described as above in terms of reactions in a restricted groundwater setting their formation is also associated both in the groundwater and wadi environments with mixing of $Mg^{2+} - 2HCO_3^-$ and $Ca^{2+} - 2OH^-$ waters (Table 2 and ref 4). This occurs in the aquifer during flash flood recharge. Subsequently during drought and recession, equilibrium is progressively restored within the host rock thereby regaining normal alkaline groundwater conditions. As $Ca^{2+} - 2OH^-$ waters are progressively added to $Mg^{2+} - 2HCO_3^-$ solutions $CaCO_3$ initially precipitates, and CO_3^{2-} becomes the dominant carbon species in solution. When the CO_3^{2-} is consumed by further precipitation of $CaCO_3$ the pH continues to increase until the brucite and serpentine solubilities are also exceeded. Consequently low temperature serpentinisation processes in the mantle sequence may take place by at least two distinct mechanisms.

6) The stable H and O isotopic compositions of the rain, surface and spring waters, as referred to earlier in the paper (Fig 3) attests to; the modern "meteoric" character of the water and the importance of evaporative processes in concentrating the major ions in the surface waters. Nevertheless the groundwaters are consistently lower in deuterium and, in general, in ^{18}O than the surface waters (Table 10, Fig 3). Since the δD and $\delta^{18}O$ isotopic data for serpentine from the Semail mantle sequence are -65 and +7.84 respectively (12), which are ^{18}O rich and D depleted relative to the surface waters, then the differences in isotopic compositions between the surface and spring waters cannot be explained simply in terms of a low temperature serpentinisation model. Thus if serpentine, the only major hydrous phase present, were to precipitate with the above isotopic composition, groundwaters should be D rich rather than D depleted relative to the surface waters as observed.

Since hydrogen gas release (5) and evaporation from the groundwater would result in D rich groundwaters, it seems that either the meteoric waters emerging from spring sites are not representative of present day climatic conditions,

or hydrogen isotopic exchange takes place between groundwater and hydrous phases in the rock. However hydrogen exchange seems to be the most probable process since:
a) Surface waters and groundwaters from other lithologies throughout the region are D enriched (for a given $\delta^{18}O$ value) relative to hyperalkaline waters (3). b) Hydrogen isotopic exchange between natural waters and preformed serpentine and clay minerals has been observed (19, 40-42), and c) A high rock/water ratio is involved in mantle sequence reactions.

7) The Ca(OH)$_2$ waters are highly reactive not only in the surface zone, precipitating carbonate minerals, but also at contact zones with the overlying crustal sequence, producing basic reaction rodingite minerals. In Oman this is typified by the formation of the rare mineral suolunite (Ca$_2$Si$_2$O$_5$(OH)$_2$H$_2$O) (52).

Table 10. Stable hydrogen and oxygen isotopic relationships for mantle sequence surface and spring waters.

Location	Surface Water δD	Spring Water δD	Spring-Surface Water $\Delta\delta D$	Surface Water $\delta^{18}O$	Spring Water $\delta^{18}O$	Spring-Surface Water $\Delta\delta^{18}O$
Karku	- 3.8	- 6.9	- 3.1	-0.94	-1.17	-0.23
Semail	+ 0.5	- 1.3	- 1.8	-0.66	-0.31	+0.35
Abal	+ 0.8	- 5.8	- 6.6	-0.87	-1.57	-0.70
Slayah	+10.7	- 5.1	-15.8	+2.54	+0.43	+2.11
Hawqayn	+ 1.7	- 7.5	- 9.2	-0.12	-1.75	-1.63
Tawiyah	- 1.6	- 7.2	- 5.6	-0.54	-1.42	-0.88
Khafifah	- 1.6	- 3.5	- 1.9	-1.12	-0.87	+0.25
Nidab	- 7.9	-11.2	- 3.3	-1.90	-2.09	-0.19
Najd	+ 6.3	- 4.9	-11.2	+1.87	-0.92	-2.79
Mazera	- 2.9	-10.5	- 7.6	-1.11	-1.08	+0.03

MINERALOGICAL AND PETROGRAPHIC EVIDENCE OF HIGH AND LOW TEMPERATURE SERPENTINISATION

With the exceptions of wholly serpentinised fractures the field and petrographic evidence indicates approximately uniform and pervasive hydration unrelated to either the water table or proximity of the weathered zone. Such alteration of the anhydrous phases (olivine and pyroxene) is pre- or syn- tectonic, and is contemporaneous with nappe generation, detachment and transport in the marine environment (1,6). The temperature at which pervasive serpentinisation occurs is in the range 180 to 500°C (43-45) although greenschist metamorphism at the basal thrust of the Semail ophiolite nappe implies a maximum emplacement temperature of about 340°C.

The low temperature serpentine is identified as late stage (cross-cutting), unstrained, colourless and generally inclusion free veins. Large compound veins of serpentine and carbonate deposited in open veins are uncommon but diagnostic. One such example from the Dibba ultramafic area displayed a unique petrofabric of cream coloured translucent botryoidal serpentine with orthorhomic extinction crosses, within net veined dolomite. The earlier phases of serpentinisation are characterised by replacement textures with alteration in the order olivine > orthopyroxene > clinopyroxene. Since 80% of the mantle sequence rock is composed of olivine the predominant characteristic of the early hydration (ie. higher temperature) processes is serpentinisation of olivine.

The petrographic relationship between precipitated serpentine and brucite is one of successive rather than co-precipitation, and is probably influenced by the sensitivity of the two alternative reactions to intermittent fluctuations in pH, silica introduced from the bicarbonate type water, silica released by differing olivine-pyroxene ratios along the ultramafic groundwater pathway and above all, by the kinetics of precipitation. The description of a brucite- "protoserpentine" gel from a similar alkaline groundwater environment in California (46) could therefore be explained by the flushing of such varied reaction products by partially mixed but Ca^{2+} - $2OH^-$ dominated groundwater types.

The chemical composition of the two types of serpentine do provide some evidence for contrasting modes of formation (Table 11), the late stage (precipitation) serpentine being Fe, Al, Mn and Ni depleted and Mg enriched relative to the earlier (alteration) serpentine. The low Fe, Al, Mn and Ni concentrations in the precipitated serpentine reflect the low concentration of these ions in ground and surface waters. Higher concentrations of Mn and Fe in the alteration serpentine may reflect oxide/hydroxide hydration products.

X-ray diffraction proved less useful in discriminating between serpentine of high and low temperature origin. Only the rare orthorhombic form is readily differentiable due to its low crystallinity which gives rise to broadened peaks. Otherwise serpentines of varying habit (vein or matric) and texture appear to be complex mixtures of the various polymorphs described elsewhere (47-48). However apart from a tendency for low temperature serpentine to cyrstallise as chrysotile rather than lizardite, there is no clear correspondence between origin and diffraction pattern.

Magnesian Carbonates

Magnesite is by far the most common carbonate mineral in the mantle sequence and accounts for about 80% of all the carbonate veins deposited, the remainder being mostly dolomite. Typically it occurs in mm to metre thick veins, is invariably pure white (probably iron free), and is very fine grained, almost procellanitic, massive to concretionary in texture leaving little or no petrographic evidence of formation conditions. The problem of the origin of magnesite veins is obscured by the current lack of any reliable method for the isotopic analysis of magnesite (49) and by the apparent lack of present day precipitation. Although generally undeformed, many magnesite veins are slickensided suggesting a relatively early post-nappe emplacement genesis.

Evidence from other ultramafic areas supports an inverse correlation between brucite at depth and surface exposures of magnesite. Hostetler et al (50), for example, found brucite to be typically unstable along the sheared borders and fracture zones of ultramafic masses, and regarded it as the "prime candidate" as the precursor of magnesite. Indeed a significant feature of magnesite occurrences both in Oman and elsewhere (50,51) is its close association with dunite but complete absence from pyroxene rich rocks such as websterite, ie. magnesite occurrence is restricted to rocks which hydrate to produce serpentine and brucite. In less arid areas brucite alters to the minerals : hydromagnesite [$Mg_4(CO_3)(OH)_2.3H_2O$], artinite [$Mg_2(CO_3)(OH)_2.3H_2O$], pyroaurite [$Mg_6Fe_2(CO_2)(OH)_{16}.4H_2O$], and/or coalingite [$Mg_{10}Fe_2CO_3(OH)_{24}.2H_2O$], none of which have been found in the Oman surface environment. Magnesian huntites, varying from $Mg_3Ca(CO_3)_4$ to $Mg_{3.3}Ca_{0.7}(CO_3)_4$, were found however. Since surface weathering does not satisfactorily explain magnesite formation, the most plausible mechanism for its precipitation therefore seems to be following brucite solution within the groundwater environment by the reaction:

$$Mg(OH)_2 + Mg^{2+} + 2HCO_3^- \rightarrow 2MgCO_3 + 2H_2O$$

This reaction is promoted by the solubility of brucite in most HCO_3^- type groundwaters, and is consistent with the field evidence in several ways. Firstly, the requirement of moderate pH waters in the above reaction implies complete serpentinisation in the vicinity of the groundwater pathways (though not necessarily complete serpentinisation of the inter-fractured matrix) and consequently the cessation of brucite precipitation and highly alkaline conditions. Secondly, the greater proportion of primary dunite in the basal zone and "sea water" hydration during emplacement would have been conducive to the precipitation of brucite. Thirdly, reactivation of groundwater pathways in the basal zone by mid Tertiary uplift would have facilitated the conversion of pre-existing brucite to magnesite at any early age, thus contributing to the modern scarcity of brucite in the ultramafic rock, and accounting for slickenside formation in the magnesite veins. The scarcity of brucite is notable in view of the primary minerals which, both in order of abundance and degree of hydration, are olivine >> ortho-pyroxene >>, clino-pyroxene. The mantle sequence should therefore have hydrated to produce extensive brucite (serpentine/brucite ratio ≈ 1.5), which must since have been altered or removed.

Table 11. Chemical variation between the two serpentine facies

Samples 1-6 by X-Ray fluorescence, 7 by electron microprobe

Sample	Alteration Serpentine*			Precipitation Serpentine			
	1	2	3	4	5	6	7
SiO_2	38.50	36.80	39.35	40.92	41.96	42.49	48.29
Al_2O_3	0.77	3.39	1.05	0.0	0.01	0.0	0.28
Fe_2O_3	8.26	11.59	8.84	6.02	1.34	1.57	0.05
MgO	38.22	34.15	37.04	40.11	41.21	41.52	35.08
CaO	0.0	0.35	0.16	0.32	0.19	0.12	-
MnO	0.13	0.17	0.12	0.05	0.02	0.01	0.05
Na_2O	0.76	0.0	0.0	0.03	0.83	0.0	0.0
P_2O_5	0.04	0.0	0.05	0.0	0.04	0.0	0.0
Cr_2O_3	0.0	0.0	0.0	0.6	0.0	0.7	0.05
NiO	0.29	0.20	0.34	0.05	0.10	0.07	0.05
S	0.0	0.04	0.0	0.0	0.01	0.0	0.12
L.O.I	13.34	13.30	12.99	13.97	13.88	14.49	13.60
Σ	100	100	100	101.56	99.62	100.32	97.09

* As analysed the Cr_2O_3 content varies from 0.37 to 0.48% (5 analyses) due to inclusions of aluminous magnesiochromite. In order to compare pure serpentine chemistries, the chromite phase has been "subtracted" using the ratios "Cr_2O_3 49.3, FeO 14.5, MgO 13.9, Al_2O_3 16.2, (Peters and Kramers) and the resulting compositions recalculated to 100%. K_2O and BaO concentrations are both less than 0.1%.

1   Dark green serpentinised harzburgite from the base of the Semail nappe.

2 and 3   Black serpentinised harzburgite from near the top of the mantle sequence.

4   Soft silvery coating on transparent precipitated tabular brucite. From the base of the Semail nappe.

5   Yellow "soapy" precipitated serpentine from the mid ultramafic unit.

6   Pale green "waxy" serpentine from the upper ultramafic unit.

7   Translucent cream botryoidal orthorhombic serpentine from vein in Dibba ultramafics.

Apart from the accessory chromite, the monomineralic serpentine composition of all samples was confirmed by X-ray diffraction.

CONCLUSIONS

Many features all serve to illustrate the irregularity of low temperature serpentinisation both in intensity and periodicity according to the prevailing hydrogeological conditions. eg 1) non-equilibrium gas and mineral assemblages, 2) irregularity of H_2 and air discharging at spring sites, 3) intermittent hydroxide precipitation in spring changes, 4) presence of $CaCO_3$, $Mg(OH)_2$ and serpentine in the upper aquifers of the mantle sequence (4), 5) variability of water chemistry at individual spring sites, and 6) the re-routing of groundwater flow following carbonate cementation.

Despite uncertainties concerning the details of the dominant weathering reactions, both in the surface and groundwater zones, the reactions considered are characteristic of a system with a high rock to water ratio (ie. a rock dominated system). Many of the present day processes currently observed in Northern Oman are at least partially influenced by reaction products formed during the initial nappe emplacement when higher temperature serpentinisation occurred. In these respects, the semi-arid ophiolite environment in Northern Oman with is characteristically infrequent, low CO_2 recharge, provides a window upon processes which, in other environments, would be either cryptic or inactive.

ACKNOWLEDGEMENTS

The authors acknowledge with gratitude the help of the NERC isotope facility (B.G.S. Wallingford) and in particular Drs A H Bath and G Darling. Numerous analytical services were provided by the Open University (thin section, X.R. Fluorescence, microprobe and "wet chemical" facilities), particular thanks being due to Dr J Watson for extensive X.R.F. analysis. Additional wet chemical analysis and X-ray diffraction were also performed at both Oxford University (Department of Geology and Mineralogy) and the Institute of Hydrology.

REFERENCES

1. Brown, M.A. 1982, "Chromite Deposits and their Ultramafic Host Rocks in the Oman Ophiolite". PhD. Thesis, Open University, 263 pp.
2. Browning, P. 1982, "The Petrology, Geochemistry and Structure of the Plutonic Rocks of the Oman ophiolite" PhD Thesis, Open Univeristy, 1-404 pp.
3. Sir Alexander Gibb and Partner 1976, Water Resources Surv. of Northern Oman 1-16 vol.
4. Neal, C. And Stanger, G. 1984, Min. Mag. 48, 237-241 pp.

5. Neal, C. and Stanger, G. 1983, Earth and Planet. Sci. Lett. 66, 315-320 pp.
6. Glennie, K.W., Boeuf, M.G.A., Hughes-Clark, M.V., Moody-Stewart, M., Pilaar, M.F.H. and Reinhardt, B.B. 1974, Verh. K. Ned. Geol. Mijnbouwknd. Gen. 1-423 pp.
7. Cook, J.M. and Miles, D.L. 1979, Methods for the Analysis of Groundwater, WD/ST/79/5 1-83 pp.
8. Truesdale, V.W. and Smith, C.J. 1975, Analyst 100, 111-123 pp.
9. Truesdale, V.W. and Smith, C.J. 1979, Mar. Chem. 7, 133-139 pp.
10. Coleman, M.L., Shepheard, T.J., Durham, J.J., Rouse, J.E. and Moore, G.R. 1982, Anal. Chem. 54, 993-995 pp.
11. Epstein, S. and Mayeda, T. 1953, Geochim. Cosmochim. Acta 4, 133 pp.
12. Barnes, I., O'Neil, J.R. and Trescases, J.J. 1978, Geochim. Cosmochim. Acta 42, 144-145 pp.
13. Searle, M.P. 1980, "The Metamorphic Sheet and Underlying Volcanic Rocks Beneath the Semail ophiolite in the Northern Oman Mountains of Arabia". PhD thesis, Open University, 213 pp.
14. Whitehead, D.C. and Truesdale, V.W. 1982, "Iodine: Its movement in the environment with particular reference to soils and plants". Grassland Research Report, 1-83 pp.
15. Barnes, I. and O'Neill, J.R. 1969, Geol. Soc. Am. Bull. 80, 1948-1960 pp.
16. Barnes, I., Rapp, J.B. and O'Neil, J.R. 1972, Contrib. Min. Pet. 5, 263-276 pp.
17. Wenner, D.B. and Taylor, H.P. 1971, Contrib. Min. Pet. 32, 165-186 pp.
18. Wenner, D.B. and Taylor, H.P. 1973, Amer. Jour. Sci. 273, 207-239 pp.
19. Wenner, D.B. and Taylor, H.P. 1974, Geochim. Cosmochim. Acta 38, 1255-1286 pp.
20. Nesbitt, H.W. and Bricker, O.P. 1970, Geochim. Cosmochim. Acta 42, 403-409 pp.
21. Hoefs, J. 1980, Minerals and Rocks. 9, 1-203 pp.
22. Rucklidge, J.C. and Patterson, G.C. 1977, Contrib. Miner. Petrol. 65, 39-44 pp.
23. Earley, J.W. 1958, Amer. Miner. 43, 148-156 pp.
24. Kuroda, P.K. and Sandell, E.B. 1953, Bull. Geol. Soc. Amer. 64, 879-896 pp.
25. Millra, Yasunion and Rucklidge, J.C. 1979, Proc. Geol. Assn. Canada/Miner. Assn. Canada, Abstr.
26. Kohls, D.W. and Rodda, J.L. 1967, Am. Min. 52, 1261-1271 pp.
27. Stumm, W. and Morgan, J.J. 1970, Aquatic Chemistry, Wiley Interscience, New York. 583 pp.
28. Ramdohr, R. 1967, Neues Jb. Min. Abh. 107, 241-265 pp.
29. Ramdohr, R. 1969, "The Ore Minerals and their Intergrowths" (Pergamon, 1969, N.W.).
30. Naldrett, A.J. 1965, Heazlewoodite in the Porcupine District Canadian Min. 8, 383-385 pp.
31. Deutch, E.R., Rao, K.U., Laurent, R.L. and Seguin, M.K. 1977, Nature, 269, 684-685 pp.
32. De Quervain, F. 1945, Schweiz. Min. Pet. Mitt. 25, 305-310 pp.

33. De Quervain, F. 1963, Schweiz. Miner. Petr. Mitt. 43, 295 pp.
34. Chamberlain, J.D., McLeod, C.R., Traill, R.J. and Lachance, G.R. 1965, Can. Jour. Earth Sci. 2, 185-215 pp.
35. Chamberlain, J.A. 1966, Canad. Miner. 8, 519-522 pp.
36. Coveney, R.M. 1972, Econ. Geol. 66, 1265-1266 pp.
37. Alabaster, A. 1982, "The Interrelationship between Volcanic and Hydrothermal Processes in the Oman Ophiolite". PhD thesis, Open University, 408 pp.
38. Coleman, R.G., Huston, C.C., El-Boushi, I.M. and Bailey, E.H. 1979, Occurrence of Cu-bearing Massive Sulphides in the Semail Ophiolite. Unpublished report.
39. Carney, J.N. and Welland, M.J.P. 1974, Geology and Mineral Resources of the Oman Mountains. I.G.S. report No. 27.
40. Suzuoki, T. and Epstein, S. 1976, Geochim. Cosmochim. Acta 40, 1229-1240 pp.
41. O'Neil, J.R. and Kharaka, Y.K. 1976, Geochim. Cosmochim. Acta 40, 214-246 pp.
42. Sakai, H. and Tsutsumi, M. 1978, Earth and Plant. Sci. Lett. 40, 231-242 pp.
43. Bowen, N.L. and Tuttle, O.F. 1949, Bull. Geol. Soc. Amer. 60, 439 pp.
44. Miyashiro, A. 1973, Metamorphism and Metamorphic Belts, London: George Allen and Unwin.
45. Chernosky, J.V. 1975, Am. Min. 60, 200-208 pp.
46. Luce, R.W. 1971, Clays and Clay Mins. 19, 335-336 pp.
47. Whittaker, E.J.W. and Zussman, J. 1956, Min. Mag. 233, 107-126 pp.
48. Wicks, F.J. 1969, "X-ray and Optical Studies on Serpentine Minerals". D.Phil thesis, Oxford Univ.
49. O'Neil, J.R. and Barnes, I. 1971, Geochim. Cosmochim. Acta 35, 687-697 pp.
50. Hostetler, P.B., Coleman, R.G., Mumpton, F.A. and Evans, B.W. 1966, Am. Min. 51, 75-98 pp.
51. Dabitzias, S.G. 1980, Econ. Geol. 75, 1138-1151 pp.
52. Stanger, G. and Neal, C. 1984, Min. Mag. 48, 143-146 pp.
53. Stanger, G. 1984, (in prep) The Hydrogeology of the Oman Mountains Phd thesis, (Open University).
54. Nahon, D., Paquet, H. and Delvigne, J. 1982, Econ. Geol. 77, 1159-1175 pp.
55. Golightly, J.P. 1979, (Evans, D.S.I., Shoemaker, R.S. and Veltman, H. Ed.) Int. Laterite Symp. New Orleans, 3-23 pp.
56. Stanger, G. 1984, (in press) Lithos.
57. Plummer, L.N., Jones, B.F. and Truesdell, A.H. 1976, USGS Water Resources Investigations 76-13, 1-61 pp.
58. Truesdell, A.H. and Jones, B.F. 1974, Jour. Res. USGS. 2, 233 pp.
59. Stanger, G. 1984, (in press) Hydrol. Sci. Bull.
60. Edmunds, W.M., Miles, D.L., and Cook, J.M. 1984, (in press) Proc. Symp. on Hydrochemical Balances in Freshwater Systems, (Uppsala).
61. Helgeson, H.C. 1969, Am. Jour. Sci. 267, 729-802.

MANGANESE CONCENTRATION THROUGH CHEMICAL WEATHERING OF
METAMORPHIC ROCKS UNDER LATERITIC CONDITIONS

Daniel NAHON, Anicet BEAUVAIS and Jean-Jacques TRESCASES

Laboratoire de Pétrologie de la Surface, E.R.A. 070.220
Université de Poitiers, 86022 POITIERS CEDEX (France).

Laterite profiles developed on manganiferous metamorphic rocks in the Ziemougoula area (North West Ivory Coast) permit the observation of chemical weathering of tephroites, manganocalcites, chlorites and spessartite garnets. Several stages are observed in the progress of weathering. In the lower part of profiles, tephroites and manganocalcites alter first into manganite, followed by chlorite into todorokite then by spessartites which weather into birnessite; the Al content of the parent garnet is leached out of the zone. Higher in the profile, garnets alter directly into lithiophorite; in this geochemical environment Al is no longer mobile. In such a layer, early-formed birnessite and manganite are transformed into nsutite along with minor cryptomelane. These later minerals can locally evolve into ramsdellite or pyrolusite. The associated nsutite and lithiophorite are the principal phases of the hard manganese crust capping the hillrocks of the lateritic landscape of the Ziemougoula area.

1. INTRODUCTION

The lateritic weathering of manganese-bearing rocks was recently studied in tropical West Africa (28). The Ziemougoula area, in the north west of the Ivory Coast was especially studied (25, 26, 1). A thick weathered cover, in which Mn oxides concentrate, develops at the expense of a Birrimian metamorphic parent rock made up of gondites and tephroitites. These formations were prospected by the Société pour le Développement Minier de la Côte d'Ivoire (SODEMI) by pits and by borings with core sampling.

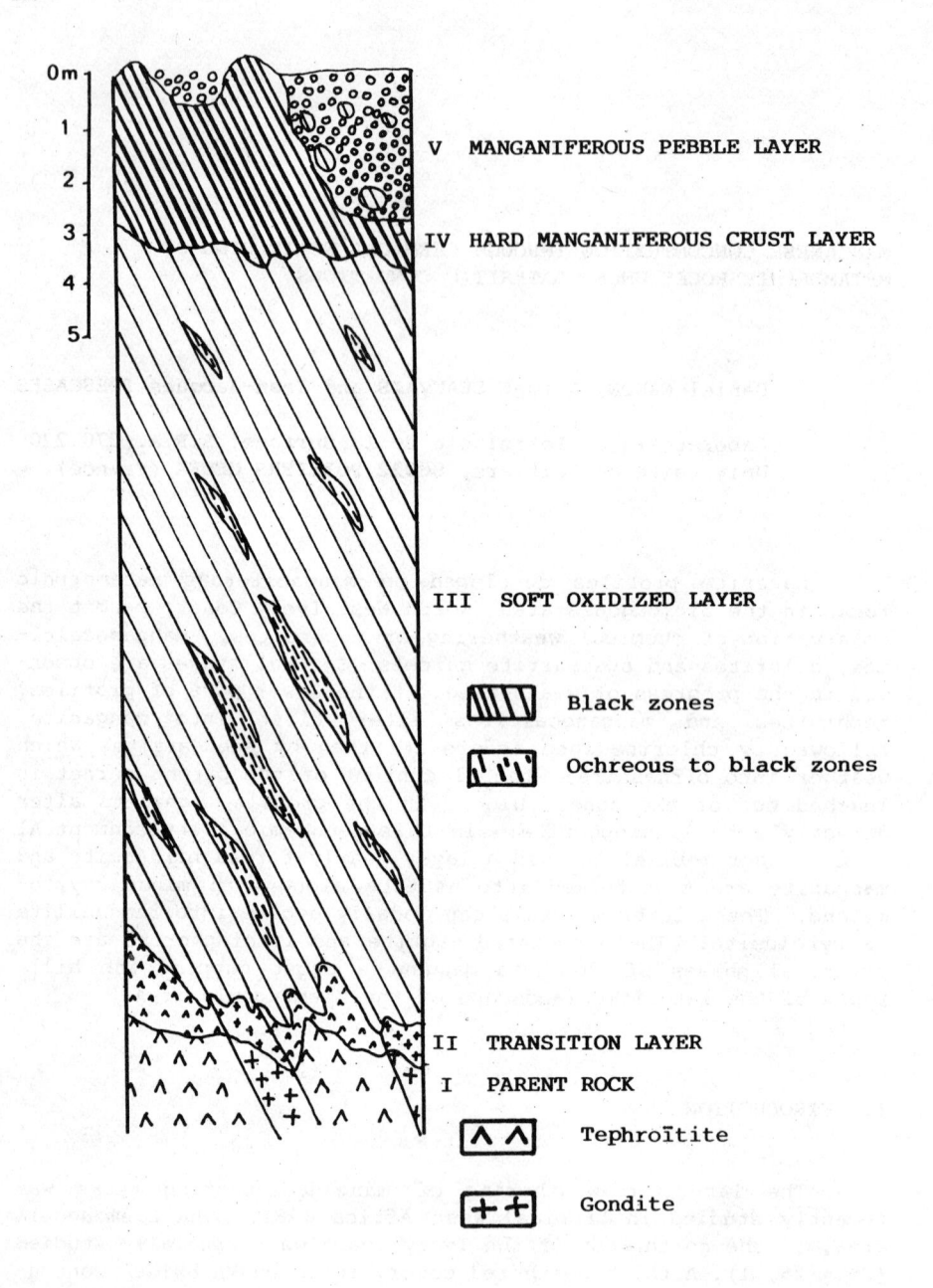

Figure 1. Sketch of weathering profile in the Ziemougoula area.

Oriented samples were collected after a detailed description of the profiles was made in the field. These samples were studied from a petrographic, mineralogical and chemical point of view by standard methods using polarizing microscope, X-ray diffraction, electron microprobe, infra-red spectrometry and scanning electron microscopy. Through pits and borings conducted by the SODEMI, a schematic profile can be drawn (Figure 1), showing main weathering layers developed along the profile. The study presented here deals with the appearance and development of different weathering stages of Mn-bearing parent minerals. The Birrimian metamorphic rock is weathered first into soft oxidized mantle (layers II and III, fig. 1) then in the higher part of advanced profiles into hard manganese crust (layer IV) which is locally dismantled into a pebble layer (layer V). The Birrimian metamorphic parent-rock consists roughly of 50 % tephroite, 35 % spessartite garnet, 7.5 % quartz, 5 % manganocalcite, 2 % Mn-chlorite and 0.5 % sulfides. These minerals respond differently to weathering. Thus, at the base of profiles, tephroite, then manganocalcite weather in this chronologic order followed by chlorite. Mn garnets are unweathered during these transformations and weather higher in the profile. But the alteration of garnets differs depending on their chemical composition; Ca-bearing garnet alters first.

2. THE PARENT MINERALS (LAYER I)

Tephroite crystals are anhedral and tightly cemented into a finegrained mosaic. The chemical compositions obtained by means of microprobe yield the following average formula for the tephroite:

$$[(Si_{0.99}Al_{0.01})O_4][Cr^{3+}_{0.001}Fe^{2+}_{0.010}Mn_{1.846}Mg_{0.136}Ca_{0.007}]$$

Nests of chlorite appear at the grain junctions of tephroite crystals; from their compositions we can note the following average formula for chlorite:

$$[(Si_{3.017}Al_{0.983})O_{10}][Al_{1.068}Fe_{0.067}Mn_{0.680}Mg_{4.140}Ca_{0.002})(OH)_8]$$

The manganocalcite evolves in patches which can locally replace the border of other parent minerals. Thus, the manganocalcite appears as the last crystallized phase of the parent rock. Here too, chemical compositions were obtained by means of the electron microprobe. The average formula is :

$$(Ca_{1.634}Mn^{2+}_{0.354}Mg_{0.012})(CO_3)_2$$

Garnet crystals are euhedral and tightly cement each other into a fine-grained mosaic arranged in thin layers or in islets or nests included in tephroite or other parent minerals. The

	1	2	3	4	5	6	7
SiO_2	2.69	0.45	7.05	9.99	0.18	0.55	0.25
Al_2O_3	0.03	0.02	3.58	0.19	23.27	0.45	0.33
Fe_2O_3	0.10	0.08	0.13	2.07	0.18	0.40	0.02
Cr_2O_3	-	0.01	-	-	-	-	-
Mn_2O_3	84.70	72.36	73.93	-	-	-	-
MnO_2	-	-	-	75.32	56.39	93.45	93.94
MgO	0.03	0.56	5.75	0.04	0.01	0.12	0.04
CaO	0.18	5.36	0.06	0.44	0.01	0.24	0.08
BaO	n.d.	n.d.	n.d.	n.d.	n.d.	-	-
CoO	0.06	0.15	0.07	0.11	0.15	0.07	0.39
CuO	n.d.	n.d.	n.d.	0.03	3.53	0.23	0.78
NiO	-	-	-	-	0.07	-	0.03
Na_2O	n.d.	0.03	n.d.	0,03	0.01	0.03	-
K_2O	-	0.03	0.01	0.06	0.01	-	-
TiO_2	-	-	-	n.d.	n.d.	n.d.	n.d.
H_2O*	12.22	20.95	9.42	11.72	16.20	4.46	4.14
Total	100.00	100.00	100.00	100.00	100.00	100.00	100.00

Table 1 - Average chemical compositions of manganiferous oxihydroxides and oxides from weathering of parent minerals in the manganese ore of Ziemougoula area.

1	=	Manganite from tephroite (7 analyses)
2	=	Manganite from manganocalcite (10 analyses
3	=	Oxihydroxide from chlorite (4 analyses)
4	=	Birnessite from garnet (5 analyses)
5	=	Lithiophorite from garnet (9 analyses)
6	=	Nsutite from birnessite (4 analyses)
7	=	Ramsdellite from nsutite (5 analyses)
-	=	Below detection limit
n.d.	=	Not determined
*	=	Obtained by difference

garnet phase was analysed by X-ray diffraction (main lines : 2.596 Å, 1.61 Å, 1.55 Å) and infra-red spectrometry (main lines : 937 cm^{-1}, 880 cm^{-1}, 645 cm^{-1}, 460 cm^{-1} and 370 cm^{-1}). Only spessartite was identified. But the chemical compositions obtained by means of the microprobe yield the following average formula :

$$[Si_3 O_{12}](Al_{1.998} Cr_{0.002})(Mn_{2.657} Fe_{0.171} Ca_{0.092} Mg_{0.080}).$$

These garnets of which the general formula is (Z) $R_2^{3+} R_3^{2+}$ are actually intermediate between spessartite (88 %) and almandine-pyrope-grossular (12 %).

3. EARLY STAGES OF MANGANESE OXIHYDROXIDE CONCENTRATION (LAYER II)

At the base of profile, tephroite, manganocalcite and chlorite weather into either manganite or into a mixture of Mn oxihydroxide + kaolinite. These transformations lead to a few-centimeters-thick weathering layer in which about half of the rock remains unweathered.

3.1. The transformation of tephroite into manganite

In some pits we have noted an ephemeral weathering of tephroite into Mn^{2+} smectites (24) which precedes the manganite stage. But generally the manganite appears as the first product of the replacement of tephroite. These modifications begin in microfissures traversing tephroite crystals and develop as a mosaic of euhedral manganite crystals easily recognized under reflected light. Table 1 gives the average chemical compositions obtained on such manganite with the electron microprobe (7 analyses). One can note that silica is not entirely leached when manganite appears. The irregular distribution of silica among manganite crystals suggests a scattering of microconcretions of SiO$_2$. In addition the pattern of distribution of Si shows that silica is more abundant near the edges of unweathered tephroite.

3.2. The transformation of manganocalcite into manganite

A well crystallized manganite mosaic develops from cross-cleavages and microfractures towards unweathered manganocalcite, progressively isolating residual fragments of the parent carbonate. Each fragment retains the crystallographic orientation of the origininal carbonate. This proves the in situ weathering of manganocalcite. The compositions of these manganite crystals (Table 1) compared to that of manganocalcite enlightens us regarding the source of the silica impurities. As the parent manganocalcite is almost depleted in this element, we can consider a silica transport from the surrounding tephroites which weather at the same time as the manganocalcite.

3.3. The transformation of chlorite into Mn-oxihydroxides

The chlorite crystals show a progressive buckling and opening up of the lamellae as a Mn-oxihydroxide weathering plasma develops at the expense of the lamellae. Analyses of hand-picked weathered chlorites from crushed rocks, by X-ray diffraction and infra-red spectrometry, show that the weathering plasma consists of a mixture of Mn-oxihydroxides (todorokite) + minor kaolinite. Microprobe analyses confirm such a heterogeneity in which minute particles of residual chlorites were probably contained (Table 1).

4. DEVELOPMENT OF THE MANGANESE OXIDE LAYERS (LAYERS III and IV)

From the base to the top of profiles, all the parent minerals weather progressively into a Mn-oxide plasma, to such an extent that few parent relicts still occur near the surface. In fact, the intense concentration of Mn oxides in this part of profiles is the result of (i) the progressive transformation of garnet crystals, (ii) the dissolution of quartz grains. Petrographic analysis reveals that the early transformation begins at grain boundaries and in microfissures traversing garnet crystals, along which sulfides can occur. From boundaries of euhedral grains and fissures, a manganiferous matrix appears and spreads out at the expense of garnet. With the advance of weathering, etching patterns (corrosion pits, cracks...) get more numerous, filling with manganiferous matrix which reaches deeply into garnet grains and cuts them into several portions. In situ alteration of garnets leads progressively to a complete replacement of garnets by the manganiferous matrix through pseudomorphic relations. However, the products of the transformation of garnets are different depending on the degree of advancement of weathering.

4.1. Transformation of garnets into birnessite

The first stage of garnet weathering consists of formation of a mixture of birnessite + amorphous silica. Birnessite is recognized under reflected light by intense pleochroism and strong anisotropy and by infra-red spectrometry (30) with adsorption bands at 1100-1000 cm^{-1}. X-ray diffraction spectra seem to indicate a poorly-crystallized birnessite, probably mixed with todorokite or Z α-manganate phase. The main peaks on X-ray diagrams are 9.567 Å (ASTM n° 18-1411, gives 9.65 Å for todorokite), 7.332 Å, 3.627 Å, 2.401 Å, with 2.401 Å best developed, and 7.332 Å, 3.627 Å, poorly developed. (5, 12, 6) show that the broadened 7-7.2 Å and 3.5-3.6 Å peaks correspond to poorly-crystallized birnessite, or to a mixture with other minerals. The average chemical analysis of such birnessites (Table 1) permits us to note the following facts.

- One can note that silica is not entirely leached when birnessite appears. The irregular distribution of silica in the birnessite matrix suggests a scattering of amorphous silica intimately mixed with cryptocrystalline birnessite. The birnessite matrix alternates with dark zones, becoming isotropic when they are well segregated. Microprobe point analyses and patterns of Si distribution demonstrate clearly that the dark isotropic zones are strongly enriched in silica. Chemical analyses have been made with the microprobe, in a zone where garnets are the dominant parent mineral and where the birnessite matrix develops in corrosion embayements of garnets. This suggests that the retention of silica is related to areas of lower permeability.

- On the other hand, Al is strongly leached during the weathering of garnets, as one can observe in comparing the birnessite composition to that of the parent garnets (Table 1). It is clear that in this part of the weathering profile, geochemical microenvironments occur in which Al is mobile and birnessite evolves. This mobility of Al is probably linked to the weathering of sulphides occurring in fissures and grains boundaries and which alter at the begining of garnet weathering, resulting in a strong decrease in pH.

4.2. The transformation of garnets into lithiophorite

Higher in the profiles one observes a second stage in the weathering of garnets. Both relics of garnets, not entirely or still not altered into birnessite, weather into lithiophorite. In fact, lithiophorite appears as the main constituent, but always along with goethite. With the advance of weathering, one can observe every stage reached in the transformation of garnets : (i) unweathered garnets in the birnessite matrix; (ii) garnets partly altered at their peripheries or in their cores into lithiophorite; (iii) patches of lithiophorite occurring in the birnessite matrix. This indicates a change in the progressive alteration of a same garnet, i.e. weathering into birnessite during an earlier stage, and weathering into lithiophorite higher in the profile. There is pseudomorphic replacement of lithiophorite after garnets which follows the early pseudomorphic replacement of birnessite after garnets. The lithiophorite phase is well characterized by infra-red spectrometry (3510 cm^{-1}, 550 cm^{-1}) and X-ray diffraction (9.478 Å, 4.68 Å, 2.356 Å, 2.122 Å and 1.868 Å). Goethite appears either intimately mixed with cryptocrystalline lithiophorite, or is well differentiated into micronodules or into septa localized principally at the periphery of altered garnet. X-ray patterns show a slight displacement of the main peaks, indicating Al substitution in the lattice. Fifteen goethites were studied. The degree of substitution is about 20 mole % of Al(OH)$_3$ in goethites. This is confirmed by microprobe analysis. The well crystallized mosaic of lithiophorite was analyzed by the electron microprobe. Average chemical composition is given in Table 1. One can note the high aluminum contents of such lithiophorite and the relative enrichment in copper. This Cu content of the lithiophorite of Ziemougoula is greater than those indicated by (9). Calcium, if present, and silica are strongly leached with respect to parent garnets; meanwhile Al and Mn keep an almost uniform distribution. The geochemical environment in which garnets weather is different from the underlying one where silica is not entirely leached and aluminum is mobile. In this latter environment a birnessite matrix develops.

4.3. Nsutite formation

As lithiophorite appears and develops at the expense of garnets one observes a destabilization of early-formed birnessite and manganite. Minute zones of nsutite + cryptomelane matrix begin to evolve in microfissures and cracks traversing the birnessite and manganite matrix. Nsutite + cryptomelane matrix rapidly develops at the expense of birnessite, progressively isolating irregular volumes of birnessite matrix and preserving zones of lithiophorite in the transforming mass. The nsutite + cryptomelane matrix consists of a cryptocrystalline phase in which the two minerals are tightly cemented, causing a strong induration on a field scale (manganese crust or "mangcrete"). Nsutite and cryptomelane have been characterized by X-ray diffraction and by observation under the reflecting microscope. The main X-ray peaks are for nsutite: 3.964 Å, 2.411 Å, 2.332 Å, 2.124 Å, 1.628 Å; and cryptomelane: 4.902 Å, 3.101 Å, 2.390 Å. Nsutite is the predominant phase and was analyzed (Table 1). The nsutite + cryptomelane matrix near fissures can be transformed into well-crystallized fan-shaped ramsdellite (Table 1). Locally, patches of pyrolusite may develop in the form of a well crystallized euhedral mosaic. Though scarce, this indicates a possible development of pyrolusite at the expense of nsutite + cryptomelane matrix.

Lithiophorite persists to the top of the profile as patches clearly separated from nsutite, or intimately mixed with it along sometimes with scarce relicts of garnets. Both oxides comprise the hard manganese crust which crops out in the Ziemougoula area. When this manganese crust breaks up to form a pebbly layer, one notes under the microscope many fissures crossing the indurated rock. Fissures are covered or filled with gibbsite crystals which are perpendicular to the walls. We chose clear crystals of gibbsite for microprobe analysis, nevertheless chemical analysis show the presence of small quantities of iron and/or manganese ($Al_2O_3 \simeq 66$ %, $Fe_2O_3 \simeq 1$ %, $Mn_2O_3 \simeq 0.7$ %). This is probably Fe and Mn substituted in gibbsite, as suggested by the slight shift of main peaks on X-ray patterns (d_{002} : 4.81 to 4.83 Å).

5. DISCUSSION

5.1. Manganite formation in the weathering environments of Ziemougoula

Manganite could be hypogene, in veins or in diagenetic stratified bodies, or supergene and generated by weathering of such minerals as rhodochrosite $MnCO_3$ and rhodonite $MnSiO_3$ (18). Through examples chosen in tropical and equatorial zones, one can note that manganite is effectively generated from the weathering of rhodochrosite as in Africa at Nsuta (28), at Moanda (27) or as in Brazil at Conseilheiro Lafaiete (19, 2). Furthermore manganite appears as the weathering product of Mn-pyroxenoids and Mn-garnets (2).

In other cases, manganite has been noted in lateritic profiles without references to the possible parent mineral (8, 31). Manganite is generally not found to be an abundant phase. In the African examples which we have studied in minute detail, manganite occurs as millimeter- or centimeter-thick layers at the base of profiles twenty meters thick. In weathering environments the geochemical conditions of manganite formation are limited. It may be preceded by hausmannite (MnO, Mn_2O_3) formation at the expense of the parent mineral, or the manganite may be followed by the formation of birnessite, cryptomelane, nsutite and pyrolusite (31). The limited stability field of manganite is corroborated by the thermodynamic data. Some values of the free energy of formation ΔF_f^o were determined for manganite, γ MnOOH, by (35, 7, 21, 3). The results obtained by different methods are all between -557.5 kj mole^{-1} and -571.5 kj mole^{-1}; (15, 16, 17, 11, 5) generally take ΔF_f^o =-557.7 kj mole^{-1} for plotting stability fields of the different manganese oxi-hydroxides in pH-Eh diagrams. Manganite appears as an intermediate mineral between hausmannite and pyrolusite, ß-MnO_2. In fact, the other "dioxides", such as birnessite, cryptomelane or nsutite are rarely considered in that type of diagram, because of their variable thermodynamic values. Manganite may exist only in a pH region higher than 7.5 (at 25°C and 1 atmosphere, for $\log [Mn^{2+}]$ = -6.00); its interval of stability, in Eh terms, is 0.15 V wide, for Eh lower than 0.55 V.

When the fugacity of CO_2 is high ($\log f\ CO_2$ = -2.5 versus -3.5 in the "normal" atmosphere), the stability field of hausmannite is completly covered by that of rhodochrosite $MnCO_3$. In a log f O_2 versus log $[Mn^{2+}]/[H^+]^2$ diagram, (4) shows that the field of manganite, between rhodochrosite and pyrolusite, is limited to between log f O_2 = -23 (for log $[Mn^{2+}]/[H^+]^2$ = 10.4) and log f O_2 = -18 (for log $[Mn^{2+}]/[H^+]^2$ = 9.3). Thus manganite appears as a transition phase between the carbonate of reducing environments and the dioxides of oxidizing environments. Chemical analyses show that: (i) the chemical composition of manganite is different for different parent materials; (ii) manganite crystallizes in microenvironments where foreign elements, such as Si, Ca, Mg, are present in weathering solutions; (iii) the purest manganite crystals are located in veins, cracks or fissures. Manganite crystals originating from manganocalcite are chemically different from those originating from tephroite. Each parent carbonate or silicate mineral is pseudomorphically replaced by manganite. This process of replacement includes both dissolution of the parent mineral and precipitation of the new Mn^{3+} phase. It is essentially congruent initially, and some of the Ca, Mg and Si (and other transition metals) are trapped and are either adsorbed to manganite crystals for Mg and Ca, or reconcentrated into microsilicifications (23, 25).

5.2. The chemical weathering of Mn-garnets

The garnets of the parent rocks in the Ziemougoula area consist essentially of spessartite (i.e. Al-Mn garnets). But some of them show an enrichment of Ca, particularly concentrated at their periphery. This slight chemical difference, probably due to compositional variations in the original sediment involved in Birrimian metamorphism, is sufficient to cause different reactivities of Ca-bearing and Ca-depleted garnets in lateritic weathering. The former begin to alter in the low part of the profile and continue to transform (if they are not entirely weathered) up the profile when spessartites, sensu stricto, commence to weather. It is clear that this behaviour of garnets plays a role in differentiating two stages in the lateritic transformation. However, the low content of Ca cannot entirely explain the evolution of the different manganese oxihydroxides that one observes. Even when garnets contain Ca they show significant Al contents ($\simeq$ 20 % Al_2O_3). It is likely that the geochemical environment changes as weathering proceeds. In this interpretation the behaviour of garnet during weathering differs in each zone and results in Al-depleted manganese oxihydroxides in one, and in Al-bearing oxihydroxides in the other. Thus garnets in a first stage alter into birnessite probably along with todorokite, and in a second stage into lithiophorite along with Al-goethite.

Birnessite is a naturally occurring poorly crystalline oxide (20), the common form of mineralized manganese in soils (33). In lateritic weathering profiles in Brazil, birnessite appears as a sparsely developed secondary product in the Conseilheiro Lafaiete area (19, 2, 22), where it originates from a possible transformation of manganite. Frondel et al. (10) have noted the presence of mixtures of birnessite and γ type MnO_2 generated from spessartite or rhodochrosite. In Tambao (Northern Upper Volta), Perseil and Grandin (28) stress that the presence of birnessite is generally linked to the first stage of rhodochrosite weathering. In Ziemougoula, birnessite occurs in weathered zones situated in the lower parts of profiles, and derives from garnets of spessartite type by pseudomorphous replacement. Some of the Ca, Mg and Si (and other transition metals) are trapped with the birnessite. Microprobe data show clearly that SiO_2 is extra-structural, and corresponds to amorphous microsilicification showing that birnessite crystallizes in environments where silica is not entirely leached, i.e., in a constricted microsystem. Moreover, in view of the phyllomanganate structure of birnessite, the incorporation of ions such Mg and Ca in the lattice is possible as attested by the regular distribution of these elements. Considering the percentage of Al in the garnets, it is clear that birnessite is formed in a geochemical environment in which aluminum is leached. This seems to characterize the lower part of Ziemougoula profiles, where sulfides alter simultaneously, and is the reason why lithiophorite appears higher in the profiles and contains Cu. Todorokite is presumably present admixed with birnessite. Todorokite is known as a supergene mineral in altered metamorphic rocks containing spessartite in Bahia (Brazil) (10). In the soil environment, however, todorokite appear to be unstable, and could be regarded as the hydrated precursor of birnessite (13, 14).

The second stage reached in the progress of garnet weathering consists of development of lithiophorite along with small quantities of Al-goethite. Lithiophorite may incorporate foreign ions and although Mn is predominantly tetravalent, the Mn^{4+}/Mn^{2+} ratio can vary. Lithiophorite is essentially of supergene origin. It is one of the major manganese compounds of soil nodules (33, 34, 36). In West Africa lithiophorite is present in lateritic weathering profiles as a pseudomorphic replacement of spessartite in the gondites of Mokta (Ivory Coast), Tambao (Upper Volta), Nsuta (Ghana) (28). For these latter authors, the lithiophorite would correspond to a transition with other Mn compounds stable in the oxidation zone such as cryptomelane, nsutite and ramsdellite. At Ziemougoula our observations are at variance with those described in other West African manganese deposits. Here lithiophorite is formed in a geochemical environment where aluminum released by weathering of garnets is abundant and not mobile. Furthermore, lithiophorite does not appear as an evolution of the early formed birnessite but appears at the second stage from the direct transformation of garnets, in a single layer of the profile, in which aluminum is not leached. Here, at Ziemougoula, nsutite (along with cryptomelane) appears only as the alteration product of early-formed birnessite or manganite, i.e. from a Mn-matrix depleted in alumina. Such an evolution can locally proceed to ramsdellite or pyrolusite, driving the oxidation process to completion (5). This transformation of birnessite and manganite into nsutite (γ MnO$_2$) was often noted in manganese deposits of West Africa, Brazil and France by Perseil and Giovanoli (29), and in Ghana by Sorem and Cameron (32).

At Ziemougoula, nsutite and lithiophorite formed in this way are associated making up the manganese crust which caps the hills of Ziemougoula landscape. At the top of profiles the manganese crust can be locally dissolved forming a discontinuous pebbly layer made up of relict pebbles of manganese crust in a red clay matrix consisting of gibbsite and Al-goethite (1). The manganese leached from the pebbly horizons can precipitates down-slope in the sequence, or lower in the profiles, and contribute to form crystallized oxides or oxihydroxides.

In conclusion, the different chemical transformations of Mn^{2+} bearing parent minerals and the secondary evolution of birnessite and manganite are isovolume transformations and could be indicated under the form of a sketch representing weathering of average parent rock (Figure 2) and through the following reactions :

(1) 2 manganocalcite + 23 tephroite + 3.86 H$^+$ + 58.32 O$_2$ + 4.5 Mn^{2+} + 68.23 H$_2$O $\longrightarrow$ 48.32 MnO(OH)$_{manganite}$ + 0.23 Al(OH)$_3$ + 0.23 FeO(OH) + 3.15 Mg^{2+} + 3.43 Ca^{2+} + 22.77 Si(OH)$_4$ + 0.30 (HCO$_3$)$^-$

(2) 4.48 spessartite + 1.54 H$^+$ + 5.28 O$_2$ + 48.43 H$_2$O $\longrightarrow$ 1.70 (Mn$_7$O$_{13}$, 5H$_2$O)$_{birnessite}$ + 8.96 Al(OH)$_3$ + 0.76 FeO(OH) + 0.41 Ca^{2+} + 0.36 Mg^{2+} + 13.44 Si(OH)$_4$.

(3) 4.52 spessartite + 2.96 Al(OH)$_3$ + 1.54 H$^+$ + 3.09 O$_2$ + 34.10 H$_2$O $\longrightarrow$ 12 (MnO$_2$, Al(OH)$_2$)$_{lithiophorite}$ + 0.78 FeO(OH) + 13.56 Si(OH)$_4$ + 0.41 Ca^{2+} + 0.36 Mg^{2+} .

(4) 48.32 MnO(OH)$_{manganite}$ + 5.37 K$^+$ + 10.74 O$_2$ $\longrightarrow$ 5.37 (KMn$_8$O$_{16}$, MnO$_2$)$_{cryptomelane + nsutite}$ + 5.37 H$^+$ + 21.47 H$_2$O.

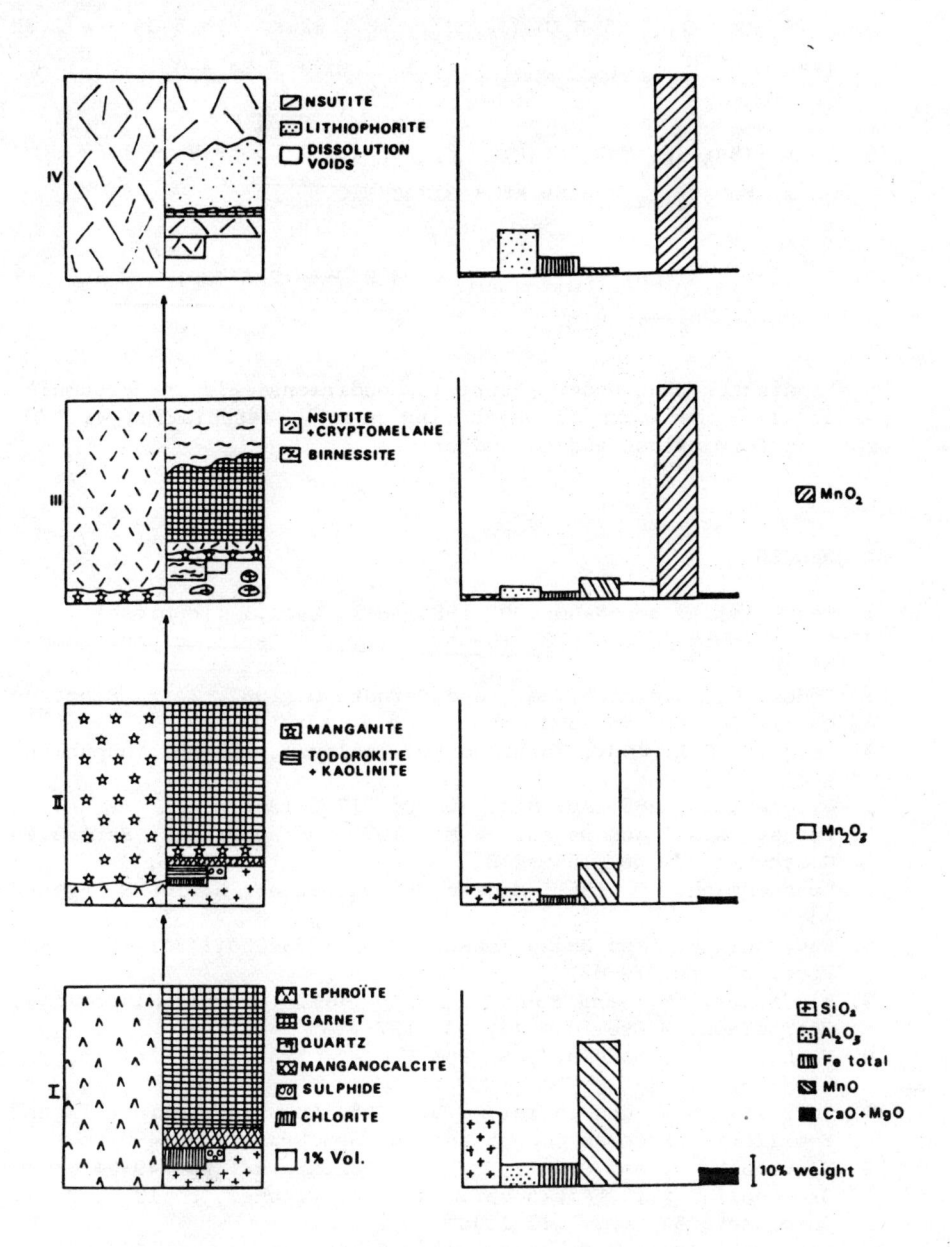

Figure 2. Sketch of mineralogical and chemical evolution of the weathering profile of the Ziemougoula area.

(5) $1.70\ (Mn_7\ O_{13},\ 5\ H_2O)_{birnessite} + 1.32\ K^+ + 0.5\ O_2 \rightarrow 1.32\ (KMn_8O_{16})_{cryptomelane + nsutite} + 1.32\ H^+ + 7.84\ H_2O.$

(6) $6.69\ (KMn_8O_{16},\ MnO_2)_{cryptomelane + nsutite} + 6.69\ H^+ + 1.67\ O_2 \rightarrow 60.21\ MnO_{2\ nsutite} + 6.69\ K^+ + 3.345\ H_2O.$

(7) $12\ (MnO_2,\ Al(OH)_2)_{lithiophorite} + 24\ H^+ \rightarrow 12\ Al(OH)_{3\ gibbsite} + 12\ Mn^{2+} + 3O_2 + 6\ H_2O.$

This indicates that under lateritic conditions, alumina accumulates in the upper part of weathering profile associating with Mn or Fe to form oxides and oxihydroxides.

REFERENCES

1. Beauvais, A. and Nahon, D. 1984, Sci. Geol., (in press).
2. Bittencourt, A.V. 1973, Thesis Univ. Sao Paulo, 81 pp. (unpublished).
3. Bode, H., Schmier, A. and Berndt, D. 1962, Zeit. Electrochemie 66, pp. 586-593.
4. Boeglin, J.L. 1981, Thesis Univ. Toulouse, 154 pp. (unpublished).
5. Bricker, O. 1965, Am. Min. 50, pp. 1296-1354.
6. Burns, R.G. and Burns, V.M. 1977, Oceanography. Ser. G.P. Glasby Ed. 7, pp. 185-248.
7. Drotschmann, C. 1951, Modern. Primarbatter. S. 10. N. Brantz ed.
8. Eswaran, H. and Raghu Mohan, W.G. 1973, Soil Sci. Soc. Am. Proc. 37, pp. 79-82.
9. Fleischer, M. and Faust, G.T. 1963, Schweiz-Mineralogy. Petrograph. Mitteilung 43, pp. 197-216.
10. Frondel, C., Marvin, U.B. and Ito, J. 1960, Am. Min. 45, pp. 871-875.
11. Garrels, R.M. and Christ, C.L. 1965, Solutions, Minerals and Equilibria, Freeman Cooper and Co. San Francisco, 450 pp.
12. Giovanoli, R. and Stähli, E. 1970, Chimia 24, pp. 49-61.
13. Giovanoli, R., Feitknecht, W. and Fischer, F. 1971, Helv. Chim. Acta 54, pp. 1112-1124.
14. Giovanoli, R. and Brütsch, R. 1979, Chimia 33, pp. 372-376.
15. Hem, J.D. 1963, U.S. Geol. Survey Water Supply Paper, 1667-A, 64 pp.
16. Hem, J.D. 1972, Geol. Soc. Am. Inc. Sp. Paper 140, pp. 17-24.
17. Hem, J.D. 1978, Chem. Geol. 21, pp. 199-218.
18. Hewett, D.F. 1972, Econ. Geol. 67, pp. 83-102.

19. Horen, A. 1953, The manganese mineralization at the Merid-Mine, Minais-Gerais, Brazil. Ph. D. Thesis, Harvard, Univ. 224 pp.
20. McKenzie, R.M. 1971, Min. Mag. 38, pp. 493-502.
21. Latimer, W.M. 1952, Englewood Cliffs, N.J. Prentice Hall.
22. Melfi, A.J. and Pédro, G. 1974, Gr. Fr. Argiles. Bull. 26, pp. 91-105.
23. Murray, J.W. 1975, Geochim. Cosmochim. Acta 39, pp. 505-519.
24. Nahon, D., Colin, F. and Tardy, Y. 1982, Clay Min. 17, pp. 339-348.
25. Nahon, D., Beauvais, A., Boeglin, J.L., Ducloux, J. and Nziengui-Mapangou, P. 1983, Chem. Geol. 40, pp. 25-42.
26. Nahon, D., Beauvais, A., Nziengui-Mapangou, P. and Ducloux, J. 1984, Chem. Geol. (in press).
27. Perseil, E.A. and Bouladon, J. 1971, C.R. Acad. Sci. Paris, Sér. D 273, pp. 278-279.
28. Perseil, E.A. and Grandin, G. 1978, Min. Depos. 13, pp. 295-311.
29. Perseil, E.A. and Giovanoli, R. 1983, Bull. Mus. Nat. Hist. Nat. Paris, Sér. C 2, pp. 163-190.
30. Potter, R.M. and Rossman, G.R. 1979, Am. Min. 64, pp. 1199-1218.
31. Roy, S. 1968, Econ. Geol. 63, pp. 760-786.
32. Sorem, R.K. and Cameron, E.M. 1960, Econ. Geol. 55, pp. 278-310.
33. Taylor, R.M., McKenzie, R.M. and Norrish, K. 1964. Aust. J. Soil Res. 2, pp. 235-248.
34. Taylor, R.M. 1968, J. Soil Sci. 19, pp. 77-80.
35. Wadsley, A.D. and Walkley, A. 1951, Rev. Pur. Appl. Chem. 1, pp. 203-213.
36. Wilson, M.J., Berrow, M.L. and McHardy, W.J. 1970, Min. Mag. 37, pp. 618-623.

RIVER CHEMISTRY, GEOLOGY, GEOMORPHOLOGY, AND SOILS IN THE AMAZON
AND ORINOCO BASINS

Robert F. Stallard

Department of Geological and Geophysical Sciences,
Princeton University, Princeton, New Jersey 08544, U.S.A.

ABSTRACT

In the Amazon and Orinoco basins, the chemistry of rivers can be related to the geology of their catchments when geomorphic factors are taken into consideration. To a first order, erosion processes are seen to occupy a continuum between weathering-limited (steep slopes, thin soils) and transport-limited (slight slopes, thick soils). Denudation rates for the former are lithology dependent, and significant removal of cation-rich, unstable, solid phases is expected. For the latter, erosion rates should depend on the regional lowering of the landscape; lithologic susceptibility to weathering should be a minor factor. The compositonal trends of river dissolved and solid load can be approximated for different weathering regimes using soil chemistry data. Preliminary results suggest that evolutionary models of saprolites produce results which are consistent with river chemistry.

INTRODUCTION

Continental denudation is a major aspect of global geochemical cycling, involving the transport of materials from bedrock and atmospheric reservoirs into the ocean, mostly via rivers. The mechanisms which control the composition of the phases coming off the continents are only imperfectly understood. The transfer is mediated by a wide variety of chemical, biological, and physical processes, often involving intermediate storage in soil, biomass, lake and ground water, and continental sediment.

Close field relationships between climate, topography, geology, soils, and vegetation have long been recognized. The understanding of the underlying causes for these relationships is central to the development of a comprehensive description of weathering and erosion. Topographic relief deserves particular attention as it is commonly accepted to be the most important factor in enhancing the erosion process. Meybeck (24) estimates that 41% of ionic inputs and 45% of silica inputs to the oceans come from humid mountainous regions, even though these occupy only 12.5% of exorheic drainage area. Data compiled by Milliman and Meade (25) indicate that the fractional contribution to the world sediment load by mountainous areas is even larger. Many of the factors which accelerate erosion on slopes are obvious; slopes are often unstable; surface runoff is more rapid, and streams and rivers are more energetic. The role of these agents in partitioning elements between dissolved and solid loads in rivers is not well understood. This paper focuses on the relationship between the chemistry of the dissolved and solid loads within the Amazon and Orinoco river systems and geology, landform development, soils, and vegetation[1].

SETTING

The Amazon and Orinoco river basins occupy most of South America north of 20 degrees south (Figure 1). In terms of discharge, the Amazon is the world's largest river ($5.5-6.2 \times 10^{12}$ m^3/yr) (30-37), about one fifth of the global river input into the ocean. The Orinoco ranks third (1.1×10^{12} m^3/yr) after the Zaire (1.3×10^{12} m^3/yr) (22,23). The Amazon ranks first or second in dissolved load (2.9×10^8 tonnes/yr), a value comparable to the Yangtze (13), and it ranks third in suspended sediment load (9×10^8 tonnes/yr) after the Ganges-Brahmaputra (17×10^8 tonnes/yr) and the Huang He (11×10^8 tonnes/yr) (21). The Orinoco ranks about sixth in terms of suspended load (200×10^6 tonnes/yr) (22), and is less than tenth in dissolved load (50×10^6 tonnes/yr) (23). Compared to a "world average", the Orinoco is a very dilute river, and its solid and dissolved loads are particularly siliceous. In the lower Amazon, the load is also dilute and siliceous; however, that in the upriver mainstem is very "average". This reflects the preponderance of dilute tributaries along its lower course.

DENUDATION MODEL

Geomorphologically, landscape development reflects those

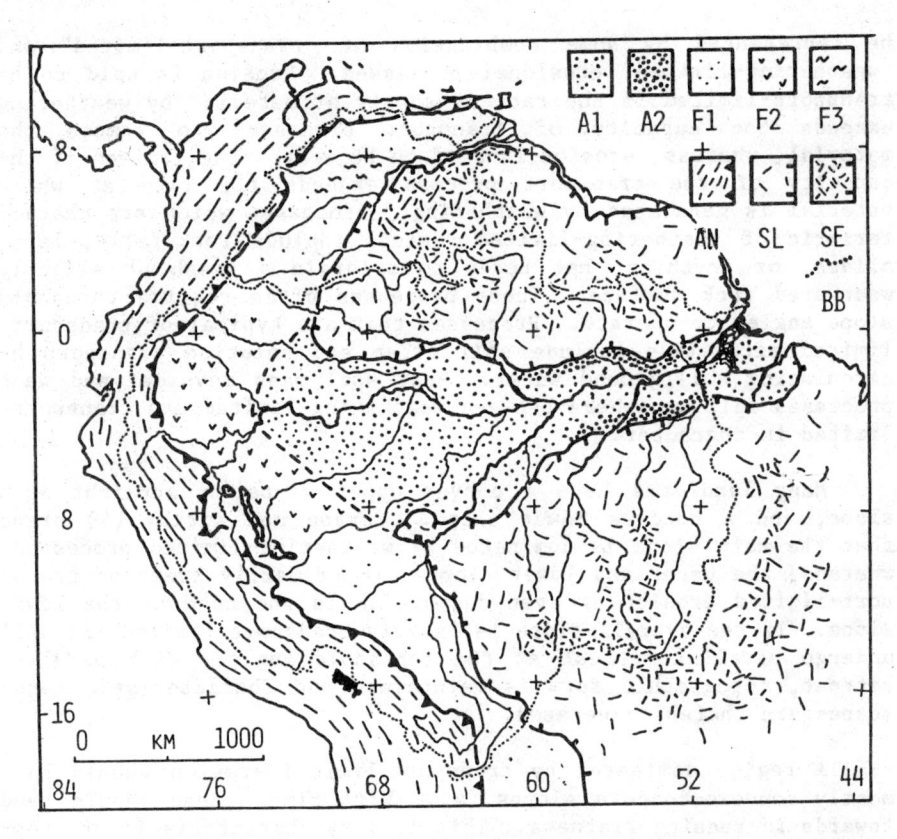

Figure 1. Morphotectonic map of northern South America. A1, Amazon Trough; A2, Paleozoic and Mesozoic sediments including carbonates and evaporites; F1, Foredeep sediments; F2, Foredeep sediments with volcanic ash; F3, Foredeep sediments with dunes; AN, Andean Cordillera; SL, low-lying shield; SE, elevated shield; BB, drainage basin boundary.

mechanisms which expose bedrock, weather it, and transport the weathering products away. Present and past tectonism, geology, climate, soils, and vegetation are all important in landscape evolution. These factors often operate together to produce characteristic landforms which presumably integrate the effects of both episodic and continuous processes over considerable periods of time.

Erosion Regimes

An approach to the study of landscapes that is particularly applicable to geochemical studies of erosion is developed by Carson and Kirkby (4). They argue that mass movement on slopes can be represented by some combination of "transport-limited" and "weathering-limited" erosional processes. Erosion is said to be transport-limited if the rate of supply of material by weathering exceeds the capacity of transport processes to remove the material, whereas, erosion is said to be weathering-limited if the capacity of the transport process exceeds the rate at which material is generated by weathering. Processes which are characteristic of weathering-limited regions include rock-falls, landslides, or anything that tends to maintain a fresh or slightly weathered rock surface. These processes often require threshold slope angles to operate. Processes that are typical of transport-limited situations include soil creep and solution transport by circulating soil/ground water. Most soil mass movement and wash processes fall somewhere between weathering-limited and transport-limited in character.

Many landforms have a convex upper slope, a straight main slope, and a concave lower slope. Carson and Kirkby (4) argue that the main slope is dominated by weathering-limited processes; whereas, the upper and lower slopes are primarily areas of transport-limited erosion or even deposition in the case of the lower slope. If the overall slope is largely transport-limited, it will undergo parallel retreat at the threshold angle. With parallel retreat, topographic form is maintained and characteristic landscapes are thereby generated.

A region dominated by transport-limited erosion should have mostly convexo-concave slopes (4). With time, these should tend towards increasing flatness. This is very characteristic of tropical rainforest areas, but less so of tropical savanna (47).

Soils, Slopes, Vegetation, and Weathering Rate

For a given set of conditions (lithology, climate, slope, etc.), there is presumably an optimum soil thickness which maximizes the rate of bedrock weathering (Figure 2) (4). For less than optimum soil thicknesses, there is insufficient pore volume in the soil to accept all the water supplied by precipitation and downhill flow. Excess water runs off and does not interact with the subsurface soil and bedrock. In contrast, water infiltrates and circulates slowly through thicker soils (especially where forested). However, if thicknesses greatly exceed the optimum, long residence times for water at the base of the profile reduce the weathering rate. The formation of impermeable soil layers can also reduce weathering rates.

There is destabilizing feedback between soil thickness and weathering rate for soil profiles that are less than the optimum thickness. Assume that a thin soil is in equilibrium such that weathering inputs balance transport losses (A on Figure 2). If

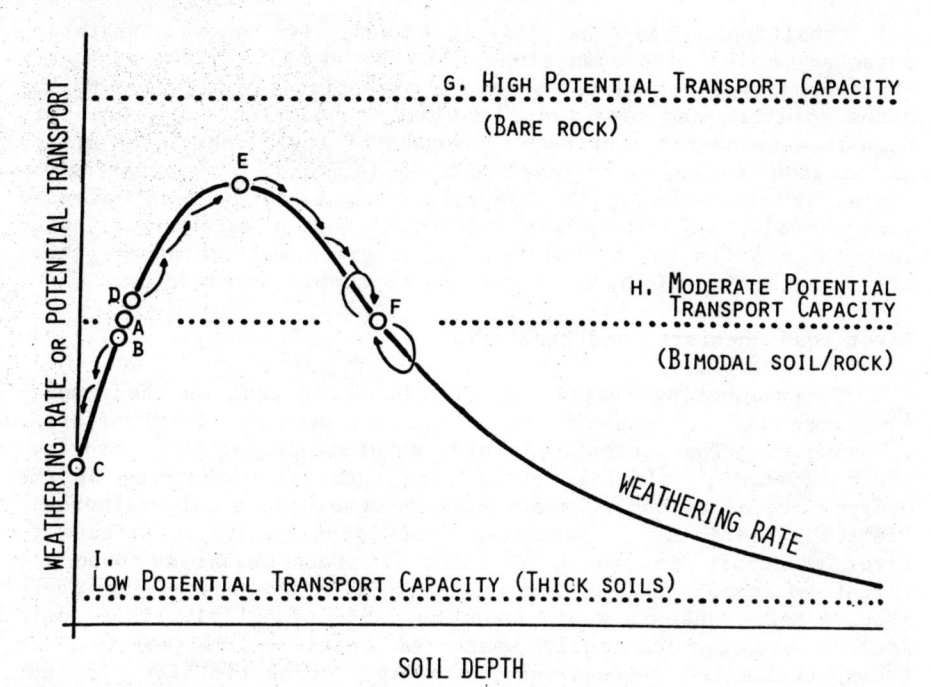

Figure 2. Solid curve portrays a hypothetical relationship between soil thickness and rate of chemical weathering. Dotted lines correspond to different potential transport capacities. For moderate capacity, C and F are stable points. Adapted from (4).

the soil is thinned, the weathering rate is reduced because less water is retained (B). The transport processes, however, are capable of removing weathered material at the same rate, and the soil would continue to thin, eventually to hard rock (B-C). This scenario would also occur if transport processes were increased in effectiveness. If the soil is thickened, or if the capacity of transport processes is decreased, the soil would tend to accumulate (D-E) Ultimately, weathering rates decrease with increasing soil thickness (E-F). Stabilizing feedback occurs, and a thick soil forms such that transport removal would balance weathering inputs. This model suggests that soil distributions should be distinctly bimodal: either thin, weathering-limited, or thick, transport-limited. For certain intermediate potential transport rates (H), either hard rock or a moderately thick soil could exist.

The effects of vegetation are complex. Vegetation reduces short-term physical erosion by sheltering and anchoring soils, but this does not necessarily reduce denudation rates. Vegetation can maintain a thin veneer of soil on steep slopes, particularly under wet conditions. As the soil thickens, it becomes unstable, detaches, and slides down slope (9,27,34,36,46). Under such circumstances, weathering rates can be exceptionally high due to the extra moisture and bioacids; likewise, denudation rates are very high because of the continuous resupply of fresh rock. The effect of erosion following fires, tree falls, and land clearing on slopes is similar (33,35). On slight slopes, over very extended time periods, vegetation may reduce weathering rates by allowing very thick soils to accumulate. For a given soil thickness, however, weathering might be faster due to supply of bioacids.

River Load Chemistry and Landforms

The weathering regime exerts a major control on the production rate and the composition of erosion products from different lithologies. The tectonic history, physical properties (porosity, shear strength, jointing, etc.), and chemical properties of the bedrock are major controls on landform development for weathering-limited situations. A particular rock type should contribute to river transport an amount of material proportional to both its extent of exposure and its susceptibility to weathering (deep groundwater transport would contribute additionally). This could include abundant partially weathered (cation-rich) solids. In transport-limited situations, however, susceptibility is not important due to the isolating effect of thick soils. The erosional contribution by a particular rock type should be related only to the area exposed. Solids would be cation-poor.

Dissolved phases are assumed to best reflect the weathering processes occurring at the erosion site, as water is not stored for long periods (many years) in soils or during fluvial transport. Weathering products, however, do undergo various reactions as they move downslope and through fluvial systems. Solids often alter when they accumulate at the base of slopes or during storage on flood plains; this obviously affects dissolved components. Soil solutions also evolve as they flow downslope. Contact with fresher materials is prolonged, and evapotranspiration can concentrate solutions (4). Both of these factors can cause the formation of different suites of clays and sesquioxides, and the precipitation of carbonates.

WEATHERING REGIMES IN THE AMAZON AND ORINOCO

The Amazon and Orinoco basins can be divided into five major morphostructural regions (Figure 1). These are the Andean Cordillera, the Andean Foredeep or Trough, the Amazon Trough, the Guayana Shield, and the Brazilian Shield. The surface lithologies of the Andean Foredeep and the Amazon Trough are quite similar to one another (mostly Cenozoic sediments). The same is true for the shields (mostly granite-gneiss terrains). Very low elevation erosional/depositional surfaces extend over the entire Amazon Trough, much of the foredeep, part of the Guayana Shield and lesser parts of the Brazilian Shield. These observations suggest that the two basins can be divided into five major erosional regions: (1) the Andes, (2) foothills of the foredeep, (3) elevated shields, (4) the Neogene depositional/erosion surface of the trough and foredeep, and (5) peneplaned shields.

Denudation rates are very different for each of these erosional regions. A simple estimate for the range of erosion rates can be made from dissolved solid concentrations for rivers draining each region. This is done in Figure 3, where the erosion scale was calculated as the product of dissolved load concentrations, mean annual runoff (1 m/yr), and a correction factor for high solid loads in the Andes and for higher than average annual precipitation in the lowland shields. The factor ranged from two for the most dilute rivers to four for the Andean Rivers. These denudation rates are in general agreement with more rigorous calculations (11,37).

The Andes and Foothills

The most concentrated samples and highest erosion rates are observed in the Andes, followed by rivers in the foothills. River chemistry is consistent with weathering-limited erosion. In these areas, sediments constitute the principal basement lithology, and the river chemistry agrees with catchment geology (37,39). For example, black shales have particularly high Mg:Ca ratios, and Bolivian rivers which drain black shales are exceptionally magnesium-rich. Rivers which drain evaporites have the highest total cation (TZ+) concentration, followed by rivers which drain carbonates, and finally by rivers which drain only siliceous rocks. This is illustrated by the ternary diagram in Figure 4, which uses (chloride + sulphate), alkalinity, and silica as input markers for the respective lithologies. Total cations increase systematically from silica to alkalinity to (chloride + sulfate). Many Andean rivers are slightly supersaturated with respect to calcite and to montmorillonite (37,40). This indicates that saturation with respect to these phases ultimately limits the supply of alkalinity and dissolved silica to the rivers. Similar limits do not apply to cations because of additional inputs from silicates and evaporites. Erosional contributions by carbonates and especially by evaporites greatly exceed those expected if inputs are proportional to the fraction of the catchment area covered by these lithologies (37,39). Much of the evaporite input is sustained by actively extruding salt diapirs (1,37,40). The presence of unstable and cation-rich minerals in the suspended and bed loads of rivers which drain the Andes (6,11,14,15,29,37,39) also suggests that extraordinarily rapid erosion is occurring.

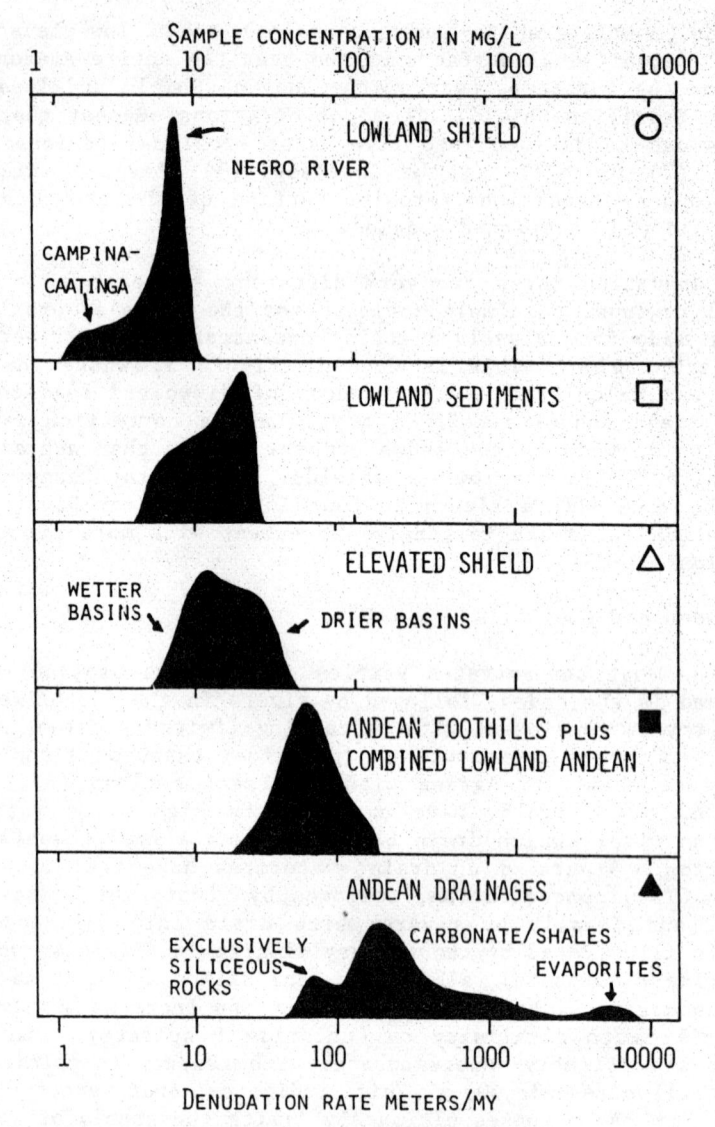

Figure 3. Histogram of sample dissolved solid concentrations with corresponding denudation rate scales. Symbols used in subsequent figures are indicated in the upper right of each panel.

The Elevated Shields

   Erosion rates in the elevated shield are much lower than in the Andes, even though elevations in the Guayana Shield reach 3000 m and exceed 2000 m in the Brazilian Shield. The topography can

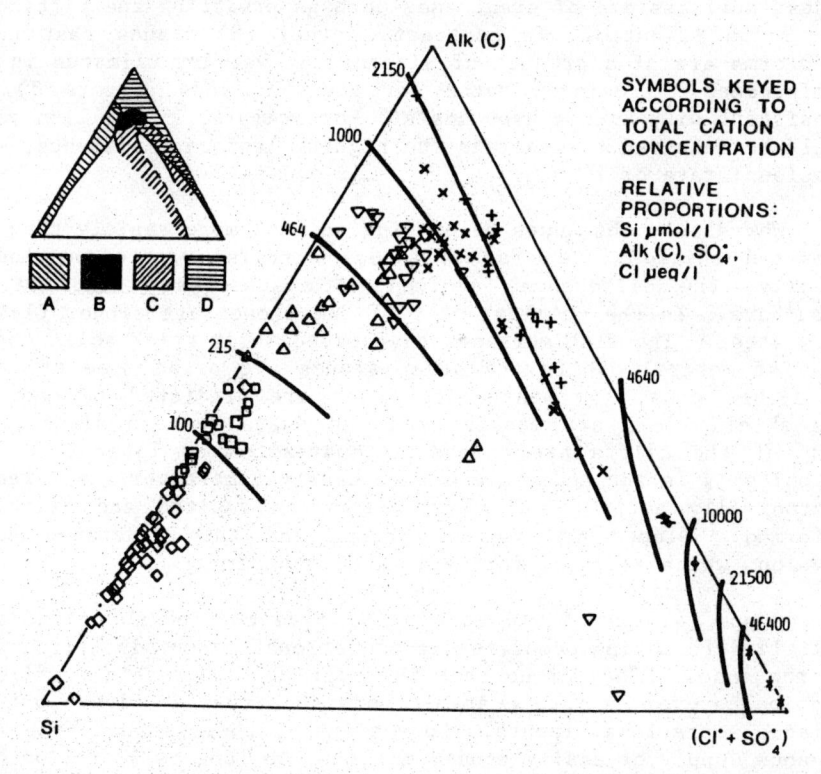

RIVER CHEMISTRY, GEOLOGY, GEOMORPHOLOGY, AND SOILS

Figure 4. Ternary diagram relating silica, carbonate alkalinity, and (chloride + sulfate) in the Amazon and Orinoco systems. Analyses are cyclic salt corrected. Curves are numbered in total cation concentration (μeq/l). The predominant symbol within each interval corresponds to samples whose concentrations fall in that interval. A - lowlands and shields; B - Amazon mainstem; C - Andean, black shales; D - Andean, carbonate and evaporite rich sediment.

be spectacular with steep, bare, rock slopes, often topped by high plateaus. Inselbergs are common in some areas. Rivers from the the shields typically carry very little solid load, and great talus piles do not accumulate below cliffs. Much of this material probably just dissolves. A karst-like topography developed on quartzite, gneiss, and granite is found on some of the higher areas (2,42). King (10) argues that such topography is formed from a series of planation surfaces separated by erosional scarps which are undergoing parallel retreat into the older surfaces. As the scarps retreat, isostatic adjustment raises the topography, inducing new scarps to form at the edge. Presumably this process continues until the landscape is flattened. In his view the oldest surfaces are of great age, perhaps predating the rifting of the South Atlantic. In contrast, Garner (8) argues that these landforms are of a product of alternating wet-dry climates in the Pleistocene. Erosional rates for these uplands (Figure 3) are consistent with King's hypothesis. For example, 150 million years would be required to generate the highest landscape elements, with erosional rate of 20 m/my.

Why should the Andes be eroding so much more rapidly than the elevated shields, even for siliceous terrains within the Andes? Clearly, the difference involves both substrate lithology and structure. In the shields, many of the slopes are either flat or very steep. The flat surfaces have exceedingly thick soils (up to tens of meters), while the steep slopes are often free of soil. In either case, low weathering rates are predicted (Figure 2). The shield rocks are massive and can sustain steep slopes, and many of the cliffs are topped by quartzites or laterites. The lithologies in the Andes are not so massive; moreover, the area is tectonically active, and rocks are often faulted and brittlely deformed. Slopes are not so steep, and thin vegetated soils develop, which is ideal for very rapid weathering.

Suppe (41) and Davis et al. (5) have modeled the effects of brittle deformation occurring in accretionary mountain belts, such as the Andes. They argue that the overall topographic profile of such belts evolves into a stable form such that erosional outputs balance accretional inputs. Consequently, there should be a continuous supply of easily eroded material so long as accretion continues. It seems reasonable that lithologic susceptibility to erosion may in turn influence the form of the entire mountain range, not just the form of the slopes. This is clearly the case when the shape of the accretionary Andes is contrasted to that of the epeirogenic shield mountains.

The Lowland Shields and Neogene Sediments

Vast tracts of the Amazon and Orinoco lowlands can be classified as having dominantly transport-limited denudation regimes. The chemistry of the dissolved and solid load of lowland rivers is consistent with erosion under transport limited conditions. Lowland soils and solid loads of lowland rivers are rich in quartz, kaolinite, and iron sesquioxides, all of which are cation-depleted phases. Most of the cation load is in solution.

In the lowlands there are three major soil regions related to zones of differing vegetation and climate (37,39): the savanna (well north and south of the equator), the Amazonian forest (central and western Amazon Basin, Andes, Guayana Shield), and the campina-caatinga forest (straddling the equator). These correspond to regions of progressively wetter climate, diminished dry seasons and more siliceous less ferrous and aluminous soils. This relationship is quite rough, as drier areas were far more extensive in the tropics during glaciations (7,32,43,44). In the savanna, rates of physical denudation are greater (31,47), and soils are definitely more aluminous than forest soils (3,37). Except in areas of unusual substrate, forest soils consist mainly of kaolinite, quartz, and iron and aluminum hydroxides. In most cases, quartz and kaolinite are overwhelmingly the dominant phases. Abundant gibbsite is reported in some savanna soils; however, classic laterites and bauxites are rare (45). The soils associated with campina and caatinga vegetation form on quartz-containing bedrock. These soils (tropical podzols) are characterized by a surface humus layer, a quartz A horizon up to several meters thick, and an underlying layer of fairly impermeable aluminous clays (no iron) cemented by humic materials (17,30). The impermeable lower layer may be a reason for the extremely low erosion rates associated with these soils. Rivers draining these regions are exceptionally brown, acidic, cation-poor, and sediment-free (17,38).

When silica is plotted against ($K^*+Na^*-Cl^*$), data fall parallel to a 2:1 trend (Figure 5) ("*" refers to cyclic salt corrected values (39); Cl^* is subracted to correct for halite). This is suggestive of a system dominated by the weathering of feldspars to kaolinite combined with background silica coming from quartz dissolution or gibbsite formation (37,40).

Dissolved solid concentrations, and by inference denudation rates, are exceedingly low. This agrees with the diminution of weathering rates in conjunction with the development of thick soils (Figure 2). Dissolved loads are lower on the peneplaned shields than on the Neogene sediments. Rocks of the shield are composed of less stable minerals than the cation-poor sediments; however, the rocks and many of the soils of the shields are less permeable than those of the sediments. The latter factor appears to dominate. Where evaporites are exposed along the lower Amazon valley (Figure 1), their contribution to the rivers is minor. This indicates that susceptibility to weathering is indeed less important in controlling lowland erosion rates.

ELEMENTAL PARTITIONING

The different styles of erosion are associated with different degrees of partitioning of elements between dissolved and solid load. There are two principal ways to partition elements between the dissolved and solid loads: by the formation of secondary phases that are enriched or depleted in certain elements, relative to bedrock, and by the selective weathering of particular phases. In the most dilute rivers on the shields, such as in the

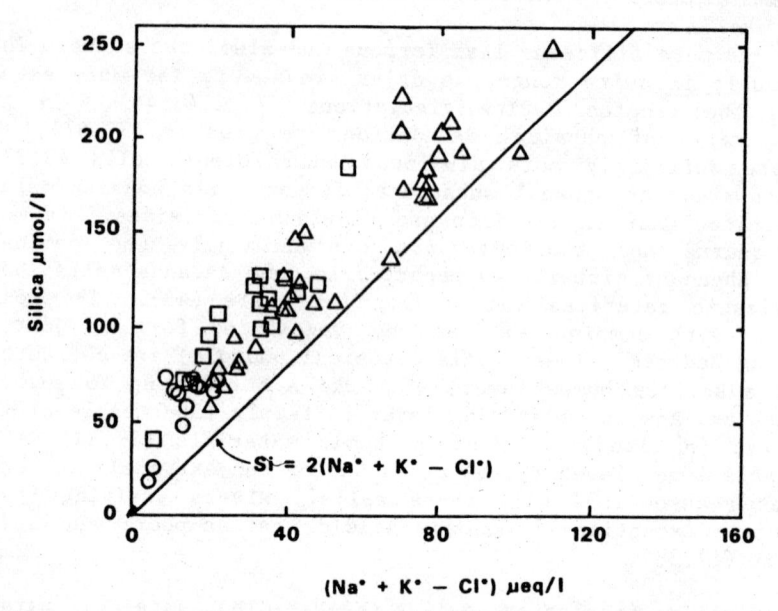

Figure 5. Si versus (Na*+K*-Cl*) for samples from the lowlands and elevated shields. Data are cyclic salt corrected. Symbols in Figure 3.

headwaters of the Negro River, elemental ratios in in the dissolved load resemble those of average shield rock (37,39), suggesting weathering by congruent solution. In other regions, silicon, iron, and aluminum are more strongly retained in the soil, while sodium, potassium, magnesium, calcium, and some silica enter solution. Partitioning of Na, K, Mg, and Ca into solid phases is seen in Andean rivers.

As rocks weather chemically, they lose their structural integrity. The structurally cohesive part of a saprolite profile is refered to here as "hard saprolite"; the less cohesive material as "soft saprolite". Primary minerals which are stable in the hard saprolite and which persist into the soft saprolite are indicated in Table 1. Solifluction, soil avalanching, and sheet runoff erosion in moist vegetated areas do not involve the hard saprolite (37). Potassium and magnesium should be enriched in solid erosion products where these processes are important, as K and Mg-bearing minerals are not strongly weathered from hard saprolite developed on acid to intermediate rocks. Na and Ca should be enriched in solution. Mg is incorporated into the lattice of many secondary clays, further accentuating its retention in the solid phase. When only kaolinite or gibbsite or other cation poor phases are forming, cation ratios in solution should match those in bedrock, unless soils are actively aggrading.

TABLE 1. Mineral Stability in Tropical Soils

MOST STABLE
Quartz* >>
    K-Feldspar*, Micas* >>
        Na-Feldspar >
            Ca-Feldspar, Amphiboles >
                Pyroxenes, Chlorite >
                    Dolomite >
                      Calcite >
                          Gypsum, Anhydrite+ >>
                              Halite+
                              LEAST STABLE

* Quartz, Micas, and K-feldspars are stable in hard saprolites developed on acid to intermediate igneous and metamorphic rocks. Quartz frequently persists through entire soil profiles.

+ Gypsum and anhydrite are observed in soils developed on salt domes in the Huallaga Basin; halite, however, is not. (see Pasquali et al., (28), López and Bisque (17), Stallard (37))

The resultant relative mobility trend is quite unlike the composition trend for common (oversaturated) igneous rocks. In the latter, as one goes from basic to acidic rocks, the Mg/(Mg+Ca) ratio decreases and the K/(K+Na) ratio increases.

The effect of partitioning is clearly seen when K*/(K*+Na*-Cl*) is plotted against Mg*/(Mg*+Ca*) for Amazon and Orinoco samples (Figure 6). The K*/(K*+Na*-Cl*) ratio represents inputs solely from silicate weathering. The Mg*/(Mg*+Ca*) ratio also includes contributions from carbonates and evaporites. Two features can be seen. First, samples from particular regions plot as distinct groupings. Second, samples from lowland rivers plot in a general field encompassing common igneous rocks, "average" shield and "average" shale compositions (no fractionation), while Andean samples are sodium and calcium enriched compared to typical igneous and shale compositions.

In the Andean samples, calcium enrichment is largely due to contributions from limestones and evaporites; however, the (Na*-Cl*) enrichment can only be due to the partitioning of potassium into solid phases of soils and river suspended load. This is particularly pronounced in rivers draining lower Paleozoic shales where water ratios are quite low compared to "average" shale and black shale ratios, reflecting the stability of potassium-bearing micas. Because of evaporite and carbonate contributions, stability trends for calcium and magnesium need to be confirmed using river-borne solids.

The effect of silica fractionation can be similarly illustrated by plotting Si/(Si+K*+Na*-Cl*) (Figure 6). Waters from most regions are silicon depleted when compared to igneous rocks, river sediments are silicon enriched. Some of the lowland rivers fall within the rock field, suggesting that in these regions silica quantitatively dissolves. The dashed line is equivalent to the 2:1 trend in Figure 5. Recall that mass balance and thermodynamic calculations suggest that cation-rich clays must be forming in the catchments represented by data falling below the trend.

The degree to which some lowland regions are dissolving is best displayed by aluminum and iron data from the Amazon system (37). It is observed that high concentrations of dissolved (< .45 um) iron and aluminum are found in the most dilute rivers. Concentrations greatly exceed those predicted by thermodynamic calculations for inorganic systems (37). Furthermore, aluminum concentrations are stable for several years in untreated samples. This suggests that aluminum and, to a lesser extent, iron exist as either organic complexes or organically stabilized colloids. Iron concentrations correlate most strongly with color (as a measure of dissolved humic materials), and aluminum with hydrogen ion concentration (37). When (Al+Fe)/(Si+Al+Fe) is plotted against K*/(K*+Na*-Cl*) very few rivers plot near the rock field (Figure 6). Those that do have the lowest dissolved loads; solid loads are negligible. Presumably, the lack of abundant cations in the soil environment allows for high levels of dissolved humic acids

---

Figure 6. Comparison of various chemical ratios in solid phases representative of the Amazon and Orinoco systems with dissolved solids in basin surface waters. Symbols for dissolved phases are given in Figure 3. For solids, circles are saturated igneous rocks from Le Maitre (18); the hexagon is average shield (12); triangles are river analyses, AM - Amazon, NG - Negro, OR - Orinoco; squares are various world averages, MC - marine clay, RV - river suspended load (20), SN - sand (29), GS - geosynclinal sediment, PS - platform sediment (12). The circles labeled "P" correspond to Parguaza granite analyses from Gaudette et al. (10) and this study. The curves with tic marks numbered 1-9 refer to a soil profile discussed later in text.

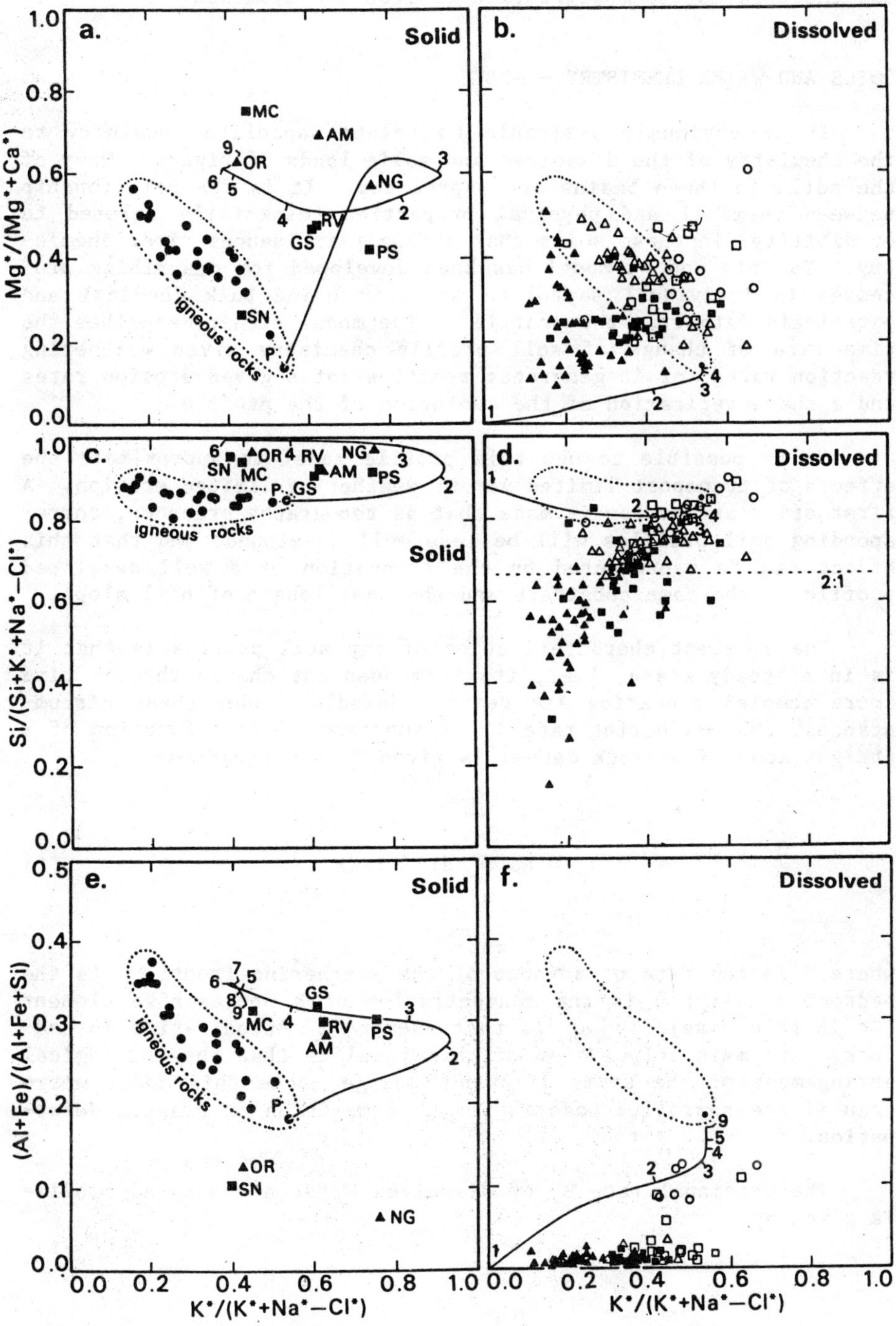
a.
Solid
b.
Dissolved
c.
Solid
d.
Dissolved
e.
Solid
f.
Dissolved
MC
AM
OR
NG
RV
GS
PS
SN
P
igneous rocks
2:1
Mg*/(Mg*+Ca*)
Si/(Si+K*+Na*−Cl*)
(Al+Fe)/(Al+Fe+Si)
K*/(K*+Na*−Cl*)

and facilitates the mobilization of iron and aluminum.

SOILS AND WATER CHEMISTRY – MODEL

It is obviously desirable to relate saprolite chemistry to the chemistry of the dissolved and solid loads of rivers. Many of the soils in these basins are saprolites. It is the relationship between chemical and physical properties (especially related to erodibility) in these soils that strongly influences river chemistry. To this end, a model has been developed for describing processes in individual saprolite profiles using bulk chemical and petrologic data from the profile[1]. The model either describes the time rate of change of soil profile chemistry given weathering reaction rates, or it generates reaction rates given erosion rates and a characterization of the evolution of the profile.

It is possible to use this profile model to approximate the effects of transport-limited versus weathering-limited erosion. A first order assumption is made that as topography steepens, corresponding soil profiles will be less well developed, and that this effect can be represented by the truncation of a well developed profile on the same substrate and the same length of hill slope.

The simplest characterization of any soil profile is that it is in a steady state, i.e., its form does not change through time (more complex scenarios are being modeled). Under these circumstances, the weathering rate J_c of substance C as a function of x (height above a bedrock datum) is given by the equation:

$$J_c = R\rho_o D_o \, d(C/D)/dx \tag{1}$$

where R is the rate of advance of the weathering front; ρ_o is the bedrock density; D is the concentration of a nonreactive element (Zr in this case); and D_o is that element's concentration in bedrock. The main stipulation of this model is that the topological arrangement of the atoms of D not change, thus this model works even if the saprolite undergoes some compaction or plastic deformation.

The erosional rate E_c of dissolved C for a truncated profile is given by

$$E_c = R\rho_o[C_o - C_s D_o/D_s] \tag{2}$$

where subscript "s" refers to the surface. The erosion rate F_c of solid C is given by

$$F_c = R\rho_o C_s D_o/D_s \tag{3}$$

An oxisol profile was sampled 20 km south of Puerto Ayacucho, Venezuela. The soil is developed on a slight ridge on a peneplain surface of very coarse-grained Parguaza granite. Streams in this region are typical of such areas, dilute and acidic. Chemistry (Si, Al, Fe, Mn, Na, K, Mg, Ca, P, Zr, Cr, Rb, Sr, LOI) and mineralogy were analyzed in bulk samples and in the clay, silt, sand, and coarser size fractions respectively. Bulk data are presented in Table 2 and Figure 7.

Smooth curves were fitted through the elemental data using a modified cubic spline regression (Figure 7), and the results are substituted into equation (1) to give weathering rates (J_c) (Figure 8). Petrographic evidence for weathering of certain minerals was used to decide whether a particular element was mobile at the weathering front. If not, the rate could be forced to be zero. In Figure 8, rates are arranged from most mobile to least mobile elements (Ca>Na≅Sr>Mg≅K≅Rb>Si>Fe>Al>Ti>Zr≅0). The surface-most sample was not included in the model as it appears to be a pisolithic lag.

The reactions agree with the soil profile mineralogy. The rapid loss of Ca, Na, and Sr is coincidental with the rapid loss of plagioclase (and modal hornblende). The delayed loss of Mg, K, and Rb is associated with the later onset of degradation of microcline, biotite, and secondary vermiculite. Quartz persists throughout the entire profile; however a substantial fraction is weathered out. This quartz weathering is associated with etching, flaking, and fracturing.

The trends for dissolved and solid phases predicted for steady state truncation of this profile are plotted on Figure 6. The numbers on the trends represent the ratio of soil Zr to bedrock Zr at that point in the profile. This is the inverse of the

Table 2.

SOIL PROFILE, 20 KM SOUTH OF PUERTO AYACUCHO

Z*	Sample Description	Quartz	Microcline	Albite	Biotite	Vermiculite	Kaolinite	Gibbsite	Goethite
90 cm	Soil surface, pisoliths	●	–	–	–	–	●	●	•
70 cm	Reddish soil	●	–	–	–	–	●	●	•
50 cm	Reddish-ochre soil	●	–	–	–	–	●	●	•
30 cm	Ochre saprolite	●	•	–	•	–	●	●	•
10 cm	Hard saprolite, ochre	●	●	•	•	•	●	●	•
3 cm	Hard saprolite, stained	●	●	●	•	•	●	●	•
0 cm	Fresh Parguaza granite	●	●	●	●	–	–	–	–

*Height above fresh bedrock in centimeters

fraction of bedrock that remains. The model trend for Mg/(Mg+Ca) versus K/(K+Na) defines the lower bound for the data in Figure 6B. This might be expected in view of the extremely Mg and Na poor nature of the granite. The trend predicted for Si/(Si+K+Na) does not define a bound to the river data field for Andean samples (Figure 6D); instead, the trend passes through lowland river data. This difference might be attributable to the limitation of silica inputs in Andean rivers through the formation of cation-rich clays. These are not important in the Puerto Ayacucho soil. The trend for (Al+Fe)/(Si+Al+Fe) also passes right through the lowland river data field (Figure 6F). These particular samples are from the most dilute rivers.

The solid phases produced by the truncation of this soil profile are quite interesting. Throughout most of the profile, the clay plus silt fraction would be mainly quartz silt, kaolinite, and gibbsite. Weak erosion into the surface layers would produce an iron stained quartz sand. Strong and continuous erosion down to the hard saprolite would generate an arkosic sand, very rich in potassium feldspar.

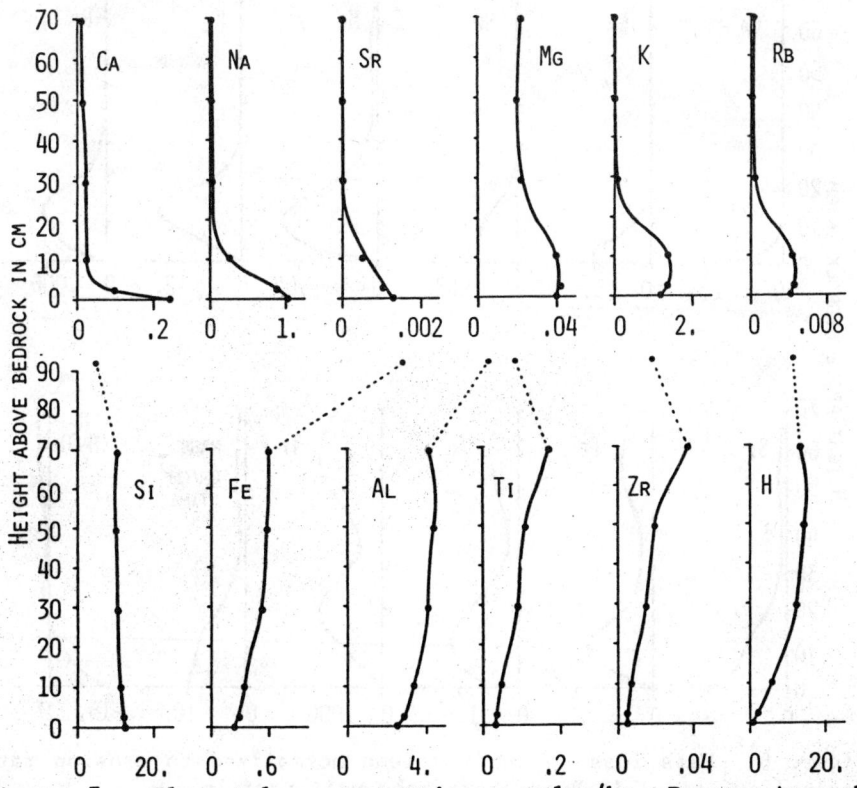

Figure 7. Elemental concentrations, moles/kg, Puerto Ayacucho soil profile.

CONCLUSIONS

In the Amazon and Orinoco basins, the chemistry of rivers can clearly be related to the geology of their catchments when geomorphic factors are taken into consideration. To a first order, erosion processes are seen to occupy a continuum between weathering-limited (steep slopes, thin soils) and transport-limited (slight slopes, thick soils). Denudation rates for the former are lithology dependent, and significant removal of cation-rich, unstable, solid phases is expected. For the latter, erosion rates should depend on the regional lowering of the landscape; lithologic susceptibility to weathering should be a minor factor. Simple weathering reactions can be used to describe compositional data for rivers in these basins. If either soils or biomass were aggrading or degrading reactions would not balance; thus, neither must be happening on a large scale.

Various other factors can complicate the model. Groundwater

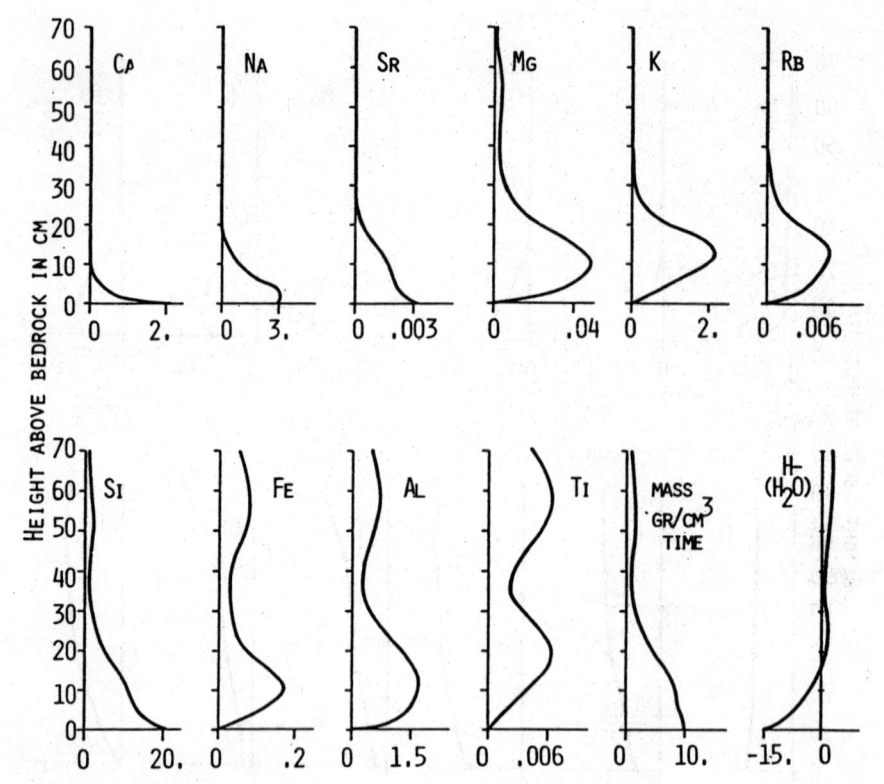

Figure 8. Mass loss per unit volume normalized to erosion rate, mmoles/(cm^3 time) in Puerto Ayacucho soil profile.

circulation, as is seen in Florida-like karst terrains, would be an obvious example. The chemical alteration of materials during transport down slope and subsequent storage in fluvial systems affects river chemistry in ways that would not be predicted from looking at local erosion. Vegetation can greatly accelerate erosion on steep slopes by anchoring soil. The anchored soil traps moisture and bioacids which facilitates further weathering and soil accumulation. This in turn leads to instability, landslides, and a new cycle of accumulation. On slight slopes, over very extended time periods, vegetation may reduce weathering rates by allowing very thick soils to accumulate.

The compositional trends of river dissolved and solid load can be approximated for different weathering regimes using soil chemistry data. Preliminary results suggest that evolutionary models of saprolites based on actual field data produce results which are consistent with river chemistry. Because the physical properties of soils (e.g., hard versus soft saprolite) are so important in determining the chemistry of dissolved and solid phases during erosion and transport, future soil work should include measurements of permeability, shear strength and cohesiveness in conjunction with chemical measurements.

The use of this approach, that of relating river chemistry to geology, landforms, and soils, offers some interesting research possibilities. On a local scale, geology and landforms should be systematically related to erosion rates and chemistry. With this, it may be possible to reconstruct the baseline chemistry of some highly human-perturbed river systems, or to describe long-term average conditions of widely fluctuating or episodic river systems. Paleosaprolites may be used to reconstruct earlier weathering conditions. Finally, to study the chemistry of erosion over the earth's surface by this approach would be ambitious. It should be possible, however, to study terrains in a suitable way through the use of remote sensing to measure slopes, perhaps space shuttle radar, coupled with geologic maps and ground studies of selected regions.

FOOTNOTE

[1] The author of this paper is currently working on a comprehensive spatial-temporal study of the Orinoco system along with researchers from M.I.T. (Edmond et al.), the U.S.G.S. (Nordin and Meade), and the Venezuelan Ministerio del Ambiente y Recursos Naturales Renobales : Projecto Orinoco-Apure (Mejia et al.) (22). This project is similar to previous work undertaken on the Amazon (21,37,38,39,40). Water analysis of the Amazon were by the author; Barry Grant at M.I.T. analyzed the Orinoco samples, courtesy John Edmond; soil and sediment analyses were by Paula Cortez, Stephanie Crane, and the author.

ACKNOWLEDGMENTS:

The most recent funding for this work comes from the author's department and includes a Dusenbury Preceptorship. A conversation in Manaus, over pizza, with geologist Victor Holm was inspiration for this paper.

BIBLIOGRAPHY

1. Benavides, V. 1968, "Saline deposits of South America". Geol. Soc. Amer. Spec. Pap. 88, pp., 249-290.

2. Blancaneaux, P., and M. Pouyllau 1977, "Formes d'alteration pseudokarstiques en relation avac la geomorphologie des granites precambriens du type Rapakivi dans le territoire Federal de l'Amazone, Venezuela". Cah. O.R.S.T.O.M. Ser. Pedol. 15, pp., 131-142.
3. Camargo, M. N., and I. C. Falesi 1975, "Soils of the Central Plateau and Transamazonic Highway" in "Soil Management in Tropical America". E. Bornemisza and A. Alvarado, ed., University Consortium on Soils of the Tropics, Raleigh, North Carolina, pp., 25-45.
4. Carson, M. A., and M. J. Kirkby 1972, "Hillslope, Form and Process". Cambridge University Press, Cambridge, 475pp.
5. Davis, D., J. Suppe, and F. A. Dahlen 1983, "Mechanics of fold-and-thrust belts and accretionary wedges". J. Geophys. Res. 88, pp., 1153-1172.
6. Franzinelli, E., and P. E. Potter 1983, "Petrology, chemistry, and texture of modern river sands, Amazon River system". Jour. Geology 91, pp., 23-39.
7. Garner, H. F. 1959, "Stratigraphic-sedimentary significance of contemporary climate and relief in four regions of the Andes Mountains". Geol. Soc. Am. Bull. 70, pp., 1327-1368.
8. Garner, H. F. 1968, "Tropical Weathering and Relief" in R. W. Fairbridge ed., The Encyclopedia of Geomorphology Reinhold, New York, pp., 1161-1172.
9. Garwood, N. C., D. P. Janos, and N. Brokaw 1979, "Earthquake-caused landslides: A major disturbance to tropical forests". Science 205, pp., 997-999.
10. Gaudette, H. E., V. Mendoza, P. M. Hurley, and H. W. Fairbairn 1978, "Geology and age of the Parguaza rapakivi granite, Venezuela". Geol. Soc. Amer. Bull. 89, pp., 1335-1340.
11. Gibbs, R. J. 1965, "The geochemistry of the Amazon River Basin". Ph. D. Thesis, Geology, University of California, San Diego, 96pp.
12. Holland, H. D. 1978, "The Chemistry of the Atmosphere and Oceans". John Wiley and Sons, N. Y., 351pp.
13. Hu, M., R. F. Stallard, and J. M. Edmond 1982, "Major ion chemistry of some large Chinese rivers". Nature 298, pp., 550-553.
14. Irion, G. 1975, "Los primeros resultos de las investigaciones de sedimentación y perfiles de erosión en la región amazónica". Universitas 12, pp., 256-257.
15. Irion, G. 1976, "Mineralogisch-geochemische Unterschungen an der pelitischen Fraktion amazonischer Oberboden und Sedimente". Biogeographica 7, pp., 7-25.
16. King, L. C. 1962, "The Morphology of the Earth". Oliver and Boyd, Edinburgh.
17. Klinge, H. 1967, "Podzol soils: A source of black water rivers in Amazonia". Atas do Simpôsio sôbre a Biota Amazônica 3, pp., 117-125.

18. Le Maitre, R. W. 1976, "The chemical variability of some common igneous rocks". J. Petrology 17, pp., 589-637.
19. Lopez Eyzaguirre, C., and R. E. Bisque 1975, "Study of the weathering of basic, intermediate and acidic rocks under tropical humid conditions". Quart. Colo. Sch. Mines 70, pp., 1-59.
20. Martin, J. M., and M. Meybeck 1979, "Elemental mass-balance of material carried by major world rivers". Marine Chem. 7, pp., 173-206.
21. Meade, R. H., C. F. Nordin Jr., W. F. Curtis, F. M. C. Rodrigues, C. M. do Vale and J. M. Edmond 1979, "Sediment loads in the Amazon River". Nature 278, pp., 161-163.
22. Meade, R. H., C. F. Nordin, Jr., D. Pèrez Hernàndez, A. Mejìa B., and J. M. Pèrez Godoy 1983, "Sediment and water discharge in Rìo Orinoco, Venezuela and Colombia". Proceedings of the Second International Symposium on River Sedimentation, 11-16 October, 1983, Nanjing, China Water Resources and Electric Power Press, Beijing, China, pp., 1134-1144.
23. Meybeck, M. 1977, "Dissolved and suspended matter carried by rivers; Composition, time and space variation, and world water balance, in". "Interaction Between Sediments and Fresh Waters". H. L. Golterman, ed., Junk and Pudoc, Amsterdam, pp., 25-32.
24. Meybeck, M. 1979b, "Concentrations des eaux fluviales en èlèments majeurs et apports en solution aux ocèans". Rev. Gèogr. dynam. Gèol. phys. 21, pp., 215-246.
25. Milliman, J. D., and R. H. Meade 1983, "World-wide delivery of river sediment to the oceans". Jour. Geology 91, pp., 1-21.
26. Oltman, R. E. 1968, "Reconnaissance investigations of the discharge and water quality of the Amazon River". U. S. Geological Survey Circ. 552, 16pp.
27. Pain, C. F. 1972, "Characteristics and geomorphic effects of earthquake initiated landslides in the Albert rainge of Papua New Guinea". Engineering Geology 6, pp., 261-274.
28. Pasquali, Z.,J., C. Lòpez E., and H. Meinhard 1972, "Meteorizaciòn de rocas del escudo de Guayana en ambiente tropical". Congreso Geològico Venezolano, Memoria IV, Tomo IV, Boletìn de Geologìa, Publicaciòn Especial n.5, pp.2245-2302.
29. Potter, P. E. 1978, "Petrology and chemistry of modern big river sands". J. Geol. 86, pp., 423-449.
30. Reichardt, K., G. Ranzani, E de Freitas Jr., and P. L. Libardi 1980, "Aspectos Hìdricos de alguns solos da Amazõnia - Regiäo do baixo rio Negro". Acta Amazonica 10, pp., 43-46.
31. Sarmiento, G., and M. Monasterio 1975, "A critical consideration of the environmental conditions associated with the occurence of savanna ecosystems in Tropical America" in "Tropical Ecological Systems". E. B. Golley and E. Medina, ed., Springer-Verlag, New York, pp., 223-250.

32. Sarnthein, M. 1978, "Sand deserts during glacial maximum and climatic optimum". Nature 272, pp., 43-46.
33. Scott, G. A. J. 1975a, "Soil profile changes resulting from the conversion of forest to grassland in the montaña of Peru". Great Plains-Rocky Mountain Geogr. J. 4, pp., 124-130.
34. Scott, G. A. J. 1977, "The role of fire in the creation and maintenance of savanna in the montaña of Peru". J. Biogeograpy 4, pp., 143-167.
35. Scott, G. A. J. 1975b, "Relationships between vegetation cover and soil avalanching in Hawaii". Proc. Assoc. Am. Geogr. 7, pp., 208-212.
36. Scott, G. A. J., and J. M. Street 1976, "The role of chemical weathering in the formation of Hawaiian Amphitheatre-headed Valleys". Z. Geomorph. N. F. 20, pp., 171-189.
37. Stallard, R. F. 1980, "Major element geochemistry of the Amazon River system". Ph.D. Thesis, Jan. 1980, Mass. Inst. of Tech. - Woods. Hole. Oceanogr. Inst. Joint Program in Oceanography, WHOI-80-29, 366pp.
38. Stallard, R. F., and J. M. Edmond 1981, "Geochemistry of the Amazon I: Precipitation chemistry and the marine contribution to the dissolved load at the time of peak discharge". J. Geophys. Res. 86, pp., 9844-9858.
39. Stallard, R. F., and J. M. Edmond 1983, "Geochemistry of the Amazon 2. The influence of the geology and weathering environment on the dissolved load". Jour. Geophys. Res. 88, pp., 9671-9688.
40. Stallard, R. F., and J. M. Edmond 1984, submitted, "Geochemistry of the Amazon 3. Carbonate and silicate weathering". J. Geophys. Res.
41. Suppe, J. 1981, "Mechanics of mountain building in Taiwan". Memoir, Geol. Soc. China n.4, pp.67-89.
42. Szczerban, E. 1976, "Cavernas y simas en areniscas precambricas del Territorio Federal Amazonas y Estado Bolivar". Venez. Dir. Geol. Bol. Geol. Pub. Esp. n.7, t.2, pp.1055-1072.
43. Tricart, J. 1974, "Existence de periodes seches au quaternaire en amazonie et dans les regiones voisines". Rev. Geomorphol. Dynam. 4, pp., 145-158.
44. Van der Hammen, T. 1974, "The Pleistocene changes of vegetation and climate in tropical South America". J. Biogeography 1, pp., 3-26.
45. Van Wambeke, A. 1978, "Properties and potentials of soils in the Amazon Basin". Interciencia 3, pp., 233-242.
46. Wentworth, C. K. 1943, "Soil avalanches on Oahu, Hawaii". Geol. Soc. Amer. Bull. 54, pp., 53-64.
47. Zonneveld, J. I. S. 1975, "Some problems of tropical geomorphology". Z. Geomorph. N. F. 19, pp., 377-392.

LIST OF WORKSHOP PARTICIPANTS

Berner, R.A.	Dept. of Geology & Geophysics, Yale University, New Haven, CT 06511, U.S.A.
Berthelin, J.	Centre de Pédologie Biologique, C.N.R.S., 17, rue Notre Dame des Pauvres, B.P. 5, Vandoeuvre-les-Nancy, France
Bikerman, M.	Dept. of Geology & Planetary Science, University of Pittsburgh, Pittsburgh, PA 15260, U.S.A.
Bourrie, G.	Laboratoire de Science du Sol--I.N.R.A., 65, rue de St. Brienc, 3500 Rennes, France
Charlet, L.	Dept. of Soil & Environmental Science, University of California, Riverside, Riverside, CA 92521, U.S.A.
Chou, L.	Laboratoire d'Océanographie, Université Libre de Bruxelles, Avenue F.-D. Roosevelt 50, 1050 Bruxelles, Belgium
Cronan, C.S.	Land & Water Resources Center, University of Maine, 11 Coburn Hall, Orono, ME 04469, USA
Drever, J.I.	Dept. of Geology & Geophysics, University of Wyoming, Laramie, WY 82071, U.S.A.
Eckhardt, F.E.W.	Inst. für Allgemeine Mikrobiologie, Universität Kiel, Olshausenstr. 40, 2300 Kiel, F.R. Germany
Eugster, H.P.	Dept. of Earth & Planetary Science, The Johns Hopkins University, Baltimore, MD 21218, U.S.A.
Fölster, H.	Inst. für Bodenkunde und Waldernährung, Universität Göttingen, Büsgenweg 2, 3400 Göttingen, F.R. Germany
Fritz, B.	Centre de Sédimentologie et Géochimie de la Surface, C.N.R.S., 1, rue Blessig, 67084 Strasbourg Cedex, France

LIST OF WORKSHOP PARTICIPANTS

Furrer, G.	EAWAG, 8600 Dübendorf, Switzerland
Gaillard, J.F.	10, rue Berthollet, 73160 Cognin, France
Graustein, W.C.	Dept. of Geology & Geophysics, Yale University, New Haven, CT 06511, U.S.A.
Helgeson, H.C.	Dept. of Geology & Geophysics, University of California, Berkeley, Berkeley, CA 94720, USA
Jones, B.F.	U.S. Geological Survey, M.S. 432, Reston, VA 22092, U.S.A.
Jørgensen, P.	Dept. of Geology, Agricultural University of Norway, P.O. Box 21, 1432 Aas-NLH, Norway
Kempe, S.	Geologisch-Paläontologisches Institut, Universität Hmburg, Bundesstr. 55, 2000 Hamburg 13, F.R. Germany
Kramer, J.R.	Dept. of Geology, McMaster University, Hamilton, Ontario L8S 4M1, Canada
Krumbein, W.E.	Geomikrobiologie, Universität Oldenburg, Postfach 2503, 2900 Oldenburg, F.R. Germany
Lasaga, A.C.	Dept. of Geology & Geophysics, Yale University, New Haven, CT 06511, U.S.A.
Lerman, A.	Dept. of Geological Sciences, Northwestern University, Evanston, IL 60201, U.S.A.
McKinley, I.G.	Eidg. Inst. für Kernforschung, 5303 Würenlingen, Switzerland
Meunier, A.	Université de Poitiers, Laboratoire de Pétrologie del la Surface, 40, avenue du Recteur Pineau, 86022 Poitiers Cedex, France
Meybeck, M.	Laboratoire de Géologie, Ecole Normale Supérieure, 46, rue d'Ulm, 75230 Paris Cedex 05, France
Monnin, C.	Laboratoire Minéralogie-Cristallographie, 38, rue des Trent-Six Ponts, 31062 Toulouse Cedex, France
Murphy, W,	Dept. of Geology & Geophysics, University of California, Berkeley, Berkeley, CA 94720, USA
Nahon, D.	Laboratoire de Pétrologie de la Surface, Université de Poitiers, 86022 Poitiers, France
Neal, C.	Institute of Hydrology, Maclean Building, Crowmarsh Gifford, Wallingford, Oxon, U.K.
Novikoff, A.	Institut de Géologie, 1, rue Blessig, 67084 Strasbourg Cedex, France

LIST OF WORKSHOP PARTICIPANTS

Paquet, H.	Institut de Géologie,], rue Blessig, 67084 Strasbourg Cedex, France
Pekdeger, A.	Geologisch-Paläontologisches Institut, Universität Kiel, Olshausenstr. 40, 2300 Kiel, F.R. Germany
Rodriguez-Clemente, R.	Instituto de Geologia, C.S.I.C., J. Gutierrez Abscal 2, Madrid-6, Spain
Schott, J.	Laboratoire de Minéralogie et Cristallographie, Université Paul Sabatier, 39, Allées Jules-Guesde, 31400 Toulouse Cedex, France
Schwertmann, U.	Institut für Bodenkunde, Technische Universität München, 8050 Freising-Wehhenstephan, F.R. Germany
Sposito, G.	Dept. of Soil & Environmental Science, University of California, Riverside, Riverside, CA 92521, U.S.A.
Stallard, R.F.	Dept. of Geological & Geophysical Sciences, Princeton University, Princeton, NJ 08540, U.S.A.
Stumm, W.	EAWAG, 8600 Dübendorf, Switzerland
Sturm, M.	EAWAG, 8600 Dübendorf, Switzerland
Tardy, Y.	Institut de Géologie, Université Louis Pasteur, 1, rue Blessig, 67084 Strasbourg Cedex, France
Trolard, F.	Institut de Géologie, Université Louis Pasteur, 1, rue Blessig, 67084 Strasbourg Cedex, France
Velbel, M.	Dept. of Geology, Michigan State University, East Lansing, MI 48824
Wieland, E.	EAWAG, 8600 Dübendorf, Switzerland
Wilson, M.J.	The Macaulay Institute for Soil Research, Craigiebuckler, Aberdeen AB9 2QJ, U.K.
Wollast, R.	Laboratoire d'Océanographie, Université Libre de Bruxelles, Avenue F.-D. Roosevelt 50, 1050 Bruxelles, Belgium
Wright, R.F.	Norwegian Institute for Water Research, P.O. Box 333 Blindern, Oslo 3, Norway
Zobrist, J.	EAWAG 8600 Dübendorf, Switzerland

SUBJECT INDEX

acid deposition 176, 197-206
activation energy 39
albite 75-94
alfisols 3
alunite 11, 121-139
Amazon basin 293-313
Andes 299-300
Appalachians 231-244
ARSAT 25
ascorbic acid 64
augite 35-52
bacteria 161-170
base supply rate 176
beidellite 3-6
benzoate 59-69
biogeochemical cycles 145, 148, 161-170
bioleaching 151, 163-167
biological mineralization 164-168, 180
biopyriboles 35
biosphere 148
birnessite 282-283
black shales 299
bronzite 35-52
brucite 249
calcite 29-31
calcium hydroxide waters 249-273
carbonates 299
cation denudation rates 181
Chad 28
chlorite 216
CISSFIT 25
citrate 59-69, 166-167
column experiments 177, 197
complex-forming ligands 55-69, 165-168
computer modeling 19-32
continental denudation 293-313
corrensite 103, 112-115
Coweeta 231-244
cyanobacteria 152-156, 161-170
diagenesis 115
diopside 35-52
dislocations 78
DISSOL 26-27
dissolution mechanisms 35-52, 55-72, 75-94, 164-168
DOC 190
end-members 23-25
enstatite 35-52
equilibrium, chemical 211-222, 231-232, 238-240

erosion 295
   transport limited 296
   erosion limited 296
etch pits 39, 220
EVAPOR 26
evaporites 299
exfoliation 147
feldspar weathering 75-94, 211-222, 233-244
ferrihydrite 119
fluidized bed 75-94
fluoride 62
forest soils 167, 175-194
fracture flow 251-2
fungi 161-170
Gaia concept 146, 149
garnet 233, 277-290
Gay-Lussac-Ostwald step rule 1-10
geomorphology 294
gibbsite 7-10, 29-30, 234-244
glacial outwash 212
goethite 29, 119-120, 234-244
granite 27-32
halloysite 3-6
Harz Mountains 199, 200
hematite 119
humic substances 190
humus layer 199
huntite 249
hydrobiotite 105-106, 234
hydrogen gas 249-250
hydrothermal alteration 136
hydrous ferric oxides 49, 119-120
hydroxy-interlayer vermiculite 7-8
hyperalkaline waters 249-273
illite 5, 216
inceptisol 175-194
intergrade minerals 106
interparticle diffraction 108-110
interstratified clay minerals 21-32, 97-116
   chlorite-smectite 103-104, 115
   chlorite-vermiculite 107-108, 115
   illite-smectite 100-103, 106-115
   kaolin-smectite 105, 107
   mica-vermiculite 103, 106, 108, 115
   random 112-115
   regular 107, 110, 115
ion exchange 80-82, 181
iron oxides 49, 119-120
isotopes 257, 262
Ivory Coast 277-290

jurbanite 11
kaolinite 3-15, 29-31, 216, 234-244
    continuum 1
kinetics 35-52, 59-72, 75-94, 232
laterite 277-290, 303
lichen 144, 167
lithiophorite 283-284
locusts 146-147
magnesite 270-272
malonate 59-69, 166
manganese 277-290
manganite 281-282
mass action law 4
mass transfer calculations 25-26
Mattigod-Sposito model 11-15
Mazzaron 135-138
microbial processes 148, 152, 154-158, 161-170, 190
mineral stability 305
nitrogen gas 249
nsutite 284-285
nucleation 70-71, 125-132
olivines 35-52
Oman 249-273
organic acids 166-167, 190
organic ligands 59-69, 221
Orinoco basin 293-313
oxalate 59-69, 166-167
oxisols 11, 309
parabolic kinetics 36, 49, 74
particle size 187
partitioning, elemental 303-304
PATHCALC 26
petroglyphs 144
pH dependence 59-69, 88-92
PHREEQE 211, 213
phthalate 59-69
portlandite 249
proton consumption rates 197-206
pyroxenes 35-52
rate laws 67-69, 78
rectorite 100, 106
river chemistry 293-313
salicylate 59-69
saprolite 232-244, 304
scanning electron microscopy 212-229
Schleswig-Holstein 212
Segeberger Forest 212
Semail ophiolite nappe 249
serpentinization 249-273
shields 300-303

smectite 3-6, 13-15, 29-31, 108-110
soil columns 177, 197
soil microcosm 177
soil solutions 175-194, 211-221
solid solution model 19-32
Solling 199
spodosol 175-194
spring water 249-273
streamwater chemistry 235-236
succinate 59-69
sulfur bacteria 165-166
suolunite 249
surface coordination 55-69
surface layers 35-36, 76-77, 85-88
surface protonation 59-61
surface reaction 77-79
temperature effects 31
tephroite 281
THERMAL 26
thermodynamic modeling 19-32, 211-222, 231-232, 238-240
topographic releif 294
tosudite 103
transition state theory 78, 89-90
tropical rainforest 293-313
ultisols 5
ultramafic rock 249-273
unsaturated zone 211-222, 234-244
variscite 10
vegetation effects 199, 297
vermiculite 7, 12-15
Vosges Massif 28
WATEQF 213
watershed budgets 183, 197, 232-244
water table 213
weathering budgets 197, 202-205, 231-232, 293-313
weathering rate 296
X-ray photoelectron spectroscopy 35, 38, 87
Ziemougoula 277-290